To Rober,

with thanks
& best wishes

[signature]

Oct. 1994

Technology Transfer and Scandinavian Industrialisation

Edited by
Kristine Bruland

During the nineteenth century Scandinavia was transformed from one of the poorest regions of Europe to one of the richest. This book deals with the technological basis of that growth; it shows, from an historical perspective, the decisive role of foreign, in particular British, technologies in the industrialisation of Sweden, Denmark and Norway. It argues that the acquisition of advanced technological capability from abroad is a key element of economic development as such, and it shows how Scandinavian industries managed to acquire such capability. This book brings together several case-studies from a wide range of industries, analysing how technology transfers took place and what mechanisms were involved.

Kristine Bruland is Visiting Fellow at the LSE (Business History Unit) and Research Fellow at the Centre for Technology and Culture, University of Oslo. She is a member of the History Research Group, Oslo.

Technology Transfer and Scandinavian Industrialisation

Edited by
Kristine Bruland

BERG

New York / Oxford

Distributed exclusively in the US and Canada by
St Martin's Press, New York

First published in 1991 by
Berg Publishers Limited
Editorial offices:
165 Taber Avenue, Providence, RI 02906, USA
150 Cowley Road, Oxford, OX4 1JJ, UK

Library of Congress Cataloging-in-Publication Data
Technology transfer and Scandinavian industrialisation/edited by
  Kristine Bruland.
    p.   cm.
  Includes index.
  ISBN 0 – 85496 – 605 – 6
  1. Technology transfer — Scandinavia.
  2. Scandinavia — Industries.
I. Bruland, Kristine
T174.3.T379   1992   91 – 36426
338.94806 — dc20   CIP

British Library Cataloguing in Publication Data

Technology transfer and Scandinavian
industrialisation.
  I. Bruland, Kristine
  338.948

  ISBN   0–85496–605–6

Printed in Great Britain by
Billing & Sons Ltd, Worcester

Contents

Contents

Maps, Tables, Figures and Appendixes

Maps

Tables

Figures

Appendixes

Notes on Contributors

Rolf Adamson is Professor of Economic History at the University of Stockholm. He has published extensively on the development of the Swedish iron industry in the nineteenth century, as well as on theoretical and methodological issues in social and economic history. He is currently working on Swedish agricultural developments in the period 1780–1850.

Rolv Petter Amdam is Senior Lecturer at the Institute for Business History at the Norwegian School of Management near Oslo. He has published a number of books and articles on the history of the Norwegian glass industry and on education, and is now working on the history of the Norwegian pharmaceutical company Nycomed.

Håkon With Andersen is Professor of History of Technology and Director of the Centre for Technology and Society at the University of Trondheim. His work covers a broad spectrum of interests in social and economic history and theoretical issues. His publications include, in collaboration with John Peter Collett, *Anchor and Balance, Det Norske Veritas 1864–1989*.

Kristine Bruland is Visiting Fellow at the LSE (Business History Unit) and Research Fellow (NAVF) at the Centre for Technology and Culture, University of Oslo. She has written mainly on British nineteenth century textile history and technology transfer. Her other publications include *British Technology & European Industrialisation. The Norwegian textile industry in the mid-nineteenth century*

Martin Fritz is Senior Lecturer in Economic History at the University of Göteborg. His main research interest has been the history of Swedish iron and iron ore exports. He also has published books and articles on the Swedish economy during the Second World War.

Fritz Hodne is Professor of Economic History at the Norwegian School of Economics and Business Administration, Bergen. He has pursued detailed research into the economic development of modern Norway for many years, and published a number of central books and articles including *The Norwegian Economy 1920–1980*.

Ole Hyldtoft is Senior Lecturer at the Institute of Economic History, University of Copenhagen. He has written extensively on urban history and Danish industrialisation. His books include *Københavns Industrialisering 1840–1914* (the industrialisation of Copenhagen), and a book on the technological development of Danish industry 1870–1930 (forthcoming).

Lennart Jörberg is Professor of Economic History at the University of Lund, and a member of the Royal Swedish Academy of Sciences. His research has dealt with the process of industrialisation in Scandinavia, Swedish entrepreneurial history and the history of prices. His major works include *History of Prices in Sweden 1732–1914*.

Even Lange is Professor of Economic History and Director of the Institute for Business History at the Norwegian School of Management near Oslo. He has published a number of books and articles on Norwegian industrial and political history. His present field of interest is the development of Scandinavian banking and finance institutions since 1850.

Svante Lindqvist is Professor of History of Technology at the Royal Institute of Technology in Stockholm. He has worked particularly on science and technology related issues in American and European history. His many publications include: *Technology on Trial: the Introduction of Steam Power Technology into Sweden 1715–1736*.

Eli Moen is Research Fellow (NAVF) at the Department of History, University of Oslo. Her work has mainly concerned issues in Norwegian nineteenth-century social and economic history. She is currently extending her research on the economic development of the Norwegian pulp and paper industry.

Gunnar Nerheim is Senior Lecturer in the History of Technology at Rogaland University Centre, Stavanger. He is the author of several books and articles about business history and the history of technology, including *Growth through Welding*. He is currently working on the Norwegian Petroleum History Project.

Trine Parmer is working as a Consultant of Education in the Norwegian Organisation of Vocational Unions, Oslo. Her work in history also includes a book about production and work-organisation in a Norwegian mechanical engineering firm after the mid-nineteenth century.

Poul Strømstad   is Curator at the Danish National Museum in Copenhagen, and co-founder of the Danish Association for Preservation of the
Industrial Environment (1979). His research has mainly dealt with technological and social aspects of artisan production, the construction and use
of old buildings, and Danish industrialisation.

Acknowledgements

'How Industrial Technology first came to Norway' by Trine Parmer is a reworked and extended version of a chapter previously published in Norwegian as: 'Hvordan industriteknologien først kom til Norge', in F. Sejersted (ed.) *Vekst gjennom Krise*, Oslo, Bergen, Trondheim: Universitetsforlaget, 1982.

'British Influence on Developments in the Swedish Foundry Industry around the turn of the Eighteenth Century' by Martin Fritz has been published in Swedish as: 'England och den svenska gjuteriindustrins omvandling kring sekelskiftet 1800' in *Polhem Tidskrift for Teknikhistoria*, 1984, 4a, Årgang 6, Stockholm.

A shorter version of 'Artisan Travel and Technology Transfer to Denmark, 1750–1900' by Poul Strømstad has been published in Danish in *Fabrik og Bolig*, 1, 1989.

An earlier version of 'The Diffusion of Technology and Industrial Change in Sweden during the Nineteenth Century' by Lennart Jörberg appeared in Swedish as: 'Teknikspridning och industriell forandring i Sverige under 1800–tallet' in *Festskrift til Kristof Glamann*. Odense: Odense Universitetsforlag, 1983.

'The Norwegian Mechanical Engineering Industry and the Transfer of Technology, 1800–1900' by Kristine Bruland was previously published in Norwegian as: 'Norsk mekanisk verkstedindustri og teknologioverføring 1840–1900' in E. Lange (ed.) *Teknologi i Virksomhet. Verkstedindustri i Norge etter 1840*, Oslo: Ad Notam, 1989.

Parts of 'The Introduction of the Bessemer Process in Sweden' by Claus Wohlert have appeared in Swedish as: 'Samspelet mellan vetenskap, teknik och yrkeskunnand' in B. Sundin (ed.) *I Teknikens Backspegel*, Stockholm: Carlsson Forlag, 1987.

'To Take Great Pains: Norwegian Wood Pulp on the British

Market in the 1870s' by Even Lange has been published in Norwegian as: 'Om at "gjøre seg en hel del umage". Norsk tremasse på det britiske marked i 1870–årene' in *Historisk Tidsskrift*, 3, 1988.

The editor and publisher gratefully acknowledge the generous financial support of NAVF (Norges allmennvitenskapelige forskningsrad) in the preparation of this work.

Scandinavia

Introduction

Kristine Bruland

The history of technology is an important component of historical studies in general in Scandinavia; not the least reason for this is the fact that in order to develop, the Scandinavian countries had to engage in substantial technology import. This book collects a number of studies by well-known Scandinavian historians on the role of technology transfer in Scandinavian industrialisation, and on the way that transfer was shaped by social, cultural and political forces.

I

The focus of this book is the technological basis of industrialisation in Scandinavia from the late eighteenth to the late nineteenth century. This historical topic is of interest for many reasons, but one is particularly simple: the Scandinavian economies are among the richest in the world, and have been for many years. They have managed to sustain high levels of productivity and *per capita* income, high growth rates of output, low inflation and unemployment rates, and – on the whole – strong positions in foreign trade. Their manufacturing performance, especially in engineering industries but also in more recently developed high technology areas such as information technology, biotechnology and new materials, is impressive. Moreover all this has been combined with social welfare policies which are perhaps the most comprehensive in the world.

This economic success stands in need of explanation, for it does not flow from any obvious natural advantages; of course, Scandinavia has important resources in terms of minerals, timber, fish stocks and latterly oil, but these resources are arguably no better than those of a number of countries in the less developed world

1

(and Norway's oil was discovered and developed at a time when it was already one of the world's richest economies). Indeed, from one perspective the Scandinavian countries share many common features with semi-industrialised economies which are far poorer: they are marginal economies on the periphery of the advanced industrial world, with generally adverse climatic and agricultural conditions. Denmark's agricultural land is good, although no better than many similar countries' or regions'; and Denmark has no other significant resources. Against Denmark's agricultural advantages must be set extremely inhospitable conditions in large parts of Sweden, and virtually all of Norway. Less than 3 per cent of Norway's land surface can be cultivated. Climatic conditions can be very difficult indeed throughout Scandinavia. The populations are small – only 6,000,000 in 1850 and 24,000,000 in the Nordic area at the present time – and dispersed over regions in which communications are difficult; this means that the economies must sustain very substantial infrastructure capital in the form of roads, railways, power supply and so on. Moreover in the late nineteenth and early twentieth centuries there was a massive out-migration, mostly by people of working age, to Canada and the USA. Only Ireland has matched the emigration rates of Scandinavia. In the area of manufacturing trade, the Scandinavian countries have always faced formidable competition from major industrial economies – in particular Germany and the UK – which are close at hand, and which maintain scientific and technological resource bases on a scale to which the Scandinavian countries can never aspire.

Against this background of disadvantage, it is not clear why the Scandinavian countries have attained such a high degree of industrial success. One commonly offered explanation is cultural: it appeals to an allegedly high degree of social cohesion and consensus which promotes economic growth by alleviating the conflicts which are produced by industrial capitalism, and which retard its growth. These factors are said to characterise the 'middle way' which is represented primarily by Sweden, but in some degree by all the countries of Scandinavia. However, there are a number of problems with this type of account of Scandinavian economic behaviour (just as there are with the various contemporary 'culturalist' accounts of Japanese economic success). On the one hand, the 'consensus culture' is arguably of relatively recent development, and could not be said to characterise the industrial climate at any stage prior to the end of the Second World War. It might also be argued that in so far as a social consensus exists, it is a product of industrial success rather than a precondition for it. It leaves aside, therefore, the important historical question of how the Scandinavian countries managed to industrialise at all, since their industrialisation is very recent, and especially since they were late-comers in an economic environment characterised by large advanced competitors. Understanding the current situation of the Scandinavian economies is very much a historical question concerning how they achieved the difficult task of industrialisation and technological modernisation.

This latter problem is the topic of this book. In the following sections of this introduction I outline firstly some of the conceptual problems (both for historical and contemporary analysis) involved in technology transfer; then a sketch of the general historical context of technology transfer into Scandinavia; and finally an overview of the contributions within this book.

II

The spread of industrialisation from the late eighteenth century, primarily from Britain, involved complex changes in economic organisation and economic life in countries which began to follow the British path. But it also involved a quite concrete diffusion of specific technologies. How has this technological diffusion or transfer problem been treated in industrial history?

I want to argue here that economic history frequently shares the well-known weakness of economic theory when it comes to technological issues and firm behaviour generally. Within economics, neo-classical production theory generally assumes that *what* firms do is independent of *how* they do it. What firms do is a capital allocation problem, guided by potential profits within an industry: firms are not constrained by the state of their technological knowledge, and can shift between industries in response to price and profit signals. How they produce, on the other hand, is a process of maximisation, which occurs in the context of available factor supplies and factor prices; feasible factor combinations are assumed to be well defined, and above all, well *known*. A technique, in this context, is defined by its capital/labour ratio. Although methods are fixed in the short term, and capital is not malleable, in the longer term firms know all potential techniques, and the problem becomes one of choosing that optimal technique which achieves an objective function of cost minimisation and hence profit maximisation. Formally speaking, the problem for the firm is to choose a K/L ratio which maximises P in an objective function such as

$$P=pq-(wL+rK)$$

where P is profits, p and q the price and quantity of output, w and r the wage and interest rates, and L and K the quantities of labour and capital used. However, in assuming the availability of all possible K/L combinations, this approach essentially erases the problem of technological obstacles to production. What kinds of emphases does this lead to? In an industrialisation or development context, the problem of industry growth is seen primarily in terms of entrepreneurship and competitive conditions: if an industry is potentially profitable, then growth depends on the supply of entrepreneurship and entry conditions. If prices are flexible and entry is free, then profit-seeking activity by entrepreneurs can be expected to take the industry to equilibrium size. Technological capabilities are in general not an issue.

In other words, this type of production theory leads to an emphasis on 'environmental' conditions rather than access to technology when discussing industrial growth. In this it has been followed by many economic historians. Of course I do not wish to suggest either that economic historians have uncritically accepted neo-classical production theory as a basis for industrial history, or that economic history consists simply of the application of economic theory to the historical past. Neither of these propositions is true. What does seem to me true, in European economic history at least, is that economic historians have concentrated on changes in the economic environment (in terms of entry conditions and de-regulation, the supply of entrepreneurship, the availability of finance, the emergence of a labour force) in discussing industrialisation, and that industrial technology has been given much less emphasis than it deserves. I have argued elsewhere that this is in large part a legacy of the great work of Alexander Gerschenkron, which has strongly influenced the history of European economic and industrial growth. In his pioneering work *Economic Backwardness in Historical Perspective*, Gerschenkron emphasised the key role played by the development of an institutional structure – institutions here being considered in the wide sense – which was in some sense appropriate to the operation of an industrial capitalist economy. These institutions included a labour market (which in turn implied the rapid or gradual transformation of peasant agriculture), a financial system, a climate of entrepreneurship and some method of generating entrepreneurs, and so on. These 'pre-requisites' for industrial production could be met in various ways, Gerschenkron argued, including the action of state policy. The latter, he suggested, was indeed historically a frequent element in state activity in Continental Europe: it was the basis of state-sponsored enterprise, for example, and of the fostering of industrial banking. Gerschenkron spoke of these institutional developments in terms of 'substitution' for the processes which lay behind British industrial success.[1] Other historians have frequently followed the emphases established by Gerschenkron. Even histories focusing directly on technology, such as Landes's *Unbound Prometheus*, strongly emphasise the 'prerequisites' for growth, and say surprisingly little about the acquisition of technological capabilities, diffusion of technologies, technological obstacles to the spread of industrialisation, and so on, especially for the really critical period in European industrial growth which I would argue begins in the mid-1840s.

However it seems clear that much of mid-nineteenth century industrialisation involved the diffusion or transfer of specific technologies, especially from Britain, to industrialising countries. Indeed the world economy is generally characterised by technological disparities, by changing leaders and followers, and by transfer between them. But the concrete details of what was involved in this remain an area where our ignorance is greater than our knowledge.

In thinking about what was involved in this process of technology diffusion or transfer, especially in early industrialisation, we have to start by considering the nature of technology itself. Technology is often defined in terms of knowledge, that is as the array of knowledges which are used in the production of a product. While this has the merit of de-emphasising hardware, with which technology is sometimes confused, it is important to remember that for firms and industries, a technology usually consists of an integration of knowledge and hardware. This 'knowledge' need not be, and usually is not, formal or codified: it is often tacit and localised, taking the form of uncodified skills or capabilities which are distributed among individuals or groups within a firm. One of the key problems for management of an enterprise is the integration of these dispersed and decentralised technological knowledges into a general 'technological capability' of the firm. What is involved in such a capability? In general, technological competition means that firms compete in terms of the performance attributes and design characteristics of products, and these are usually associated with specific process technologies. In this context, firms firstly need to understand the way in which the evolution of

1. K. Bruland, *British Technology and European Industrialization: The Norwegian Textile Industry in the Mid-19th century* (Cambridge, 1989), chap. 1.

demand is (re)shaping the desired technical attributes of products, and how their own products should adapt to this evolution. On the process side, since firms do not normally produce their own equipment, an initial problem is to develop the ability to know the range of process technology options available to the firm, and to assess them in terms of their suitability for the firm. This choice of technique problem involves assessments concerning the integration of different types of equipment, problems of setting up and maintenance, operating skills and types of labour required and so on. Even for simple production processes, this frequently involves very many different types of equipment and components. Search for, and acquisition of, a process technology therefore normally entails a complex of technical assessments and skills. Operations involve rather different skills, to do with the adjustments, modifications and adaptations which are necessary to get equipment actually to work, and then to maintain that equipment. It is almost never possible to buy even standardised items of equipment and simply run them; they require skills and learning. To sum up this brief discussion, technological capability involves knowledges and search capabilities relating to the technical development of product characteristics; an array of process hardware; and the integration of hardware with the set of unwritten knowledges which are necessary in order to operate, maintain and connect it. The technological capability to run an industrial production process is thus a matter of considerable complexity. It implies difficult processes of machine and equipment acquisition within the context of the development and integration of a large set of production skills.

Moreover all this of course happens within a social context which shapes not just the environmental conditions but also the internal operations of firms. Social changes, in terms of such things as labour recruitment, management and norms of working behaviour; the social recognition of entrepreneurs and their ability to finance their activities; the cultural role accorded to science and technology, are all very significant factors. All play a concrete role in the ability of firms to set new technologies to work. Against the background of these considerations, international 'technology transfer' is far more than a matter of the import of equipment. Much more important is the process by which search and operations capabilities are acquired by firms, and the ways (social, intellectual and technical) in which these capabilities are developed. On this topic, especially in terms of empirical studies, we remain ill-informed in European economic history.

III

What was the economic and political context in which Scandinavia began to industrialise? In the mid-nineteenth century, as Lennart Jörberg has pointed out, 'Denmark excepted, the Scandinavian countries must have been among the poorest of Western European nations.'[2] The issue of how these countries began the path towards prosperity is therefore not a trivial one: the transition from poverty to wealth occurred against the background of extremely unpropitious events and resources. What kinds of processes were involved in 'late' industrialisation in nineteenth-century Scandinavia?

The main problems were those of adjustment to the new economic and technological regime initiated primarily in Britain and France. The eighteenth century was a period of very significant technological advance in both these countries, an advance which moreover accelerated towards the end of the century. In a range of manufacturing industries, in services such as transport, and in general scientific and technical investigation, major advances occurred; these were moreover associated with sharply growing output in particular regions of Britain, France and Germany. In terms of information flows and movement of people, Europe had of course been highly integrated for many years, indeed many centuries, and the sharpening pace of technological change was therefore no secret. In the Scandinavian countries responses were quite rapid, and it was not long before technical societies – along the lines of those in Britain and France – were established, state-sponsored industrial visits (especially to Britain) began, and attempts actually to use the new technologies commenced.

For Europe generally, the long-term impact of this period of technological advance was strongly affected by the Napoleonic Wars. The collapse of the 'Continental System', which was the basis of French political and economic policy, left, as Sidney Pollard has pointed out, a major economic heritage:

> It derived from the growing isolation of the continent in twenty-two years of war from the new technology which was then developing at great speed in Britain, particularly in the textile and iron industries and in engineering. Normally, contact with other manufacturing countries would have been sufficiently close to allow them either to copy, or to learn to protect themselves against, the more advanced or the cheaper

2. L. Jörberg, 'The Industrial Revolution in the Nordic Countries', in C. Cipolla (ed.), *The Fontana Economic History of Europe: The Emergence of Industrial Societies (2)*, pp. 375–6.

British imports as they developed stage by stage. As it was, when the shipping lanes and ports were once more opened after the war a flood of British products burst over the unprepared markets.[3]

These points imply that for 'late industrialising' economies (and also, in rather different ways, for the defeated French economy) the process of industrialisation from the early nineteenth century was not simply a growth process but one of rapid adaptation in an exceptionally turbulent and complex economic environment. Broadly speaking, two types of problems had to be solved. On the one hand, there was a process of social and institutional change associated with the industrial economy; in some cases this change was driven by economic forces, while in others it followed from conscious decisions of public policy. On the other hand there was the task of acquiring and implementing specific new technologies, and marketing the products of these technologies.

The underlying social and institutional changes in turn had two dimensions, which are sometimes run together. Firstly there was the long-run historical process through which market-oriented and ultimately capitalist economies were created. This development was by no means coterminous with the industrial economy itself, for capitalism – in the sense of markets in products and labour, economic calculation, and so on – came into existence slowly and many centuries before the transition to industrialisation. Capitalism in this sense coexisted with a world of feudal institutions and constraints: toll systems, entry constraints into industries and so on. This brings us to the second dimension of social change, which is more specifically related to institutions. Alexander Gerschenkron, whose work was briefly described above, emphasised the importance of 'substitutions', meaning institutions or processes – such as state-sponsored industrial banking – which substituted for the less directed processes which had driven British industrialisation. From a technological perspective, however, substitution was not an obviously available option, especially for areas such as Scandinavia. In facing British and later German competition, it was necessary at least to approach the technological level of the more advanced economies, and this in practice meant that it was necessary to acquire and transfer directly the technologies which were used within those economies: technology transfer became in key respects the core of the industrialisation process.

3. S. Pollard, *The Integration of the European Economy since 1815* (London, 1981), pp. 24–5.

One of the factors which made an extended process of technology diffusion or transfer possible was the fact that British industrialisation was heavily oriented towards innovation in process technologies of a rather generic kind. Steam power, spinning and weaving techniques, metal manufacture and engineering, were all areas in which the basic innovations concerned changes in processes which were potentially widely applicable. Some of the key technologies were developed by specialised capital goods producers, who had a clear economic incentive to diffuse the new techniques as much as possible, including internationally. In my earlier book, *British Technology and European Industrialisation*, I showed how British textile engineers in effect marketed complete 'packages' of technology – involving equipment, design, construction and operating skills, raw material supply and so on – which significantly eased the acquisition problems of emerging textile entrepreneurs outside Britain.

However that particular process of textile technology transfer was not generally applicable. Technologies are complex and highly differentiated, both among themselves and through time, and it follows that transfer and diffusion processes for technologies are likely to be highly heterogeneous. The papers which are collected in this book are intended to throw some light on this general problem of heterogeneity and complexity in foreign technology acquisition.

IV

Scandinavian economic and social historians have always accorded a strong place to technological factors in economic growth and the development of their societies, and the history of technology is a flourishing discipline in Scandinavia. Of course such things cannot be measured, but it is very likely that in relative terms the history of technology is stronger in Scandinavia than in any other part of the world. This does not mean that there is a uniform approach to the subject. If there is any connecting theme within Scandinavian historiography in this area it is probably that technological change is far more than an economic process: it involves, and is shaped by, social and cultural factors, and by political decisions. The papers collected here exemplify recognition of this broad point, but otherwise represent a varied set of studies in terms of time-periods, industries and technologies. This variety is, I hope, a strength for it points up the diversity and complexity of the transfer process. Not all of the attempts at technological diffusion discussed in the following pages were successful, but this too is a central element in the acquisition of general technological capability.

The papers are arranged broadly in chronological order, spanning the period from the late eighteenth to the early twentieth century. The structure is also a result of a specific analytical concern; it is meant to explore whether the process of technology transfer to Scandinavia in some sense reflects the changes in technology supply during the period. The supply of technology was altered firstly in the sense that it became the concern of a machine-producing sector separated from consumption goods producers, and secondly a matter of international competition. What impact, if any, did the institutionalisation of capital goods production, first in Britain, have on Scandinavian industrialisation and the transfer of technology? And are the subsequent changes in international technological leadership reflected in the case of Scandinavia? Can we observe an increased role of British machine-making enterprises after the emergence of a separate capital goods industry in Britain, followed by subsequent decline as German, and later American, competition challenged Britain's position as 'the workshop of the world'? This book is not meant to provide answers to these question, which must be seen as subsidiary, but it is meant to reveal changes over time in the process of technological diffusion, and to ask whether the Scandinavian experience can contribute to theories of technological change and development.

The first six chapters all deal with what might be called the early phase in Scandinavian industrialisation, until about 1820. These chapters show how central technological developments within a number of sectors diffused to Scandinavia – often as a result of joint efforts between individual entrepreneurs or artisans and the state. Transfers were effected by extensive search for information, often involving industrial espionage, and the use of foreign skilled workers. However, Svante Lindqvist shows how one of the attempts, namely the introduction of Newcomen steam technology to the Swedish iron industry, failed as a result of social and cultural factors which prevented successful adaptation of the technology to its new environment. The British influence on the Swedish iron industry resulced in successful transfer and adaptation of new methods and products, as seen in both Martin Fritz's and Rolv Adamson's studies which deal with slightly later periods. The adoption of advanced techniques was swift and involved extensive visits to Britain as well as the employment of British labour. Both in Sweden and the joint kingdom of Denmark and Norway, the state actively encouraged technology transfer. Rolv Petter Amdam's study of the transfer of British technology to the German-influenced Norwegian state-run glass industry illustrates the 'mercantilistic' ideas and efforts behind the process, and documents some of the problems foreign industrial spies faced when caught red-handed in Britain. Active encouragement by the authorities took different forms in the course of the nineteenth century, which is evident in Poul Strømstad's chapter on artisan travel and technology transfer to Denmark. The Scandinavian states did not confine financial support to developing the old, staple industries like iron and glass; the introduction of modern industrial textile production in Scandinavia rested, *inter alia*, on state-initiated industrial espionage and finance, demonstrated by Trine Parmer in her chapter on the establishment of the first mechanised textile factory in Norway in 1813.

Chapters 7 and 8 are more general. Fritz Hodne discusses the transfer of technologies to a number of Norwegian and Scandinavian sectors, integrating results from many of the studies presented in this book, in the light of diffusion models. This central chapter thus provides a synthesis, which sums up much of the Scandinavian experience of nineteenth-century technology transfer and analyses its nature. Likewise, Jörberg's chapter discusses general aspects of the application of foreign technology in the course of Swedish industrialisation, where he stresses the interplay between indigenous resources and foreign technology.

Chapters 9, 10, 11 and 12 are more specific and all deal with the mid-nineteenth century, the period in which the Scandinavian economies seriously started growing and the pace of diffusion increased. Ole Hyldtoft's study of the Danish brick and tile industry reveals a complex process of learning and technology transfer from many European areas, and argues that the British influence has been underestimated. My own paper explicitly relates the development of a Norwegian engineering sector and technological capability to the activities of the British capital goods producers seeking markets abroad. Claus Wohlert describes the introduction of the Bessemer process in Swedish steel making.

Chapter 12, by Håkon With Andersen, discusses the course of Norwegian shipping in relation to theoretical approaches to technology diffusion. Andersen stresses structural changes in ship ownership and management strategy, which determined the diffusion (or failure to diffuse) of advanced marine technology developed abroad; here, as with other contributions, diffusion is seen not just within its social context, but as actively shaped by that context.

11

Chapters 13 and 14 cover the later period, the late nineteenth and the early twentieth centuries, and throw some light on the relative positions of Britain and Germany in the capital goods export market. Gunnar Nerheim's paper on the diffusion of water turbines in Europe shows the diverging paths taken by British and Continental makers of power sources, where in particular German firms rise to predominance. It also refers particularly to Norwegian engineering capability in turbine construction, or the lack of it, and the entry of American firms as suppliers of electrical power generation. British and German presence is evident also in Eli Moen's study (chapter 14) where technology transfer is accorded a decisive role in the industrialisation of Norway's paper and pulp industry.

Chapter 15 is concerned with the development of a modern Norwegian wood-processing industry, namely the production of pulp, from a historiographical perspective. Even Lange's study can, for our purposes, also be seen as an instance not only of acquired technological capability but a reversal of the direction, or flow, of technology. It argues for the important role played by the supply side, in this case Norwegian entrepreneurs, in developing a new raw material, pulp, in the face of insignificant demand pressure in the 1870s. This entailed the successful creation of a market for pulp among British paper makers. We observe, thus, how inputs to modern production methods now, in the 1870s, move the opposite way, from Norway to Britain.

The general picture which emerges from the chapters below, however, is one of overall success in acquiring, adapting and implementing relatively advanced foreign technologies. Because these studies are focused on Scandinavia, however, there is one important element which cannot be considered but which is of interest for further research. This concerns the specific social conditions which enabled Scandinavia to import and develop these technologies. In the long run, Scandinavia succeeded reasonably well in developing an advanced technological capability, while similar 'marginal' economies – Ireland, for example, or some of the Mediterranean economies – clearly found things much more difficult. There are important comparative issues here which may be of interest not simply for historical studies, but for modern issues in the economics of development, or even for industrial strategies within the advanced economies. What were the factors which shaped receptivity to British, and later German and American, technologies? Why were some countries of peripheral Europe able to absorb these capabilities, and ultimately become some of the richest economies in the world, while others remained for over a century in relative poverty? The chapters presented here suggest that these large questions depend of course on the analysis of the acquisition and diffusion at firm level, but that this is something which is strongly shaped by social conditions of a complex character. The role of social factors as a central feature in shaping the rate and direction of technological advance is increasingly recognised in modern research, but important comparative challenges still lie ahead.

1
Social and Cultural Factors in Technology Transfer

Svante Lindqvist

1 Introduction

A general definition of 'technology transfer' is 'the movement of a technology from one country to another or from one industry or application to another'.[1] This definition covers three different phenomena, i.e. three different kinds of movement:

1. Established or new technologies from industrialised nations to developing nations.
2. New technologies from one industrialised nation to another.
3. New technologies from one industry or application to another within a nation.

The last of these phenomena is more often designated 'diffusion of innovation', and the term 'technology transfer' will be used here to describe the first two, namely the movement of technologies across national boundaries. Technology transfer has been a matter of interest to economists for a number of years, and many policy studies have been devoted to an appraisal of the effectiveness of efforts to stimulate economic growth by technology transfer. This reflects a growing awareness that the process of transfer is as

1. This article is an attempt to generalise and expand the conclusions in my book *Technology on Trial: The Introduction of Steam Power Technology into Sweden, 1715–1736*, Uppsala Studies in History of Science, no.1, Uppsala, 1984. Earlier versions of this article have been given as papers at the Program in the History and Philosophy of Science and Technology at Cornell University on 4 February 1987; at the Office for History of Science and Technology at the University of California, Berkeley, on 2 March 1987; and at the Department of History and Sociology of Science at the University of Pennsylvania on 23 March 1987. For valuable comments and suggestions on this article, I am grateful in particular to John Law.

15

important to technological development as – or even more important than – the actual innovation.

Most empirical studies are case studies of technology transfer in the vaguely defined period from, say, the end of the Second World War to the present day that we regard as our own time. Historical studies of technology transfer, on the other hand, have mainly attracted the interest of other historians, and much remains to be done to relate the findings of these historical studies to the work of the economists and policy makers of today. Some may object that historical studies are irrelevant to the urgent problems of economic growth and social change facing modern societies. They may argue that historical studies are of no practical value, since conditions in the past were different. But that is, in fact, just what makes them so useful.

Technology transfer across national boundaries often involves moving a well-established technology to a totally different environment. There are several variables – technical, economic, social, cultural, political, geographical etc. – that may influence this process. Modern studies of technology transfer take into account only a limited number of these factors. Many factors are taken for granted without discussion, even though they may be highly relevant to the case being studied. Historical studies of technology transfer have their empirical value in that they increase our knowledge of this wide range of factors and their relative significance. Such studies deepen our understanding of the complexities that are involved when a technology is moved across cultural boundaries. To take into account the findings of historical studies when considering the nature of technology transfer is thus merely to increase the number of parameters that are examined. The conclusions drawn from such a wide variety of empirical data will be of more universal value than those drawn from studies limited to our own time.

In the following pages I shall try to show the significance of social and cultural factors in the process of technology transfer. My case study may seem too far off in time, and even too particular to be of any relevance for policy makers of today. But even if the conditions in this case were distinctly different, that is in accordance with what I have said above, it may have some usefulness for illuminating parameters rarely considered.

2 The Introduction of the Newcomen Engine into Sweden

The first attempt to build a Newcomen engine in Sweden was made in the late 1720s, only some ten years after it had been successfully introduced in the collieries of Britain. This was the engine built by Mårten Triewald at the Dannemora Mines, and well known today from his book *Kort beskrifning* (A short description of the fire and air engine), first published in 1734.[2] This book is one of the few sources of information concerning the invention and early history of the Newcomen engine, and it has often been quoted by historians. The engraving of the Dannemora engine in Triewald's book has been reproduced almost as often as the wood-cuts in Agricola's *De re metallica*.

The initial stage in the process of technology transfer was a result of the institutional structure of the Swedish civil service rather than of individual initiatives. Basic knowledge about the new technology was brought to Sweden c.1720, only a few years after its introduction in the British mining industry, as a matter of routine by the Swedish civil service, i.e. by the regular industrial reconnaissance abroad by the Board of Mines, the government department which exercised ultimate authority over the mining industry.

Foreign entrepreneurs played a major role in the second stage of technology transfer. They competed for monopolies of the new technology at the front line of its diffusion from Britain to the Continent. In so doing, they pushed the front line further forward and assisted the spread of the Newcomen engine across Europe. Each tried to get the better of his rivals by claiming greater theoretical understanding of the underlying scientific principles. Thus the Swedish Board of Mines was led to believe that scientific knowledge was essential to anyone intending to build a Newcomen engine, but this was a subtlety deliberately introduced by the foreign entrepreneurs in a competitive situation.

But science had a social value in early eighteenth-century Sweden. It was an interest shared by the higher strata of society, and not yet thought of as a profession. If science was a prerequisite for building Newcomen engines, it followed that only gentlemen were able to build them. Craftsmen, such as the master builders who had

---

2. Mårten Triewald, *Kort beskrifning, om eld- och luft-machin wid Dannemora grufwor*, Stockholm, 1734. Also published as: *Mårten Triewald's Short Description of the Atmospheric Engine. Published at Stockholm, 1734. Translated from the Swedish with Foreword, Introduction and Notes*, Newcomen Society, Extra Publication No.1, London, 1928.

successfully built and operated Newcomen engines in Britain and on the Continent, belonged to a lower social group, and were thus not thought to possess the necessary scientific knowledge. High social status *and* a proven knowledge of science were the prerequisites for anyone who wanted to build a Newcomen engine in Sweden in the 1720s.

In 1726, Mårten Triewald returned to Sweden from a ten-year sojourn in Britain possessing these qualities. He was the son of a German blacksmith in Stockholm, and he had left Sweden in 1716 to seek his fortune abroad. In London he attended lectures on 'the new natural philosophy' – i.e. Newtonian physics – and was attracted into the outer orbit of the Royal Society. He became a faithful admirer of Newton, and was to remain an ardent Newtonian for the rest of his life. By chance Triewald was hired as supervisor of the first Newcomen engines to be put up in the Newcastle collieries. He worked in the Newcastle area as a mining engineer for some eight years, and during this time he acquired the practical knowledge of building and operating Newcomen engines. But his main ambition was to follow the examples of J. T. Desaguliers and other members of the Royal Society. After years of self-education he began to give public lectures in physics illustrated with experiments. He began as an unaffiliated entrepreneur who taught in rented rooms in Newcastle, but in 1725 he gave one or two well-advertised series of lectures in Edinburgh. He was able to attain a certain standing in Britain and became affluent enough to buy books and a comprehensive collection of scientific instruments.

When he returned to Sweden in 1726, he possessed some of the essential qualifications of anyone wanting to build a Newcomen engine in Sweden: practical know-how together with high social status and a proven knowledge of science. One prerequisite was lacking, however, namely the backing of a private group with the means of realising the project. A few months after his return, however, he succeeded in finding support for his idea from such a group, the Partners of the Dannemora Mines.

The Dannemora Mines are situated some 50 km north of Uppsala. The deposits had been worked since the fifteenth century, and probably much earlier. In the early eighteenth century some twenty mines in Dannemora were owned by seventeen surrounding ironworks, the Partners of the Dannemora Mines. The operation of the mines was highly decentralised. Each mine was owned by several ironworks jointly, and the ironworks held differing numbers of shares in the various mines. The ironworks were entitled to as many weeks' mining as they had shares in the mine concerned.

Each one had its own mine bailiff, who paid the workmen to dig ore during these weeks. The next mine bailiff then took over the work in the mine and dug ore for the duration of his employer's weekly shares. When all the Partners in the mine had worked the mine for a period corresponding to the number of their shares, the first mine bailiff began the cycle again.

The draining of water from the mines was, however, a common concern of the ironworks, and one that for long had been a major problem. The traditional sources of power – water, wind and muscle power – had by the turn of the century proved insufficient, and in 1709 one of the mines had to be abandoned and was subsequently flooded. The Partners were therefore positive to Triewald's suggestion to build an engine 'which by the power of fire' would keep all mines free of water. A contract was drawn up, and work on the engine began in 1727. It was first put into operation in 1728, but the engine was never able to operate continuously for sufficiently long to drain all the water from the mine. Triewald had left Dannemora as soon as the engine was completed, and entrusted its maintenance to his assistant, Olof Hultberg, a young student from Uppsala University. The Partners' irritation grew over the delay and increased cost, and in 1730 they began a heated correspondence with Triewald, which eventually led to a lawsuit that was to continue for the rest of Triewald's life (he died in 1747). The engine, however, had already been shut down for good in 1736.

Why did this attempt to build a Newcomen engine in Sweden fail? We shall discuss a number of factors responsible for the failure, and our purpose is to examine what was specific and what was of general validity in this case of technology transfer. The huge bronze cylinder of the Dannemora engine will serve as our mirror, and in its convex surface we may find reflected not only an image of early eighteenth-century Sweden but also some general characteristics of this phenomenon called technology. My thesis is that social and cultural factors were decisive for this 'failure' of technology transfer. These factors, however, are often neglected in the transfer of modern technologies to developing countries. The reason for this may be that they are impossible to express in the traditional quantitative terms of efficiency and economy, or perhaps even more because we do not tend to think of technology as a cultural and social phenomenon.

3 Technical Factors

Swedish mechanical technology during the eighteenth century was based mainly on wood. Strictly speaking, it was a composite wood-and-iron technology, as iron was used for shaft ends, bearings and brackets, but wood was still the main material for machine parts. Triewald had counted upon casting the pumps in iron, having been familiar with this practice in Britain, but no Swedish ironworks could manage this. Instead he had to build the pumps from bored-out elm logs in the traditional Swedish manner. However, the logs proved too weak for his original design, which was based upon the assumption that iron would be used, and this was a problem that took about a year to overcome. There was a difference in *technical expertise* between Britain and Sweden, and it necessitated a period of modification that both delayed and increased the cost of the project. That this is a general problem in technology transfer is supported by Nathan Rosenberg, who wrote in an article on the transfer of technology to America in 1800–1870:

> It required considerable technical expertise to borrow and exploit a complex foreign industrial technology. This should hardly be a surprising proposition, but it seems to be worth repeating it in view of the vast number of foreign aid and economic assistance programs in the years since the Second World War which have come to grief because of the absence of the appropriate skills in the receiving country.[3]

There was, however, another less obvious but more important technical factor. An indirect sign of this factor was the problem of maintaining a sufficient supply of water in the tank above the cylinder to condense the steam. Much effort was devoted to sealing the tank with sheet lead and mending the leaking valves. But the crucial factor was probably to be sought elsewhere. Triewald had not been content to build an engine of normal cylinder diameter when he transplanted the Newcomen engine to Sweden from its British matrix. He cast a 36-inch cylinder instead of the one of about 26 inches that was the average in Britain at the time when he left. He thereby increased not only the volume of the cylinder and thus the power of the engine by over 90 per cent, but also the

3. Nathan Rosenberg, 'Selection and Adaptation in the Transfer of Technology: Steam and Iron in America, 1800–1870', in *L'Acquisition des techniques par les pays non-initiateurs* (Colloques Internationaux du Centre National de la Recherche Scientifique, no.538), Paris, 1973, p.68. Also published in Nathan Rosenberg, *Perspectives on Technology*, Cambridge, 1976, pp.173–88.

volume of steam to be condensed. This meant that almost twice as much water was now required to condense the steam with the same rapidity. The attempts to seal the tank and the injection valve were probably nothing more than symptoms of a major defect in his original design.

But one cannot increase just *one* parameter of a technological system without being confronted with a series of difficulties. The Newcomen engine, with its many connected and interdependent components, constituted such a system. In Britain the average cylinder volume used in the engines grew gradually – almost organically – along with the dimensions of all the other components of the system: the boiler capacity, the depth and diameter of the mine pumps, the robustness of the automatic valve mechanism, the capacity of the pump supplying water to the tank, the tubes and fittings, the valves etc. The sizes of all these components were mutually dependent, and they had to be chosen accordingly. This could be done only by the slow process of trial and error, step by step. The growth in the volume (or diameter) of the average Newcomen engine cylinder therefore shows the same exponential character as the growth of any critical parameter in any technological system.

When Triewald quite drastically increased the cylinder diameter to 36 inches, he strained all the other components of the system to the limit. That is why he ran up against problems such as the inadequacy of the water supply for condensing the steam. This was probably also the problem underlying the engine's frequent breakdown, even when it seemed to be due to other causes. It was often out of service during its first few years, and to overcome these problems a longer period of modification would have been needed – a technical development that would have taken time and money. Triewald's design was inspired by his ambition to build not only the first Newcomen engine in Sweden, but also the biggest ever. In a letter to his former teacher in Britain, the natural philosopher John Theophilus Desaguliers, he wrote in 1728: 'I have erected the first & largest Fire Engine for drawing Water & Oare in this kingdom, the Cylinder being 2 lines more than 36 Inches in Diameter.'[4]

Even a sixth of an inch was worth mentioning when Triewald wanted to boast of his achievement to his friends in the Royal Society! The reasons for the operational unreliability of the

4. Royal Society, London: letter from Mårten Triewald to John T. Desaguliers dated 20 November 1728.

Dannemora engine are therefore ultimately to be found in Triewald's personal ambition. This in turn may stem from his background: the blacksmith's son who had emigrated to Britain for better or for worse wanted to show when he returned that he was second to nobody; the colliery engineer in Newcastle, who had worked for many years in a subordinate position, wanted to show that he could now do better than anyone in Britain. The crucial technical factor in this failure of technology transfer was thus due to the personal ambition and social mobility of an individual in a specific historical context.

There is a parallel between the transfer of technology from Britain to Sweden in the 1720s and the transfer of technology from industrialised to developing countries today. The level of technological development in a nation has great symbolic importance: it is an expression of the nation's affluence, military strength and culture in general. A developing country will therefore often acquire the most modern technology possible, and regard as patronising any idea of introducing an older technology that is less than modern in the donor country. The recipients are not willing to put up with crumbs from the rich man's table. Also, there are usually strong reasons in the donor country for exporting the most modern technology. This is because it is principally the largest companies – with a heavy investment in modern technology – that are in a position to sell on distant and difficult markets. These companies make their profits from modern technology and have no interest in a return to earlier, obsolete technology. The attitude of the receiving country and the economic interests in the donor country coincide, and the result is that the technology introduced into developing countries is often far too sophisticated.

In addition, there are the usual problems of adaptation and modification, and in consequence the projects often fail. On the other hand, there are examples of successful technology transfer when the technology exported is one that has already been partially replaced by a new, more advanced technology in the donor country. The exported technology is only 'unmodern' in a relative sense, and its great advantage is that it has been well tried and tested. It has been used for a long time, and its teething troubles have long since been overcome. There is also ample knowledge about the way to use it, the most common problems arising and the solutions to these problems.

The complex relationship between the donor country and the receiving country was in the case of the Dannemora engine embodied in one person: Mårten Triewald. His reasons for choosing an advanced level of technology was, as earlier suggested, his personal ambition, which was in turn to be explained by his changing social position in early eighteenth-century Britain and Sweden. If Triewald had been content to build a Newcomen engine of average size it would probably have worked much better. In fact the best thing he could have done would have been to dismantle an old engine that he himself had helped to erect and operate for several years, and then bring it over and re-erect it at the Dannemora Mines. Even this would have called for modification in a new environment, but it would not have resulted in so many new technical problems if the level of the technology had been the same as, or preferably below, the average in the donor country. *The tendency to export far too advanced a technology* (in comparison with the general level in the *donor* country) may thus be regarded as a mistake commonly made in technology transfer.

4   Geographical Factors

The severity of the winter in Sweden by comparison with that in Britain upset the original schedule for the construction of the Dannemora engine. As construction work could be undertaken only during the summer half of the year, the building of the engine fell almost a year behind time. Once it was completed, the engine was out of action for long periods when necessary repairs were prevented by ice in the mine or frozen pumps. The delays increased the Partners' irritation, and the *difference in climate* thus contributed to the failure.

The major geographical difference was in *natural resources*. In Britain, Newcomen engines were coal-fired, but Sweden was to all intents and purposes devoid of fossil fuel. To run a Newcomen engine on wood required considerable quantities, because the heat content of wood is far less than that of coal. Wood was also a natural resource that was at a premium in the mining districts, where it had many important uses: fire setting, ore calcination, in blast furnaces and forges etc. It was, however, not an actual unavailability of wooden fuel that was the direct cause of the engine's failure. The problem lay rather in the fuel economics, which will be discussed below.

5 Economic Factors

Triewald had put the cost of building the engine at 24,000 copper-daler, but the cost of adapting the technology to the new environment raised the total cost to over 50,000. Despite this outlay, the engine still did not provide continuous operation. The Partners were reluctant to invest more in the project, but it is possible that the engine could have been used successfully to drain water, if only more time and money had been spent on development. However, this project was financed on a very different basis from later schemes to introduce the Newcomen engine into Sweden. The next such attempt was not made until the Swedish Ironmasters' Association, a private credit institution for the mining industry established in 1747, agreed to underwrite them. The costs were thus not borne by the individual ironworks, which would not have ventured for a second time to participate in these experimental projects without a financial guarantee. As all transfers of new technologies give rise to additional costs for adaptation to a new environment, the absence of *external financial backing in the initial stage of technology transfer* appears to have been a contributory cause of the failure of the Dannemora engine.

But the crucial question is of course whether the engine would have been deemed economic by comparison with the traditional technologies. According to Triewald's original estimate of the fuel consumption, the engine would consume only 300 cords of wood fuel per year. Had this been correct, the engine's running costs would have been less than 50 per cent of the costs for the traditional sources of power. By 1730 it was, however, evident that the engine consumed about 9 cords of wood per day. This meant that the engine's running costs *exceeded* the existing costs by 50 per cent. Triewald had underestimated the fuel requirements by a factor of 10 when inferring the Newcomen engine's consumption of wood fuel from its consumption of coal. The Newcomen engine was economic in the context in which it had been developed, i.e. the British coal industry, but it was, as Graham J. Hollister-Short has written, so symbiotically linked to the mining of coal that 'it could not in any significant or lasting way break clear of the technological matrix in which it had first come to maturity'.[5] The basic economic reason for the failure of the Dannemora engine was thus *the difficulty of reproducing the cost-effectiveness of the technology in a new environment*.

5. Graham J. Hollister-Short, 'A New Technology and its Diffusion: Steam Engine Construction in Europe 1720–c.1780', *Industrial Archaeology*, vol.13, 1978, p.122.

6 Social Factors

The Newcomen engine was a centralised prime mover in that it produced high power and could replace a large number of the traditional, smaller power units in the mining industry such as horse-whims, windlasses and water wheels. But since work in the Dannemora Mines was decentralised, in other words production was organised on a system of weekly shares in the different mines, the adaptation of the new technology to the existing social order presented a major problem. If the engine was unreliable as far as continuity of operation was concerned, it might unfairly favour or penalise the different ironworks by keeping the mines free from water in some weeks but failing to do so in others. It thus threatened to undermine the whole system that formed the basis of the social order.

One of the disadvantages of the early Newcomen engines lay not in their technical design but in the scarcity of experienced persons to keep them in service. In a paper entitled 'The Transfer of Power and Metallurgical Technology to the United States, 1800–1880', Brooke Hindle stressed the role of immigrant mechanics in the initial transfer of steam engine technology.[6] Their importance in general is confirmed by the case of the Dannemora engine, and to this we may add a social dimension.

There were only two men in Sweden who could keep the engine working: Triewald and his assistant Olof Hultberg. The reasons for Triewald's reluctance to spend his time in Dannemora will be reviewed below, but in addition Hultberg left Dannemora at the first possible opportunity. Why did Hultberg, the 'immigrant mechanic', not remain in Dannemora? The answer lies in his status, or lack of it, at the Dannemora Mines. He was an outsider, a student at Uppsala University, who had suddenly been entrusted with a responsible post in a traditional mining community. His position in relation to the head mine bailiff and the mine bailiffs of the Partners was ambiguous. (It was as if an undergraduate engineering student were made responsible for the maintainance of a Pittsburgh rolling mill with no clearly defined authority to match his new responsibility.) But Hultberg had to work and live within the mining community, and he was dependent on the head mine

6. Brooke Hindle, 'The Transfer of Power and Metallurgical Technologies to the United States, 1800–1880: Processes of Transfer, with Special Reference to the Role of the Mechanics', in *L'Acquisition des techniques*, pp.407–28. Cf. David J. Jeremy, *Transatlantic Industrial Revolution: The Diffusion of Textile Technologies between Britain and America, 1790–1830* Cambridge, Mass., 1981, esp. pp.254–7.

bailiff and the rest of the community for wood fuel, materials and labour for the engine, as well as for his salary, food and housing. They for their part were also dependent upon him, for his efforts to keep the engine in regular service affected the work of all who hoped to be able to exploit the mines drained by the steam engine. It must have been obvious to them that the reliability of the engine would determine how much ore each ironworks could mine during its allotted weeks, and they held Hultberg responsible for the engine's failings.

Hultberg had been appointed *Konstmästare*, a title usually given to a person responsible for all the mechanical contrivances used for draining water and hoisting ore from a mine. But Hultberg was *Konstmästare* 'at the engine', which meant that his authority extended only to the engine and its pumps. However, the consequences of the engine's performance reached beyond the stone walls of the engine house and the wooden pump pipes. This led naturally to friction between Hultberg and the mine bailiffs. Olof Hultberg had to represent the entire social basis of the new technology when Triewald was away, but he had not been properly integrated in the social order in terms of authority and prestige. The only position in the hierarchy of the Dannemora Mines that Hultberg could occupy, as the engineer responsible for a centralised power system, was below the head mine bailiff, but above the mine bailiffs of the various ironworks. The head mine bailiff must have realised this, because he told Hultberg on one occasion in 1729 that he would 'have him dance to a different tune and teach him better ways' once he came under the authority of the Dannemora Mines.

It is understandable that Hultberg preferred to leave Dannemora for what was sure to be a considerably more peaceful existence as a maker of scientific instruments in Uppsala. After he had left, the engine remained silent: all the hardware was there, but there was nobody to operate it. The case of the Dannemora engine illustrates the fact that the hardware components are only a part of a technology, and thus only a part of a successful technology transfer. Practical know-how, experience and skill must also be transferred, and these are all human qualities. The engine may have worked by vacuum, but its engineer could not work in a social vacuum. Thus this case shows the importance of *integration of immigrant technicians in the existing social order*. This is not only necessary for successful operation in the initial stage of technology transfer, but a precondition for the diffusion of knowledge into the new environment. The engine house that still stands in Dannemora is a silent reminder that technology is more than hardware, more than precision-bored cylinders, pressure-resistant boilers and so on. It is also a matter of human beings.

The problems of the Dannemora engine, and their contribution to this failure in technology transfer, discussed so far may be summarised by a quotation from the paper by Nathan Rosenberg mentioned earlier:

> New techniques frequently require considerable *modification* before they can function successfully in a new environment. This process of modification often involves a high order of skill and ability, which is typically underestimated or ignored. Yet the *capacity* to achieve these modifications and adaptations is critical to the successful transfer of a technology – a transfer which is too frequently thought of as merely a matter of transporting a piece of hardware from one location to another.[7]

Apparently neither Triewald nor the Partners realised what this need for modifications demanded of them. If the Partners had realised it, they would have made sure that Hultberg enjoyed the authority and prestige appropriate to his responsibilities. They would also have been more understanding about the delays and increased costs that the modifications entailed. They would have been wise to apply the old engineering rule, 'Multiply all estimated costs and time schedules by π.' But after a few years of unreliable operation they were probably not very interested in investing more time and money in attempts to make the engine run properly. As an alternative to the traditional methods of draining the mines, the engine would at best be equally expensive, but besides this there was the specific problem of its unreliability as a centralised power system, and the effects of this on the decentralised system of working the mines. If Triewald, on the other hand, had realised the need for modifications, he would have spent more time at Dannemora trying to make his engine run properly than he did. Triewald's absence from Dannemora once the engine had been built calls for an explanation, and in order to provide this we have to examine a further set of critical factors in technology transfer.

7   Cultural Factors

In his paper 'Technology Assessment from the Stance of a Medieval Historian', written in 1970, Lynn White Jr. showed some

---

7. Rosenberg, 'Selection and Adaptation', pp.55ff.

of the different levels of complexity that may be found in the assessment of new technologies.[8] He concluded: 'My thesis is that technology assessment, if it is not to be dangerously misleading, must be based as much, if not more, on careful discussion of the imponderables in a total situation as upon the measurable elements. Systems analysis must become cultural analysis, and in this historians may be helpful.'[9] So far we have been able to distinguish a number of measurable elements: *technical* (technical expertise, the level of the technology), *geographical* (climate, natural resources), *economic* (external financial backing, reproducing cost-effectiveness), and *social* (adaptation to the existing social order, integration of the immigrant technicians). Finally we must take into account the imponderables in a specific cultural situation.

Triewald asserted that he had introduced 'this glorious invention' to his native country by building the Newcomen engine at the Dannemora Mines. This was an expression of the utilitarian ideals that came to dominate Sweden during the eighteenth century, but which did not make their breakthrough as a politico-economic programme until after the change of regime in 1738–9. In Britain the conception of science and technology as a public utility had long been institutionalised in the Royal Society: this was a line of thought that could be traced back to Francis Bacon. It was from the circle around the Royal Society that Triewald had derived the inspiration for his work, and his values had been shaped during his years in Britain. When Triewald claimed that he had built the engine in Dannemora for the 'public utility', this was an early and private attempt to apply utilitarian ideals as a politico-economic programme. This led to something of a cultural conflict in Dannemora, for the Partners' values had been moulded by a harder reality: the technical and economic situation in a traditional Swedish mining society. When Triewald spoke of the 'utility' of his engine, he used the term in a broader and more general sense than the Partners. Throughout the project the parties applied different criteria of 'utility', and this brought them into conflict. In court, the Partners argued that their criteria should apply, namely that the literal wording of the contract should be observed, and here Triewald's more vaguely formulated utilitarianism came off second best.

But Triewald also had a more personal ambition. He wanted to

---

8. Lynn White Jr., 'Technology Assessment from the Stance of a Medieval Historian', in idem, *Medieval Religion and Technology*, Berkeley, 1978, pp.261–77, originally published in *American Historical Review*, vol.79, 1974, pp.1–13.

9. Ibid., p.276.

be recognised by the Royal Society, and he succeeded in this by being elected a Fellow in 1731 and having several of his letters published in the *Philosophical Transactions* in the 1730s. The Dannemora engine was only one route to his goal, but he omitted neither to mention it in his first letter to the Royal Society in 1728 nor to send a copy of his book on the Dannemora engine to the president of the society in 1734.

The role of technology in realising his ambitions is illustrated by a closer look at the dates of his letters to the Royal Society after his return to Sweden and before the founding of the Royal Swedish Academy of Sciences in 1739. During the period 1728–38 he wrote ten letters to the Royal Society,[10] and eight of these were written on Saturdays, or on the Sabbath or other holidays. A natural rhythm was superimposed on the division of his life between science and technology by the alternation of working week and weekend. Technology was his breadwinner, and in the long term a way to improve his social position; for the time being science was something to which he could devote himself only when his weekly work as an engineer was over.

The social dimension of science and technology, set in a specific historical context, goes some way towards explaining Triewald's lack of interest in keeping his engine in operation once it had 'demonstrated its great effect'. His main ambition was to realise his British ideals at a time when science was not yet a profession but an interest shared by the higher strata of society. It was through technology that he succeeded in acquiring the position he sought. When he came back to Sweden in 1726 he was referred to as *Mechanicus*, a general title applied to anyone who had studied technology. But at Dannemora and elsewhere he demonstrated the knowledge he had acquired in England, and in 1727 he was appointed *Directeur* of mechanics in the Board of Mines. With his many books, among them the one on the Dannemora engine and especially the two volumes of his lectures on natural philosophy, he showed that he was able to impart this knowledge to others, and in 1735 he was made Head of Education in the Fortifications Corps. As *Capitaine-Mechanicus* he had a well-paid position with high social status that gave him the opportunity to pursue the role of gentleman scientist, and by the end of the 1730s he had succeeded in attaining the goals formulated during his years in England. He lived in an elegant house that he had built in Stockholm, and on the

10. Full references to these letters are found in Lindqvist, *Technology on Trial*, p.366, n.18.

edge of the city he had a summer house with large grounds. In his garden he cultivated exotic fruits and experimented with crops found in Sweden. He was a member and indeed founder of an academy of sciences enjoying royal patronage in the capital city. In its proceedings he wrote in his native language on all kinds of experiments and observations that might be of 'public utility'. The erstwhile Newcastle colliery engineer had come a long way.

There is a lesson to be learned here: what is a technological 'failure'? We tend to measure it in the traditional terms of efficiency and economy. The introduction of the Newcomen engine in the Dannemora Mines was a failure, but then only in the simple sense that the objectives of the Partners were not realised. The introduction was successful as far as Triewald was concerned because it promoted his own ambitions. We, too, at times may be too quick to define contemporary attempts to introduce modern technologies to developing countries as 'failures'. They may be failures as far as efficiency and economy are concerned, if sufficiently narrowly defined. But they may serve other purposes in the recipient country, for example in terms of national prestige and personal ambitions, or perhaps because of secondary effects that will be of greater and more long-lasting value, such as international contacts, technical education of the domestic work force, the establishment of an infrastructure or the introduction of various supportive technologies. For the *donor* nation even a 'failure' in technology transfer may be useful in terms of access to raw materials, a new export market, or political and military-strategic gains in developing friendly relations with a developing country.[11] However, none of these qualities can be expressed in terms of efficiency and economy, and they will not show up in a cost-benefit analysis of an isolated case of technology transfer.

Finally we shall touch on yet another cultural factor that may have contributed to this failure in technology transfer, namely culturally conditioned ideas on the nature of technological development. Mårten Triewald did not employ a systematic method in his scientific work. Scientific discoveries were to him just that – 'discoveries' in the strict sense of the word, connections that were suddenly revealed to the attentive observer. There are several examples of this in his letters to the Royal Society. A new explanation of aurora borealis dawned upon him when he was travelling

---

11. Melvin Kranzberg, 'The Technical Elements in International Technology Transfer: Historical Perspectives', in John R. McIntyre and Daniel S. Papp, *The Political Economy of International Technology Transfer*, New York, 1986, pp.31–45.

on a sledge across the ice one winter evening and saw the northern lights glowing as the moon rose.[12] Another letter to the Royal Society, published in the *Philosophical Transactions* under the title 'An Extraordinary Instance of the Almost Instantaneous Freezing of Water', was the description of how he entered his lecture theatre at the Palace of Nobility in Stockholm on a winter's day and happened to see the (supercooled) water in a glass tube containing Cartesian divers suddenly turn to ice.[13]

Triewald himself formulated his method in a discussion in the newly established Royal Swedish Academy of Sciences in the autumn of 1739. At one meeting a Fellow had presented a divining rod, with the aid of which he believed it was possible to detect metals. Triewald's friend and co-founder of the Academy, the botanist Linnaeus, was sceptical, but another Fellow thought that the Academy ought not to reject uncertain observations, for in many cases they had turned out to lead to important discoveries. This view was supported by Triewald, who, according to the record, said, 'Yes, indeed, most things in the world have been discovered more by chance than a priori.'[14] This opinion was not the expression of a superficial approach to science, but of a deep religious conviction. Like many of his contemporaries, Triewald was influenced by physicotheology.[15] The English naturalist

12. British Museum, London: Sloane MSS, 4,053, pp.293–96. Letter from Mårten Triewald to Hans Sloane dated 25 July 1734. Copy in Royal Society, London, Letter Book Copy, vol.21, pp.269–75.
13. British Museum, London: Sloane MSS, 4,025, pp.293ff. Letter from Mårten Triewald to Hans Sloane. Copy in Royal Society, London, Register Book Copy, vol.16. Mårten Triewald, 'An Extraordinary Instance of the Almost Instantaneous Freezing of Water; and giving an Account of Tulips, and such Bulbous Plants, Flowering much Sooner when their Bulbs are Placed upon Bottles Filled with Warm Water, than when Planted in the Ground', *Philosophical Transactions*, vol.37, 1731–2, pp.79–81.
14. E. W. Dahlgren (ed.), *Svenska Vetenskapsakademiens protokoll för åren 1739, 1740 och 1741*, Stockholm, 1918, vol.1, p.45.
15. Bengt Hildebrand, *Kungl. Svenska Vetenskapsakademien: Förhistoria, grundläggning och första organisation*, Stockholm, 1939, pp.208, 304ff., 322ff. Hildebrand remarks that Triewald, Linnaeus and Anders Johan von Höpken – the three foremost of the founders of the Royal Swedish Academy of Sciences – shared exactly the same physicotheological belief. The Academy may in this perspective be seen as an institutionalisation of their religious belief. For the relationship between religion and economy in eighteenth-century Sweden, see Tore Frängsmyr, 'Den gudomliga ekonomin: Religion och hushållning i 1700-talets Sverige', *Lychnos*, 1971–2, pp.217–44 (English Summary, pp.243ff.). For physicotheology in general, see A. D. Atkinson, 'William Derham, F. R. S. (1675–1735)', *Annals of Science*, vol.8, 1952, pp.368–92; Clarence J. Glacken, *Traces on the Rhodian Shore: Nature and Culture in Western Thought from Ancient Times to the*

William Derham was one of the foremost proponents of this philosophy, which sought to show the existence of God through the purpose and beauty of Nature. The catalogue of the library left by Triewald includes several of Derham's works, in English, German and Swedish editions. It is reasonable to assume that Triewald had bought and been influenced by Derham's *Physico-Theology* (1713) while in England, because he quotes it several times in his published lectures of 1735–6, which in all other respects reflect what he had learned in England.[16]

At the time of his death he also possessed at least four different works by Christian von Wolff, all dating from before his return to Sweden in 1726. Among these was Wolff's *Vernünfftige Gedancken von Gott, der Welt und der Seele des Menschens* (Rational thoughts on God, the world and the human soul), and he seems likely to have owned at least this one in the early 1730s, as he quotes from it in 1734 (see below). Moreover it was in about 1730 that Wolff's ideas began to gain followers in Sweden.[17] However, Tore Frängsmyr has pointed out that there were two kinds of Wolffianism, since Wolff himself 'made a distinction between learned men and common people. In his Latin work the physico-theology was not to be found at all; he rejected its value before strict scientific readers, but he accepted it for a wider public.'[18] This intellectual double-dealing by Wolff was a conscious social stratification of his philosophical arguments. Triewald's disdain for the academic world and his campaign against Latin as a language for 'the learned' is well known. It is not surprising therefore that Triewald, the son of a German blacksmith, came into contact with Wolff's philosophy in its more popular form, i.e. through books in which Wolff put forward physicotheological arguments in German for the common people. (All the books by Wolff in Triewald's library were in German.)

Triewald's physicotheological views had an important bearing on his views on technological development. Mechanical inventions

End of the Eighteenth Century, Berkeley, 1976, pp.375–428, 504–550; Wolfgang Philipp, 'Physicotheology in the Age of the Enlightenment: Appearance and History', *Studies on Voltaire and the Eighteenth Century*, vol.57, 1967, pp.1,233–67.

16. For references to Derham, see: Mårten Triewald, *Föreläsningar öfwer nya naturkunnigheten*, Stockholm, 1735–6, vol.1, pp.38, 61, 63, 64, 106, 196, 198; vol.2, 'Oration', p.85, note x.

17. Tore Frängsmyr, *Wolffianismens genombrott i Uppsala: Frihetstidens universitetsfilosofi till 1700-talets mitt* (Acta Universitatis Upsaliensis, no. c.26), Uppsala, 1972, with an English summary, 'The Emergence of Wolffianism at Uppsala'.

18. Tore Frängsmyr, 'Christian Wolff's Mathematical Method and its Impact on the Eighteenth Century', *Journal of the History of Ideas*, vol.36, 1975, p.666.

were not 'inventions' to Triewald in the true sense of the word: creations of something that did not exist before. They were more like discoveries: findings of something that had existed before but had remained unknown. 'To make an invention' was just another way of revealing God's ingenuity in creating this world. God was in a sense the true inventor of all mechanical constructions, and man a mere 'discoverer'. Triewald expressed this view in the first chapter of his book *Konsten at lefwa under watn* (The art of living under water), in 1734:

> The Great God is the only and true originator of all things; whereas we wretched beings can but now and again become aware of a part of the infinite wisdom in God's creation; and turn the same to our use when we put our minds to a particular end, having previously made ourselves fully aware of all the laws of nature, which the Lord of Nature has so laid down that they can never be upset or changed by us; so we cannot credit ourselves with any more, when we discover what is useful, than having after contemplation come upon the way in which God's work has been revealed to us.[19]

For this passage, Triewald quoted a passage from Wolff's *Vernüfftige Gedancken von Gott* in a footnote, and the illustration at the top of the page showed a symbolic sun, containing a motif signifying God. Wolfgang Philip has written that this was the symbol favoured above all others in the physicotheological books of the Enlightenment: 'Its rays fill nature. They wrap men who "contemplate" the objects of nature and become "enlightened" in this contemplation.'[20] The continuation on the next page of Triewald's book makes it clear that these 'discoveries' also included mechanical inventions. He states that all new inventions 'rebound to the glory of God and the utility of mankind', and that the inventors 'have by their discoveries brought forth the glory of God, and thus remark-ably served the human race, to alleviate the troubles that exist in this wretched life'.[21] A similar passage appears in the first volume of his lectures, published the following year.[22] God has, Triewald writes, rightfully imposed the punishment for man's sins saying 'in

---

19. Mårten Triewald, *Konsten at lefwa under watn eller en kort beskrifning om de påfunder, machiner och redskap hwarpå dykeri – och bärgnings-societetens privilegier äro grundade, hwarmed de anstält profwen under twenne riks-dagar för Sweriges rikes högl. ständers herrar deputerade*, Stockholm, 1734, p.1.
20. Philipp, 'Physicotheology', p.1,266.
21. Triewald, *Konsten*, p.2.
22. Triewald, *Föreläsningar*, p.285n.

the sweat of thy face shalt thou eat bread'. But he has also given man several means to facilitate this work, as well as rationality to 'invent' them and the ability to apply them through human muscle power or other forces of nature.

Triewald's view of the act of invention is expressed clearly in his well-known story of how Thomas Newcomen invented the steam engine. This appears in his book *Kort beskrifning*, which was published in the same year as his book on the art of diving and thus written in the same spirit:

> For ten consecutive years Mr Newcomen worked at his fire-engine which never would have exhibited the desired effect, *unless Almighty God had caused a lucky incident to take place.* [Triewald then reports the story of how Newcomen discovered the method of condensing steam by the injection of cold water. He continues:] Though somebody might think that this was an accident, *I for my part find it impossible to believe otherwise than that what happened was caused by a special act of Providence.* To this conclusion I – who knew personally the first inventors – have been brought more than ever when considering that *the Almighty then presented mankind with one of the most wonderful inventions which has ever been brought into the light of day.*[23]

The italics are mine, but they could just as easily have been Triewald's. This passage has often been quoted by historians, as it is the only account of the moment of invention of the steam engine. Unfortunately it is doubtful whether we can regard this as a historical source in the sense that it tells us anything of what actually took place in that historic Dartmouth workshop some time during the first years of the eighteenth century. It would be more correct to regard it as simply another illustration of Triewald's religious beliefs.

This physicotheological view of the act of invention ('technotheology' we might call it) also implied that the existence of the steam engine gave us knowledge of God. At a time when the whole mining industry of England was threatened by the problem of flooding, God had in his wisdom revealed to mankind the steam engine to enable the riches and bounty which he had laid down in the earth to be brought up. Like the wonderful and wise arrangements of nature, the steam engine bore witness to its divine origins and to the fact that everything was arranged in the most wonderful way in this best of all worlds. There are some celebrated lines by

---

23. Triewald, *Kort beskrifning*, pp.2ff.

Linnaeus, expressing his belief in the miracle of creation as a proof of the benevolence and omnipotence of God. I saw, Linnaeus wrote, 'the eternal, all-knowing, all-powerful God from the back when he advanced, and I became giddy! I tracked his footsteps over nature's fields and found in each one, even in those I could scarcely make out, an endless wisdom and power, an unsearchable perfection.'[24] Triewald had witnessed the impressive power of the Newcomen engine and its usefulness to the mining industry of England. When he saw these engines, 'which by the power of fire and air drain an incredible amount of water from the deepest shafts', he became giddy! He saw in their mechanical design an endless wisdom and power, an unsearchable perfection. This explains Triewald's indignation at the Partners' complaint to the Board of Mines. The Partners had not only impugned his good faith ('Laudanda Voluntas') when he wanted to introduce the Newcomen engine into Sweden 'for the public good', they had done something far worse. The Newcomen engine was, to Triewald, like natural science to Linnaeus, a proof of an almighty and benevolent God. How dared they question this? Had he not revealed to them the glory of God in Dannemora?

There is a lesson to be learned here, too, as well as in the previous discussion on technological 'failures'. It is not the obvious fact that an early eighteenth-century view on the nature of technological innovation happened to be influenced by a contemporary philosophical tradition. Rather, that something we take for a fact – i.e. that inventions are true creations by individuals – can be a sentiment, and as such culturally dependent and thus subject to changes over time. This conclusion recoils upon ourselves, and then in the sense that follows from this we must accept the notion that our own views on the nature of technology may be just that, i.e. views. We have no logical ground for assuming that Triewald's ideas on the nature of technological innovation were sentiments whereas our own should be facts. Our views on the nature of technology may be as relative as his, and they too may depend on our own specific cultural situation. That major political decisions of today on science policy and education might be based upon ideas on the nature of technology that are as capricious as Triewald's is a distressing thought. This should, at least, make us more humble when we encounter different views on the nature of technology in

24. Quoted from the English translation in Sten Lindroth, 'The Two Faces of Linnaeus', in Tore Frängsmyr (ed.), *Linnaeus: The Man and His Work*, Berkeley, 1983, p.12.

other cultures, and these differences may be the most fundamental and yet imperceptible obstacles in the process of technology transfer.

Summary of Factors

Factors influencing the process of technology transfer, and the specific subset revealed in this case study of the failure to introduce the Newcomen engine into Sweden, 1726–36

1. Technical factors
   (a) The available technical expertise in the recipient country
   (b) The tendency to export far too advanced a technology (in comparison with the general level in the *donor* country)
2. Geographical factors
   (a) The difference in climate
   (b) The difference in natural resources
3. Economic factors
   (a) External financial backing in the initial stage of technology transfer
   (b) The difficulty of reproducing the cost-effectiveness of the technology in a new environment
4. Social factors
   (a) The adaptation of the new technology to the existing social order
   (b) The integration of immigrant technicians in the existing social order
5. Cultural factors
   (a) Reasons for the transfer that cannot be measured in the traditional terms of efficiency and economics of the new technology
   (b) Culturally conditioned ideas of the nature of technological development

2

How Industrial Technology First Came to Norway

Trine Parmer

1 Introduction

The introduction of new technologies and the modernisation of traditional crafts and trades first began in Britain. From the middle of the eighteenth century the British economy experienced unprecedented growth. The textile industry was the first sector to introduce machinery in the many interrelated production processes in yarn and cloth manufacture. New industrial technology was adopted first of all in cotton, and modernised cotton production efficiently and brutally out-competed the widespread cottage industry in this field.

It did not take many years before the new industry, via different channels, reached the Continent. From small centres it subsequently spread from country to country and continually conquered new markets. It also reached Norway, and when and how this happened is the main theme of this chapter.

Until around 1845–50, we can characterise Norway as an industrially underdeveloped country. Even so we find a small number of factories in operation before this date which stand out as notable exceptions among the more traditional enterprises spread through Norway. These enterprises were engaged precisely in cotton production, and operated with a view to expanding the market for cotton goods in Norway. As was the case in Britain, yarn production was the first to be mechanised. In this chapter we shall focus on the earliest of these factories, Mads Wiel's Cotton Factory, which was established in 1813–15, and we shall examine more closely how it was possible to start modern industrial production at that time.

Seeing that we are concerned with the first instance of foreign textile technology being transferred to Norway, we shall begin with a short presentation of the technical equipment which was acquired by Mads Wiel. After that we shall go back in time to trace the underlying causes for the migration of the new technology to Norway. Our main emphasis is on the latter. Finally, we shall take a closer look at Mads Wiel, the man behind the new industrial venture, and try to explain the aims and motives behind his pioneering role.

2 Mads Wiel's Cotton Factory

In the spring of 1813, building started on a new factory near Fredrikshald (Halden today), in the Tistedalen area on the southeasternmost coast of Norway. It was to be a cotton factory with spinning as the most important production process. The location of the factory was chosen with care – just by a suitable waterfall of the Tiste. A concern of this kind was no exception in itself, but certain features of the factory were novel.

The man behind the new venture, Mads Wiel, was a merchant and lumber trader in Fredrikshald. In addition, he ran a tobacco plant and a rope factory. Wiel was an exceptionally active man, and in 1812 he decided to set up a cotton factory for both yarn and cloth production. In Wiel's time a factory could not be built in the twinkling of an eye. The war in 1814 caused some delays, and it was not until 1815 that the building was completed. By then also the inventory was ready to be installed.

In a fire insurance evaluation from 1817, we find the following description of the building:

> 762B, cotton factory A building of 3 floors constructed of half-timber and boards and painted, roughcasted and painted on the inside, foundations on granite rock with 4 stone pillars, 29 ells [old Norwegian measurement = 2 feet] long, 13 ells in width. Ground floor: 1 large room with $11\frac{1}{2}$ windowbays also 1 smaller room with $5\frac{1}{2}$ windowbays, both rooms having 2 brick heating stoves. On the 1st floor, a large room with 12 windowbays, and a chamber with 7 windowbays and brick stoves in each room. Second floor: large attic in the middle, 2 side rooms with two brick heating stoves, 8 windowbays. Roof in sections and tiled, and two water pumps . . . 5,000 specie daler.[1]

The building was, thus, 18.2 m long and 8.15 m wide. In a

---

1. Fire Insurance Records A, Halden, Litra B, fos. 164–6, State Archives, Oslo.

valuation of 1846 we get the additional information that its height was 7.5 m. It was a small factory by our standards, but even so large enough for the purpose. The largest room on the ground floor was about 107 m², and the smallest approximately 31 m². This gave a floor space of 139 m²; for all three floors it was 414 m². Unfortunately we have no description of how the technical equipment was arranged. It would have been of great interest to know how much machinery and equipment there was in each room and on each floor. However, what we do know is how this inventory was designed. This is very important for establishing the kind of production that was carried out in the factory.

The technical equipment was extremely varied – from technically advanced machines to simple hand tools. The machines required to be powered mechanically, and for this purpose Wiel installed a water wheel of the overshot type in the basement of the building. This overshot wheel constituted the sole source of mechanical power. At the time two types of artificial power were used in industrial production, water wheels and steam engines, in principle performing identical purposes. In the textile factories in Britain both were used; in his choice of power technology Wiel's factory was therefore not lagging behind.

Next followed the transmission equipment between the water wheel and the machinery. We know that Wiel managed to obtain wooden models of all the cast iron wheels which were needed. These were made at Eidsfoss Works, a Norwegian iron foundry, in 1815. The transmission equipment is listed in a fire insurance evaluation of 1815, which, *inter alia*, is the only complete description of the inventory: 'A water wheel, a cam wheel with axle, a vertical axle with two gears [or cogwheel pinions?], a horizontal axle of 20 ells in length with two distributors [?] and an iron pinion [gear?], all supplied with necessary equipment.'[2] The machines were the most expensive equipment in the building, and they were ready for operation from the autumn of 1815.

For the preparation of the flax used for spinning, Wiel did not have a devil, but cleaned the flax by use of combs, leaving only the finer fibres left for spinning. At this point there was a marked difference between Wiel's factory and the British textile enterprises, where this process was performed mechanically. Manual combing would, presumably, give a product of comparably inferior quality.

For carding, Wiel had 'one single carding machine with six

2. *Besiktelses- og taksasjonsprotokoll 1769–1826* (valuation protocol), fos. 187b–188, Halden City Archives.

rollers'. He was, however, in the process of getting a second, double carding machine made. (This was built in the factory, and thus it has been possible to reconstruct exactly what it looked like.) The wear and tear on the wire teeth was probably quite severe, since the cotton was only cleaned manually in advance. Even so they must have lasted throughout the first year and a half that the factory was in operation, unless cards were also made at the factory: the first order for cards was dispatched to Copenhagen in the summer of 1816. Wiel's carding machines functioned in exactly the same way as in the British factories, although some parts may have been made from different materials.

According to the fire valuation, the process of drawing was performed by a 'complete drawing machine with necessary fully complete high rollers with attached cylinder'. The machine was rapidly extended. In January 1816 eight new rollers – two front, two back and four top rollers – arrived from Copenhagen. The head of a drawing machine consisted of one front, one back and two top rollers, which means that this dispatch added two new heads to the machine. For slubbing, the cotton factory at Tistedalen had 'a complete slubber with necessary roller wheels'. Even though Wiel had only one machine for drawing and one for slubbing, they were presumably so advanced in design that they had enough capacity to prepare the cotton for the spinning.

The market which Wiel was hoping to exploit comprised Fredrikshald and the neighbouring towns in south-eastern Norway, and his customers would normally come from the middle class. In addition, Fredrikshald was a centre for Swedish–Norwegian trade, and Wiel may have envisaged new possibilities for cotton yarn in this connection. One would immediately consider water twist to be the most likely yarn made at Wiel's, seeing that calicoes were in general demand.

For spinning, Wiel had acquired '4 complete spinning machines with the necessary bevelled wheels and top rollers of 108 spindles size'. These machines were most probably throstles. The four throstles constituted advanced and costly production equipment in spinning. The yarn which was produced in these machines was a genuine industrial product and bore no comparison to the hand-spun yarn process. Reeling took place at Wiel's on '4 hanks, each 16 spindles'. These reeling machines were not large, but they undoubtedly performed the required task. Wiel also bought one winding machine. For weaving, there were '6 looms with accessories' and '3 stocking frames'. Supposedly these were all of ordinary type and were used for both wool and cotton, since both

Table 2.1 Fire insurance valuations of Mads Wiel's cotton factory and machines (in riksdaler: 1 riksdaler = 2 Danish krone)

The building, without inventory	3600
1 single carding machine	1500
1 drawing frame	1000
1 slubber	1500
1 spinning machine	1375
1 winding machine	1500
1 reel	50
1 loom	13.2

Source: *Besiktelses- og taksasjonsprotokoll 1769–1826*, fos.187b–188, Halden City Archives

types of yarn were produced in the factory.

In order to comprehend the value these machines represented, it is useful to relate the fire insurance valuation for each machine to the value of the building. Table 2.1 clearly illustrates that the machines, excepting the loom listed last, were not ordinary handicraft or workman's tools but modern and expensive machinery.

We have seen that the machinery which Wiel had installed lacked only one link in the chain to be complete in comparison with contemporary British cotton mills. This missing link, the devil, belonged to the preparatory process, and was probably the part which could most easily be replaced by manual work without serious consequences for the rest of the production. Wiel's technical equipment was not quite, but almost, equal to comparable contemporary machinery in more advanced economies.

The spinning machinery which has been described was of a very high technical standard. Within spinning, it was unique in Norway. It was not only powered mechanically, but also built on extremely advanced technical principles. The machines were of the same type as the British cotton spinning machines, and were used in the same type of industrial production. There is little doubt that contemporary Norwegian textile manufactories operated with considerably less advanced equipment.

It is, however, necessary to point out that the remaining activities in Wiel's factory – namely cotton weaving, hosiery production, wool production, dyeing – were all carried out manually, with traditional methods and equipment. This part can safely be defined as manufacturing activities.

Wiel hired '20 women and children' to work in his factory. This

41

indicates that he not only needed skilled labour, but also unskilled workers who could be placed at the various links in the production process. It should therefore be apparent that the factory was run on both modern and traditional methods side by side, but that cotton spinning was clearly the most labour- and capital-intensive activity. We have seen that the technical equipment in Wiel's cotton factory was more advanced than one would expect in a period when it was still illegal to export machines from Britain to other countries. How was this possible, and by what means had Wiel managed to acquire these machines?

3 Charles Nordberg and the Diffusion of Technology

The cotton spinning which Wiel started in his factory in Tistedalen probably had the most advanced technical equipment which could be found in Norway at that time. But where did it come from? The Norwegian iron foundries had no possibility of producing such machinery. Britain had an export ban on new machinery until 1843. We must therefore direct our gaze towards another, closer country.

Copenhagen was the metropolis of the North in Wiel's time. In the latter half of the eighteenth century all kinds of enterprises were started up there, and the government both encouraged and gave economic support to the various ventures which saw the light of day. This applied to Denmark as a whole, but Copenhagen was naturally enough the centre. Through an acquaintance, Wiel made contacts with a gentleman in Copenhagen named Nordberg, from whom he made inquiries about cotton production and the purchase of machinery for a complete cotton spinning mill. At the time, Wiel was a young, unknown Norwegian aged 21, and it is therefore quite surprising that he should take this initiative with a business connection in Denmark whom he had never previously dealt with. Wiel's first letter, written in Norwegian, reads:

Fredrikshald, 24 August 1812

Mr Nordberg, Copenhagen.

From one of my friends residing where you live, Mr J. Kofoed, I have learnt, that you will undertake to supply me with the machinery required for a cotton spinning mill. From your information [I understand that] 4 spinning, 2 carding and 2 other machines belonging thereto, [can be] installed complete for 26,000rd and that you want to know my decision regarding this as soon as possible. You must, however, allow me first to receive your kind opinion about the following. How much, or how many lb spun yarn in total, do these machines produce per day, as I find it best to supply myself with raw cotton for one year's work? What is the wage cost per lb, how many people are needed for running the machines, and can the machines be run by water power? I have, in my opinion, a suitable waterfall near town, and if so the wages could be cut considerably. What size of rooms are required? And for what wage do you think you could get me a competent master, if the machines could be transported here by land, whereby the risks of transport would be eliminated? And if so, how many carts would be necessary?

The yarn would be used for clothing and stockings, and I therefore also require the necessary looms. Please also let me know at what price you would be able to supply me with a loom for cloth, and one for stockings. To run these I want the same master who supervises the spinning, if he can manage all the work, and if it is possible to get such a man. In anticipation of your esteemed opinion, which I hope to receive soon, I have the honour to sign myself,

Yours respectfully, M. Wiel.

This is quite a list of items that Wiel wanted prepared and sent. Nordberg sent a positive reply, and the machinery was, as we know, made and dispatched to Fredrikshald.

We shall here take a closer look at Nordberg. Who was he and what was his profession? At the turn of the nineteenth century the prime source of advanced textile technology and skills was Britain. Furthermore we know that Britain was unwilling to share the benefits of its inventions with others, and therefore made sure that neither experts in this field nor skilled craftsmen or machines disappeared out of the country. None the less it did happen that skilled workers migrated to other countries with their expertise, but this was basically a form of escape which the authorities attempted to prevent by all means. The government enacted a number of new laws to prevent the export of machines, equipment and craftsmen. Foreign recruiting agents who were caught could be sentenced to several years in prison. Some, however, were successful and returned safely to the Continent with both equipment and skilled workers.[3]

On the other hand, there were many foreigners who came to Britain to study, *inter alia*, textile manufacture according to modern

3. O. W. O. Henderson, *Britain and Industrial Europe 1750–1870*, Leceister, 1972, pp.10–26.

principles. According to J. R. Harris, it was possible to do this quite openly. In some places foreign guests were received with courtesy, since their hosts intended to build up a market for their products outside Britain.[4] These visiting observers had varying intentions, anything from smuggling machines and parts to procuring drawings and receiving oral or written information. All this could subsequently be used in domestic production, in an attempt to compete with the British imported goods which flooded the markets. The majority of west European countries carried out this type of industrial espionage actively, but the British did not make it easy for them. Most of the time foreigners were received with the greatest suspicion, and they were not allowed to make tests or construct models. If somebody managed to gain an insight into how a machine worked, the usual outcome was that it turned out to be extremely difficult to get it constructed properly at home. In addition, it was difficult to get the country's own inexperienced workers to operate the machines efficiently. The machines were often large and heavy, and needed trained hands.

The Nordic countries carried out industrial espionage too; at least the Danish and Swedish governments sent hand-picked agents to Britain with this in mind. Around 1760 a young Swede of about twenty years of age left for Britain to specialise in various forms of production and technology. His name was Charles Axel Nordberg. He remained in Britain for fifteen years. We know little about where he travelled and what he studied, but when he returned safely to Sweden about 1775, he founded a 'Manchester' factory where both cotton spinning and weaving were carried out. This indicates quite clearly that he had unravelled the secrets of the textile industry in England.

But Nordberg did not remain long in Sweden. He was not satisfied with the terms he was given by the Swedish state. The efforts of the Swedish King Gustav III to introduce a national costume had deprived him of a large part of his market. Thus, in 1799, when he was invited by the Danish Chamber of Commerce to come to Copenhagen to establish a 'Manchester' factory there, he accepted and left Sweden.[5] It seems that the Danish state developed a policy of active support to newly established enterprises earlier than the Swedish state. Nordberg was far from the

---

4. J. R. Harris, 'Industrial Espionage in the Eighteenth Century', *Industrial Archeological Review*, vol.7, 1985, pp.3–4.
5. Carl Bruun, *København. En illustrert skildring af dets historie, mindesmærker og institusjoner*, Copenhagen, 1901, vol.3, pp.509ff.

only Swede from whom the Danish state benefited; Jøns Mathias Ljungberg, professor in Mathematics and Philosophy at the University of Kiel, was another. Ljungberg was enticed by all the new inventions in Britain and left for England to study them at close quarters, roughly at the same time as Nordberg returned home. Ljungberg stayed in Britain for a year and a half, and he filled several books with detailed drawings of the most extraordinary appliances, but especially spinning and carding machines. Ljungberg also travelled directly to Stockholm, where he offered to put both his knowledge and labour at the disposal of the Swedish Chamber of Commerce. They were not interested, and in 1778 Ljungberg therefore left for Copenhagen.[6]

Thus in the year 1779 both Ljungberg and Nordberg were in Copenhagen, and a meeting between the two was arranged by the Chamber of Commerce. Together they took the initiative in establishing the new cotton factory. Ljungberg was responsible for equipping the mill with the required spinning machinery, and Nordberg the weaving and printing equipment. In 1780 the factory started to operate, but it was soon obvious that there was not enough space; additional buildings had to be erected and adjoining sites bought up.

By 1784 the following machinery was installed in the factory: a carding mill with two carding machines, a rolling mill for finishing by calenders, both hot and cold (to press and smooth out woven cloth), and a reeling mill. In addition there were a wash house, pump house, dye-works and a bleaching house. For the actual spinning process, there were thirty spinning machines, and the weaving mill had 123 looms. There was a separate building for printing, where the cotton material had its pattern imprinted. There was also a public finishing establishment on the spot, which Nordberg managed. Nordberg was also in charge of the finishing process, and saw to it that woven material was given the correct surface treatment and shine. This was a complicated process, and inferior finish was regarded as the main defect in late eighteenth-century Danish wool and silk products.[7] An Englishman, Joseph Dalton, was appointed as dye and print foreman. To carry out these processes it had thus been possible to employ a specialist who had learnt the technical skills without resorting to spying.[8]

The 'Manchester' factory was a large enterprise, with a total of

6. *Biographisk Lexicon övfer namnkunniga svenska Män*, Uppsala, 1842, vol.8.
7. C. Nyrop, *Niels Lunde Reiersen. Et Mindeskrift*, Copenhagen, 1894, p.188.
8. Bruun, *En illustrert skildring*, p.506.

800 employees. We have reason to believe that it operated with a combination of processes similar to those Wiel was establishing, i.e. mechanical spinning and manual weaving side by side. It is somewhat unclear how the 'Manchester' factory was powered, but it seems likely that it was run by water power. Seeing that Denmark's first steam engine was introduced some time between 1785 and 1790, there was probably no alternative before that time.[9]

Another important factor as regards the activities of the factory was that there should be continuous employment of apprentices, but never more than thirty at any one time. The apprentices received a fixed wage, and the factory was given a premium by the government for every apprentice who qualified, the idea being that the factory should act as a sort of school for spinners and weavers, who after their apprenticeship could go out and practise their expertise.

The directors of the factory had a clear policy as regards acquiring raw materials for weaving. No foreign yarn was imported. It was partly produced in-house and partly by Norwegian spinning mills, especially the state-run woollen manufactory at Kongsberg. This had been established by the General Storehouse (Generalmagasinet) in 1778, but was later conveyed to the 'Manchester' factory and the woollen manufactory at Blaagaard in Denmark.[10]

As we can see, the Danish-Norwegian state had a definite commitment concerning textile production in both countries; the 'Manchester' factory, the Royal Cotton Manufactory in Copenhagen, and the woollen manufactory at Kongsberg in Norway were all established on the initiative of the government.

Nordberg's and Ljungberg's factory had now got well under way, and they were acquiring practical experience in cotton manufacture of all types. The government was extremely interested in promoting industrial activities which had a promising future, but wanted them to be run privately. Thus the enterprise was restructured into a partnership in 1782. The government gave financial support, but otherwise generally left matters in the hands of the entrepreneurs. Economic liberalism began to take root in Denmark–Norway, and the basic principle of this policy was that both trade and industrial activities should develop and compete freely without much interference from the state. Even though the enterprise was changed into a partnership, it remained in fact the

9. Axel Linvald, *Kronprins Frederik og hans Regering 1797–1807*, Copenhagen, 1923, vol.1, p.360.
10. Bruun, *En illustrert skildring*, pp.513ff.

property of the government, which all the time held the majority of the shares. The authorities did not wish to cut themselves off from the enterprise because they considered it to be of great economic importance.

Only five years after the establishment of the factory, significant progress had taken place in Britain in further developing the machines and processes which had formed the basis for the factory. In 1785 a model of a new spinning machine was successfully smuggled to Denmark. This machine inspired Nordberg to invent a spinning machine for both wool and cotton. He was commissioned by the government to construct the machine, which he did, and was paid by the authorities. We glimpse here the beginnings of his later career as machine-maker. It soon transpired, however, that the machine did not measure up to expectations, and in 1787 Nordberg was sent back to Britain to study the new inventions more closely.[11] In Wiltshire and Somerset he had a good look at the clothing factories and smuggled out drawings of various carding and spinning machines. In Manchester he was especially on the look-out for cotton manufacture; in Birmingham he was interested in machines for production of steel and iron, and in Sheffield mills of various kinds.[12] Presumably the trip must have been of great significance for Nordberg's further career as machine-maker.

The same year he was appointed member of the General Factory Commission, as was the Norwegian commercial assessor Haagen Mathiesen. Nordberg was also appointed Perpetual Director of the Royal Danish Cotton Manufactory.[13]

When Nordberg returned home, he immediately set about constructing new machines, of better make and more efficient than those which had been running during the previous seven years. A large sum of money was provided for the purpose. The construction of these machines, however, caused great difficulties. The different machine parts had to be made by carpenters, turners and smiths following Nordberg's instructions. The workers were quite inexperienced in such tasks, and there was much trial and error with the more intricate metal parts.[14] In order to solve this problem Nordberg again travelled to England on a spying mission in 1791.

11. O. J. Rawert, *Kongeriket Danmarks industrielle Forhold fra de ældste Tider indtil Begyndelsen af 1848*, Copenhagen, 1850, p.507.
12. Linvald, *Kronprins Frederik*, p.360.
13. Nyrop, *Niels Lunde Reiersen*, p.167.
14. J. O. Bro Jørgensen, *Industriens Historie i Danmark*, Copenhagen, 1943, vol.2, p.175.

On his return, and, with the help of an English mechanic called John Smith, he finally succeeded in producing the machines at the beginning of 1792.[15] The new carding, slubbing and spinning machines were in operation at the factory until 1795, when both the buildings and inventory were sold by auction. The government had changed its attitude regarding industrial activities and was no longer so interested in being directly responsible economically for the factory. Nordberg's two trips to England were the most important industrial espionage missions (until 1848) which were initiated and paid for by the Danish authorities.[16]

The year before, in 1794, Nordberg together with some partners had bought a property with water power at Usserød near Hørsholm just outside Copenhagen. They established a textile factory with machine-spinning, dyeing, weaving and printing. The factory employed 131 workers who worked altogether eighteen machines connected with the spinning process and twelve power looms. There were also plans to put up twelve stocking frames. From the information available about this factory, there seems to be a clear parallel to Mads Wiel's cotton factory, which was established twenty years later. The government gave strong encouragement to the enterprise at Usserød, and provided economic support. In 1798 the factory was sold to another partnership, which continued to run it until 1810. All the time it was difficult to run the business profitably, and the new owners had problems, particularly in getting new cards for the carding machines and qualified people to carry out repairs on the machines. A few years later Mads Wiel had exactly the same problems; he had to stop production at his cotton factory for a couple of months in the spring of 1816 through lack of usable cards.

Very few people in Denmark possessed the skill required for making proper cards for the big carding machines. In the course of the eighteenth century cards and heckles were, of course, made domestically, but these were intended for home crafts manufacture. The factories had to import cards from Holland, Belgium or Germany. To solve this problem, the government in 1788 invited a German named Collard to come to Denmark; he was given funds to establish a carding factory in Fredericia in return for providing reserve stocks of cards. It was from the mechanic Collard that Wiel obtained all his cards for the double carding machine which was constructed at Wiel's factory in 1816. This indicates that Collard

15. Ibid.
16. Rawert, *Kongeriket Danmarks industrielle Forhold*, p.507.

produced carding equipment for a period of at least thirty years. The same year as the factory at Usserød was sold, in 1798, Nordberg started his own workshop for machine production in Copenhagen, aided by a government grant of 5,000rd. From now on his main occupation was to make machines for carding and spinning rather than supervising their operation. This did not mean out-competing Collard's manufacture; the market was large enough for both of them.

It rapidly became clear that some of Nordberg's machines were defective. This prompted a fourth trip to Britain in 1799 and serious machinery-making began on his return. In this, Nordberg was helped by John Smith, the same mechanic who had assisted him in 1792. At Nordberg's works machines were not factory-made; the individual parts were made by different craftsmen outside the works, while the parts were assembled at Nordberg's workshop. This splitting up of the production process resulted in low efficiency and high production costs.[17]

Despite a second grant from the state in 1804, Nordberg went bankrupt in 1806 and was forced to move to Fredericia, a refuge for bankrupts. Here Nordberg established another mechanical workshop which he ran until his death in 1812. He was by that time seventy-two years old. His previous workshop was taken over by a carpenter called Krebs, whom Nordberg had trained.

The above is an outline of Nordberg's career, but the picture is not complete yet: there is an additional aspect which concerns the role of his son, Charles Nordberg Jr. in the development of Danish machine-making capability and Norwegian textile mechanisation. The younger Nordberg had been trained by his father and was in 1803 commissioned by royal decree to construct a complete set of cotton machinery based on the latest models in order to demonstrate his technical qualifications. This he did; the machines were approved and purchased for the new cotton spinning mill which the prime minister, Count Schimmelmann, had decided to establish near Helsingör.[18]

From what we now know about Nordberg senior, it is easier to assess the importance of the contact between him and Mads Wiel. It is reasonable to assume that Nordberg was the best Danish business contact Wiel could have had at the time. There is no doubt about Nordberg's knowledge about mechanised cotton production, and he had long experience in the management of both manual and

17. Jørgensen, *Industriens Historie i Danmark*, Copenhagen, 1943, vol.2, p.178.
18. Rawert, *Kongeriket Danmarks industrielle Forhold*, p.511.

mechanised factory operations. Furthermore, Nordberg had improved his knowledge of machine-making during a period of several years, and last, but not least, he was experienced in training employees in both textile production and textile machine manufacture.

Nordberg agreed to make the machines for Wiel's factory following the request of August 1812. Unfortunately Nordberg died the following November, which could easily have led to postponement for Wiel. However, the connection was continued without a break by his son. The younger Nordberg made the machinery and also supplied a 'skilled master foreman' for the factory in Tistedalen; this man was named Theim, and he too had been trained by Nordberg senior. Theim got hold of a smith he had met during his apprenticeship to join him in Norway.

Wiel's factory near Fredrikshald was not the only Norwegian destination for workers previously in Nordberg's employment. A mechanic named Gellertsen, with exactly the same background at Nordberg's as Theim, was sent to the woollen manufactory at Kongsberg in 1812 to supervise the technical management. As already mentioned, contact between this enterprise and the 'Manchester' factory in Copenhagen had early been established. In 1818 Gellertsen left Kongsberg for Drammen, situated just south-west of Oslo, to build the textile machines that were needed for the newly established Solberg cotton spinning mill. In 1820 the mill was moved to just outside Drammen to exploit the water power of the Solberg river, and in 1821 started operating.[19]

We can identify a number of links of communication between Britain and the rest of Europe in the transfer of technical skills and knowledge during Europe's industrial revolution. The central role played by Britain has been treated, among others, by Herbert Heaton, who writes: 'The early stages of the new-style industrialization in other countries benefited from the entry of British machines, artisans, managers, entrepreneurs, and investors, as well as from the visits, fleeting or prolonged, of continental observers.'[20] As examples of British emigration of expertise and machinery, Heaton mentions two pioneers who came to play a significant role in the industrial revolution on the Continent. The cotton manufacturer John Holker fled to France in 1750 and was instrumental in the

19. *Solberg Spinneri 100 år 1818–1918. Jubileumshefte*, pp.5ff.
20. Herbert Heaton, 'The Spread of the Industrial Revolution', in M. Lranzberg and C. W. Pursell Jr. (eds), *Technology in Western Civilization*, New York, 1967, vol. 1, p.505.

modernisation of the French textile industry. William Cockerill went to Belgium in 1799, where he and his descendants established machine workshops with a wide range of products. From these workshops machines were subsequently spread all over Europe. Through British emigration and the smuggling of new technology, drawings and models, new centres of dissemination were created outside Britain.

The spread of technology took place, perhaps even more through the last group mentioned by Heaton, namely Continental observers. Heaton remarks, 'Foreigners snooped around factories, iron works, and mines, and frequented taverns in search of artisans who might give them information, smuggle them into industrial plants, or be willing to emigrate,'[21]

This was plain industrial espionage, and Nordberg was, as described above, one of the foreign spies. Nordberg was an important channel in the spread of modern technology to Scandinavia; he returned from Britain to Stockholm, then moved to Copenhagen, where his operations acted as a point of departure for spreading the new technology to other parts of Denmark and subsequently to Tistedalen, Kongsberg and Drammen in Norway. The latter transfer, from Denmark to Norway, was legal, but was otherwise very similar to the transfers of technology by the emigration or escape of British artisans.

The transfer of technology from Britain to Norway via Denmark described above falls into two parts. The 'package' of technology Nordberg brought to Denmark consisted primarily of elements which were easy to carry out of the country, without risking detention and imprisonment. It appears that Nordberg spent most of his time in Britain making detailed studies of machine construction. The outcome was a collection of drawings which subsequently formed the basis for his own machine-building in Copenhagen.

All in all, Nordberg made four visits to Britain with the purpose of industrial espionage. He was twenty years old the first time he came, and 59 on his last visit. Although he remained in Britain for fifteen years after his first arrival he did not, presumably, achieve the level of skill and workmanship which was required to succeed as a builder of modern textile machinery. In his middle age he had to make another three trips across the North Sea to seek information about new technology under development. Seeing that Nordberg did not succeed as well as one would expect as a machine

21. Ibid., p.504.

manufacturer, it is reasonable to look for at least one of the causes for this in the 'smuggling-package' he brought back. It might be the case that if Nordberg had managed to return to Denmark with machine parts and skilled workers, it would have proved easier to overcome the domestic production problems.

The other part of the diffusion process, the transfer of technology from Nordberg's workshop to Wiel's cotton factory in Norway, was of a different character. It is important to note that the package Wiel received in 1815 included both complete machinery and expertise – namely two of Nordberg's apprentices, Gellertsen and Theim. Even so, the transfer process entailed certain difficulties. One of the problems was caused by a combination of inadequate means of transport and the political situation at the time. As we have seen, Wiel ordered his machinery in August 1812. During 1813 the roads through Sweden were closed, and in order to transport Theim and the machinery which was ready in 1814 to Fredrikshald, Wiel arranged for one of his ships, the *North Star*, to sail for Copenhagen under the Danish flag and equipped with both Danish and Swedish papers. Nevertheless during the voyage back to Norway the machines were discovered by the Swedish authorities, declared to be contraband and duly taken to Sweden. In August 1814 the machinery was released and brought across the border to Norway – two years after having been ordered, and one year later than the delivery date agreed with Nordberg.

The next problem was that the dispatch was incomplete; several machines as well as parts were missing. Wiel had received no written information about this. There was, however, only one possible solution to the problem, namely to send the ship back to Copenhagen. Theim went along in order to check the remaining equipment, and almost refused to return to Norway because of his recent experience of the dismal conditions during the stormy voyage to Denmark.

Furthermore, when it soon became clear that more machinery was needed than the items Wiel by now had received from Copenhagen, the additional order of one carding machine resulted in similar delays and problems. It took a year and a half before the machine arrived and enabled the spinning machines to operate at full capacity. The demand for machine parts was equally difficult to meet. Norwegian producers were unable to make parts of sufficient strength, and most parts had to be made in Copenhagen. Such problems delayed the process of transferring technology from the centre, in this case Nordberg's workshop in Copenhagen, to Wiel's cotton mill in Fredrikshald at the periphery in Norway.

Nordberg's activities as industrial spy, textile manufacturer and machine builder played an important role in the development of his assistants' and colleagues' competence and capabilities. Many of them subsequently founded enterprises within the same fields, and via Nordberg this capability spread also to Norway, first of all to Fredrikshald, Kongsberg and Drammen. Nordberg's activities were typically Scandinavian: he started his career in his country of birth, Sweden, then moved to Denmark where his most important contribution was as a skilled mechanic and entrepreneur. Soon after his death, Nordberg's skills and knowledge were spread by his apprentices to Norway – the last of the three Scandinavian countries to enter the industrial age.

4 Mads Wiel, a Norwegian Industrial Pioneer.

Mads Wiel was the first Norwegian entrepreneur to establish contact with one of Scandinavia's first centres of industry in Copenhagen. Who was he, and what attracted him to the new industry? In particular, what qualifications did Wiel have, at the age of twenty-one, which enabled him to take the initiative and introduce to Norway modern technology developed abroad, at the start of the nineteenth century?

It has not been possible to establish the exact origins of the Wiel family, but it seems that they came to Norway from Denmark. In 1671 the sources show that Mads Jensen Wiel was active in timber exports and shipping in Drammen, activities which were continued by relatives after his death.[22] In 1750 Mads Jensen Wiel's grandson, Peder Wiel, moved to Fredrikshald; his brothers Claus and Truels followed a couple of years later.[23] Truels had two sons, Johannes and Mads (Mads was born in 1791); Johannes, at the death of their father, inherited the family estate, while Mads carried on the merchant activities. Mads Wiel also ran a rope works, a saw mill, a grocer's shop and flour mills, and started up a tobacco manufactory. The trade in timber was central, and Wiel exported timber first of all to England and Denmark.

At twenty years of age Mads Wiel was not particularly well qualified when his father died in 1811. However, in 1812 he travelled to Copenhagen; during his month-long stay he established important business contacts and befriended Kofoed, later a

22. H. K. Steffens, *Slægten Wiel på Strømsø og Fredrikshald*, Kristiania, 1903, p.1.
23. H. P. Norløff, *Familien Wiel på Fredrikshald*, Oslo, 1942, p.5.

High Court judge and chamberlain. In identifying a connection between Wiel and modern textile technology, it is interesting to note that Kofoed's grandfather, Hans Hansen Kofoed, was the director of the military woollen manufactory in Copenhagen – the equivalent position to Nordberg's in the 'Manchester' factory. It may be that information and insights into modern textile production were passed on to Mads Wiel through his acquaintance with the Kofoed family. Another possible source of information would be Wiel's cousins in Copenhagen, Peder and Magdalena Wiel; Magdalena was married to Niels Lumholtz, the son of the bookkeeper at the 'Woollen Manufactory' Mathias Lumholz.[24] If this is the same as the Cloth Manufactory at Blaagaard, which was Copenhagen's largest woollen manufactory, it is likely that Wiel heard about Nordberg through Lumholtz.

Whichever channel led Wiel to Nordberg, what was important was that Wiel, through his friends and family, received information about what kind of production was carried out in Copenhagen and elsewhere in Denmark. The advanced techniques which were in operation in Denmark from the end of the eighteenth century were unknown in Norway, and Wiel's experience and knowledge of modern textile techniques gave him a unique position within the local community of Fredrikshald.

We do not know the reasons behind Mads Wiel's decision to start textile production. It seems likely, however, that he was aware that the repeal of the ban on the import of raw cotton to Norway was imminent, which may have played a role in his decision to order textile machinery only four months after his return to Fredrikshald.

Considering all the factors we see that Wiel had several advantages over most of his contemporaries. His financial background was strong, and he was experienced in a broad range of manufacturing activities and active in extensive trade. Before 1810, a typical aspect of the composition of the east Norwegian merchant class was the dominant share held by merchants who had immigrated from Britain, the Netherlands and, in particular, Danish and north German trade centres.[25] Such connections opened up closer contact with the much more varied and advanced economic life and developments abroad, through travel, family kinship etc. Foreign travel was, furthermore, frequently regarded as an integral part of the general education of the immigrants who now sought Norwegian

24. Steffens, *Slægten Wiel*, p.62.
25. Odd Halvorsen, *Øst-norsk trelasteksport og trelastborgerskap 1760–1810* (Cand. Philol. thesis, University of Oslo, autumn 1979), p.179.

citizenship.[26] Thus the merchant immigrant community enjoyed certain advantages over other groups. Mads Wiel, on the other hand, never had the opportunity to spend extended periods abroad, and his journey to Copenhagen in 1812 remained his sole foreign trip. Although his visit was short, it nevertheless decisively influenced Wiel's ability to exploit the possibilities which existed at the time.

The entrepreneurial function played by Wiel and his likes was to act as intermediaries, linking local Norwegian communities with the advanced European economies.[27] They provided new technology because they usually also procured the necessary technical expertise from abroad. These skilled workers were the real carriers of technology. Not until much later in the nineteenth century did Norwegian entrepreneurs receive technical education and act as managers.

According to Fredrick Barth the position which the entrepreneur stakes out for himself is, seen in relation to resources, competitors and customers, his niche.[28] This niche will rapidly change if the entrepreneur, in his capacity as an intermediary, travels to more advanced economies, enabling him to exploit two different markets – the one he has left and the one he enters. This is an advantageous situation which constantly provides him with a better niche as a point of departure. The entrepreneur will find, through his travels, new opportunities to exploit. Wiel found exactly such an opportunity. He gained access to modern technology in an advanced market, and cheap labour in the underdeveloped domestic market. The combination of these two factors provided the basis for the industrial concern in which he invested in 1812. As we know, the technology included the new textile machines from Copenhagen, while labour was provided locally from Tistedalen. There were a number of men employed at Wiel's sawmills in Tistedalen, at a time when female family members had few opportunities for paid work. A textile factory employing women would naturally provide a welcome opportunity.

Not to employ the women would, from an economic point of view, be directly detrimental to Wiel's interests; the provision for the whole family would nevertheless be borne by Wiel, directly or indirectly. Wiel's expenses were not significantly increased, since the female workers' wages were, after all, used to repay the credit

---

26. Ibid., p.180.
27. Frederik Barth (ed.), *The Role of the Entrepreneur in Social Change in Northern Norway*, Oslo, 1972, p. 4.
28. Ibid., p.9.

Wiel had extended to their husbands, who were employed by him. The women and children who made up the available labour capacity were, in addition, cheap labour compared to the men. Thus, their employment was both profitable and increased the family's potential purchasing power. The main share of the family's added income would return to Wiel, which in turn strengthened the ties between employer and employees.

In our attempt to identify the reasons behind Wiel's decision to start cotton spinning production it may be useful to look at some of his other manufacturing activities. As already mentioned, he owned a tobacco manufactory and a rope works. Rope-making is a type of textile production based on hemp for raw material. Hemp is an annual plant, and the fibres were extracted by a method similar to that used in the production of linen. After cleaning, the fibres were spun, on spinning wheels, and then made into rope.[29] Tobacco production also involved processing plant fibres by the use of spinning wheels; this process was managed by a 'spinning master'. This means that Wiel was familiar with the process of spinning, and this may have played some role in his decision to invest in cotton, and later wool, spinning. We may be observing a case of technical convergence acting as an agent for change.

The most important features of Wiel's new undertaking were the machinery and the mechanical power. One may ask why he did not choose to expand and modernise one of the product lines he had already established. The answer to that may be that it was probably easier to mechanise a new industry than older branches based on established, traditional methods. In addition the market for cotton yarn would appear promising; the demand for yarn imported from Britain was considerable, and this may have encouraged domestic production.

Our knowledge about the market in cotton goods in the Fredrikshald area is, however, regrettably limited. Still, Wiel's involvement in trade probably meant that he was well informed about existing demand. Because of the blockade, which was not lifted until 1814, imports were banned, although considerable smuggling took place. Naturally, this makes it difficult to estimate the size of imports. As late as 1834, however, 5,994 lb of cotton goods (excluding cotton yarn) were imported to Fredrikshald, and in 1844 the figure had reached 21,513 lb.[30]

29. *Gamle Bergen Årbok*, 1975.
30. H. P. S. Krag, *Underretning om Fredrikshald By og dens Krigshistorie*, Christiania, 1848, p.9.

It is probably the case that a combination of different factors gave Mads Wiel the idea to start cotton manufacturing. Wiel's general activities and his trip to Copenhagen together provided him with the broad overview which was required in order to transform the idea of a cotton factory into reality. As regards the question why Mads Wiel in particular became Norway's first industrial entrepreneur, the answer must be sought in his character; his initiative demanded an unusual amount of courage and enterprise. In this respect we can compare him with the Norwegian industrial entrepreneurs or bourgeoisie of the 1840s and 1850s. On the basis of what we know, we see that Mads Wiel in most ways was a traditional merchant who lived in a period and society which lacked the impulses which characterised the 1840s and 1850s. The fact that he nevertheless took the initiative of starting up a completely new form of economic activity a whole generation before the emergence of the new entrepreneurs, suggests that he was a marked exception. He was young, daring and untraditional in his thinking; the latter was, perhaps, the most important precondition for the role he came to play as an intermediary, linking non-industrialised Norway with the industrialised world.

3

British Influence on Developments in the Swedish Foundry Industry around the turn of the Eighteenth Century

Martin Fritz

1 Introduction

The decades before and after 1800 can be seen as a period of reconstruction and renewal in the history of Swedish iron casting – as indeed in that of western Europe – following some centuries when the structure of production had remained unchanged.[1] Traditionally, the casting of cannons and cannon balls had played a predominant role in foundries in many countries but there also existed – though on a smaller scale – the casting of various household articles, such as cauldrons, pots and pans, weights, oven hearths etc.

Casting of this kind had been carried out since the Middle Ages and was intimately linked to the blast-furnace: instead of casting pig iron for further processing into malleable forge iron, the molten iron was poured into moulds in casting pits. Apart from the manufacture of cannons, which in Sweden was concentrated in a number of specialised ironworks, called 'styckebruk' or ordnance works, the production of iron casting was more in the nature of a temporary, sideline activity. Furthermore, the supply of water for blast-furnaces was limited, so that they could be kept going only during parts of the year.

1. This article contains material used in a lecture given at the Nordic Symposium on The History of Technology on 14–16 June 1988 and is based on a study yet to be published on Swedish iron casting during the nineteenth century, in collaboration with Bengt Berglund (Department of Economic History, University of Göteborg) and the Swedish Historical Foundry Association.

For the quality of the finished castings, the mixture and the proportions of the various types of iron ore used were very important, as was the time during the blasting period when casting took place. 'A blast-furnace might easily be disturbed by a multitude of circumstances', reads a contemporary account.[2] Knowledge concerning these circumstances remained for a very long time incomplete, and quality control left much to be desired, again with the exception of the manufacture of cannons, where by gradual experience the right proportions of the different iron ores were learnt. The iron's brittleness or hardness, along with the difficulties of achieving thin-walled, that is lighter, products, were all factors which restricted the usefulness of the material. Of all pig iron produced in Britain at the beginning of the eighteenth century, only 5 per cent appears to have been used for casting.[3]

Towards the end of the eighteenth century, however, we find a growing demand for iron castings. (In what follows, I exclude the manufacture of cannons.) This was a result of a slow but general increase in purchasing power creating a demand for iron tools in agriculture and iron castings for household use, such as stoves and kitchen utensils, but it was also a result of the onset of industrialisation, with a growing demand from mechanical engineering works for iron castings which were more durable than wood and cheaper than wrought iron products. In addition, castings were, during the nineteenth century, increasingly put to new uses, such as canal locks and iron bridges, drain pipes, staircases and so on. Iron castings also began to be utilised for articles with ornamental functions like railings and fountains, or for everyday use as umbrella stands and spittoons.

In Britain Abraham Darby succeeded, in the early years of the eighteenth century, in producing pig iron by using coke as fuel in his blast-furnace. It took until the middle of the century, however, before it proved possible to process the pig iron produced by the coke method into malleable forge iron. Once this was achieved, there was a gradual change-over in Britain to coke blast-furnaces. In contrast, the new coke-produced iron rapidly proved excellent for casting purposes, and allowed for the casting of products with thinner walls, and therefore lighter products than castings produced with charcoal. Since lighter products required less iron, less fuel and less time, the coke blast-furnace yielded more profit. Thus

2. K. Karmarsch, *Handbok i mekanisk teknologi*, Stockholm, 1862, vol.2, p.114.
3. C. K. Hyde, *Technological Change and the British Iron Industry*, 1700–1870, Princeton, 1977, p.127.

it was in the field of casting that Darby's invention initially gave the best returns. In addition, Darby succeeded in casting in sand moulds, which proved cheaper than the clay moulds previously in use.

In spite of these improvements in Britain, blast-furnace casting still had its limitation. As mentioned earlier, quality control was still primitive, and production possible during only part of the year. It was difficult to produce large castings such as cannons – hence the characteristic double-shafted blast-furnaces in Swedish ordnance works – and, somewhat later, steam engine cylinders and other large articles. Furthermore production took place at the blast-furnace and not near the markets, i.e. the larger towns with mechanical workshops. Finally, the production of iron castings was a subsidiary activity at the blast-furnace, where the process of pattern-making and moulding had little connection with the main production. As a background we have the growing demand for cast iron products referred to earlier.

2 New Melting Methods

In Britain during the eighteenth century two different types of furnaces were developed, which both made it possible to melt pig iron and scrap iron for production of castings. This offered several advantages. One was that these furnaces could be located next to existing blast-furnaces and thus be of complementary use when large castings were manufactured. When the furnaces were tapped simultaneously, the casting process was quicker, which improved quality. Another advantage was that the new furnaces could be set up separately and freely located in the vicinity of the market, towns and densely populated areas. These furnaces were based on the use of coal and coke, which were cheaper than charcoal. They further made it easier to control the quality, provided that the properties of the melted pig iron could be regulated.

These new furnaces were of two different types, air furnaces and shaft furnaces. The former type was generally called reverberatory furnaces and were known in Britain during the first half of the eighteenth century. This type of furnace was heated with coal, had a long vault with a hearth at one end, and a tall chimney – 18–24 m high – at the other end, which created a powerful draught and intense heat, rendering a blast superfluous. These furnaces produced strong and hard cast iron used primarily for the casting of cannons and for other large castings, in combination with direct casting from blast-furnaces. According to the German historian of mining and metallurgy, Ludwig Beck, it was the need for large castings which prompted the construction of the new reverberatory furnaces.[4] Around the turn of the eighteenth century there existed, however, a number of foundries in Sweden for the manufacture of cauldrons, which were based on the use of reverberatory furnaces.

The second type of furnace – the shaft furnace – had a predecessor in a smaller, portable tip furnace on wheels, primarily using scrap iron. This was described by Réaumur in the 1720s and referred to at some length in *De ferro* (On iron), a treatise published in 1734 by Emanuel Swedenborg. During the last decades of the eighteenth century the so-called cupola furnace was developed in England – one source refers to a patent taken out by John Wilkinson in 1794. The furnace looked like a small blast-furnace, 6 m high at the most, but more often much smaller; for instance the height of the cupola furnace at the Motala Verkstad (mechanical engineering works at Motala, Sweden) in 1830 was specified as beeing 2.25 m high. The cupola furnace was built on fire-proof clay and might be either circular or square in shape. Just as iron ore and fuel were placed in layers in the blast-furnace, pig iron and coke, or sometimes charcoal, were fed into the cupola furnace layer by layer.

Forced blast was necessary in this type of furnace. The following comment referred to the cupola furnace in Hällefors, Södermanland, in 1820: 'This cupola furnace, notwithstanding that its volume represented but one twentieth part of the common blast-furnace shaft, none the less required close upon as much blast as the blast-furnace, as is also the case in England, where cupola furnaces are heated with coal.'[5] For the production of larger castings, two or more cupola furnaces could be placed side by side, and both could be tapped simultaneously, the molten iron being channelled to a common mould. Cupola furnaces were used mainly for casting machine parts which were to be finished with file and chisel. As skills improved, different types of pig iron were mixed in predetermined proportions in order to achieve desirable properties in the finished product.

The iron was conveyed from both reverberatory and cupola furnaces in a clay-lined channel from the tap hole of the furnace to moulds, which were arranged in front of the furnace or dug in a pit; or else the iron was first tapped into clay-coated ladles, which were carried by hand, or moved with a swing-crane, and the iron poured into the mould.

---

4. L. Beck, *Die Geschichte des Eisens*, Brunswick, 1879, vol.3, pp.380, 746.
5. *Jerkontorets Annaler (JKA)*, 1820, pp.5ff.

These melting furnaces thus proved to have several advantages in the field of iron casting: there was greater freedom as to the location of the foundry, and engineering works frequently grew up around it; buying from different sources gave better opportunities to control the quality of the iron; larger castings could be produced more quickly; and the fuel (coal and coke) was cheaper than charcoal.

Reverberatory furnaces were introduced in Britain during the first half, and the cupola furnaces towards the end, of the eighteenth century. Soon after the middle of the century reverberatory furnaces 'in the English manner' were first erected in Sweden. The first cupola furnace in Sweden was built as early as 1805, which indicates a swift technology transfer.

British influence on early Swedish engineering industry was quite extensive, as has been pointed out in the literature.[6] Since an iron foundry as a rule was an integrated part of any engineering works, if not its very nucleus, this is obviously also true for the casting of iron. In the following account we shall describe with practical examples the transfer of knowledge concerning the new reverberatory and cupola furnaces from Britain to Sweden. Two fairly obvious aspects are dealt with: 1) British foundrymen in Sweden; 2) Swedish travellers in England.

3 British Foundrymen in Sweden

The first attempt at remelting pig iron in a coal-heated reverberatory furnace was made in the 1760s at an 'art foundry' in Klarabergsgatan in Stockholm with the purpose of casting rollers, on the initiative of Jernkontoret (the Ironmasters' Association). This attempt, based mainly on knowledge gained from abroad and on general experience, was a failure, mainly because the heat was too intense for the bricks. Towards the end of the 1760s we can see the beginnings of casting of general household goods. Gustaf Broling mentions in his report on this new furnace that 'an experienced foundryman was procured from Amsterdam, an Englishman by birth named Evans. This foundry, so it is said, operated but a few years, whereafter the said Englishman was obliged to return to London with his family.'[7]

6. T. Gårdlund, 'Teknik och tekniker i den tidiga svenska verkstadsindustrin', *Ekonomisk Tidskrift*, 1940.
7. G. Broling, *Anteckningar under en resa i England åren 1797, 1798 och 1799*, Stockholm, 1811–17, vol.3, p.184.

We know a good deal more about the next attempt to work with a reverberatory furnace 'in the English manner', namely a cauldron foundry at Röda Sten just outside Göteborg, granted privilege in 1764. The owner, Johan Cahman, who was of German descent, intended to melt pig iron with imported coal 'for the casting of various smaller and finer items of iron'.[8]

In order to recruit skilled labour, Cahman used agents in Britain – according to his own records he was also there himself – even the year before the granting of the privilege, to recruit skilled workers from the large foundry of Carron and Co. in Scotland. British industry offered tough resistance to the exportation of machines and skilled labour. A very proficient foundryman called Thomas Lewis did, however, leave his British workplace in 1765 together with a fellow worker with the intention of moving to Göteborg and there developing Cahman's foundry at Röda Sten, but both men were caught. Lewis tried again and finally arrived in Göteborg, probably in 1766. The following year, the British chaplain in Göteborg passed on the information that he had met with five fellow countrymen, all employed at Cahman's foundry, whom he in vain had tried to persuade to return to their native country.[9]

Apparently, Cahman's travelling expenses were paid by the Manufacturers' and the Ironmasters' Associations, which considered it a matter of importance to support this production, even though it entailed considerable expenses in the import of coal and the recruiting of foreign labour. The Cahman foundry manufactured mainly cauldrons of various types, pots and pans, and also weights, stoves and some machine parts.

As for Lewis, he left Göteborg after a quarrel with Cahman and moved to Stockholm, where in January 1769 he received a privilege for a foundry, also 'in the English manner'. With the support of Robert Finley, of British descent, and of the Ironmasters' Association, which continued to show an interest in the new foundries, Lewis was able to establish the Bergsund foundry on Södermalm in Stockholm. In 1771, however, Finley went bankrupt and Lewis was ruined too. Nevertheless Lewis found himself a new supplier

---

8. G. Bodman, *Fabriker och industrier i det gamla Göteborg*, Göteborg, 1925, pp.321ff. This location can have depended partly on the opportunity there was of buying scrap iron from the weighing of iron in Göteborg.

9. Lewis's activities in Sweden have been studied by C. Sahlin, 'Thomas Lewis och hans insatser i den svenska gjuteriteknikens utveckling' *JKA*, 1928. See also Ashton, *Iron and Steel in the Industrial Revolution*, 2nd edn, Manchester, 1951, pp.202ff., and W. O. Henderson, *Britain and Industrial Europe, 1750–1870*, Liverpool, 1954, pp.5ff.

of credit in another business firm into which he had married. The same year Lewis himself acquired another British foundryman, William Wilde, who later played an important role in the history of iron casting in Sweden, and after the death of Lewis was employed by several firms to set up new foundries.

Production at Bergsund was also quite diversified – though mainly specialising in household goods and stoves, which were sold in a small shop in Gamla Stan (the old part of Stockholm). It seems that Lewis had difficulty in getting hold of suitable types of pig iron for his casting. This was, moreover, a problem for Swedish foundries throughout the early decades of the nineteenth century. In an attempt to deal with these difficulties, and to supervise the running of the blast-furnace personally, Lewis in 1780 bought the blast-furnace at Nyhyttan and placed Wilde there as foundry master.

Thomas Lewis showed a lively interest in new techniques, made several attempts at introducing technical improvements at Nyhyttan, corresponded with his former employer at Carron and Co. and followed the evolution of Watt's steam engine. Lewis is also given credit for having improved the vital moulding techniques, in so far as he, together with Wilde, succeeded in replacing the expensive clay moulding with moulding in sand, also for smaller and more delicate articles. At the turn of the century this method was generally known in Sweden.

Lewis died in 1783, but Bergsund continued to operate and later developed into one of the country's leading engineering works, from 1806 under the ownership of a single owner, Gustaf Wilcke. In the same year, Wilcke employed as foreman the very man who perhaps more than anyone else has come to be regarded as the founder of the Swedish engineering industry, Samuel Owen.

Owen's life is known in broad outline. After a thorough education in Britain, he came to Sweden, and it was on his second visit in 1806, undertaken in order to set up a steam-engine, that he was employed by Wilcke. During his brief sojourn at Bergsund – it lasted for three years – production was extended to include steam-engines, threshing mills and rollers for a mill at the Klosters iron works, the first to function in Sweden.

In 1809 Owen set up an engineering works and foundry on Kungsholmen in Stockholm, for the manufacture of steam-engines, flour and threshing mills and iron steam ships, and the firm expanded to become the country's leading engineering works, until Motala took the lead in the 1830s. Initially the going was slow, however. In his autobiographical notes he wrote: 'When first

I cast on Kungsholmen I had not a single man who had ever seen iron melt; hence it was no easy matter to establish an engineering factory.'[10] There were, nevertheless, a few British skilled workers whom Owen had summoned to his side. His machines 'witnessed to a creative spirit different indeed to that which has hitherto prevailed and certain of them were far removed from the cumbersome nature which with few exceptions had characterised Swedish castings hitherto'.[11] Broling's assessment of Owen's engineering works was that it must 'be considered one of the most fortunate acquisitions made in later years for the improvement of Swedish crafts and manufactures'.[12]

Running a foundry and engineering works during the first half of the nineteenth century was no sinecure, however, and in 1843 Owen was obliged for financial reasons to leave his workshop. At the age of seventy and with the title of ironmaster, he began to work at the Åkers ordnance works, where he energetically participated in casting operations, among other things in the construction of a cupola furnace.

Before the 1830s, which saw the beginning of an expansion in Swedish engineering industry, there was besides Bergsund and Owen's works only Motala. The Motala engineering works were set up in connection with the construction of the Göta Canal in the early 1820s. The head of the project, Baltzar von Platen, engaged during a visit to Britain in 1822 a British engineer called Daniel Fraser on the recommendation of the well-known Thomas Telford. Fraser stayed on in Motala as technical manager for twenty-one years, until 1843. A few years later, a British foundry master, Andrew Malcolm, was appointed to supervise the casting from the cupola furnaces – he too stayed until 1843, after which he ran an engineering works in Norrköping together with his brothers. Other departments at Motala were also managed by British experts summoned from abroad.

The Motala works were designed not only to serve the needs of the canal but also to function as the 'central workshop' for the whole country and as a training centre for engineers and technicians. It was for the explicit purpose of 'establishing an engineering works by the aid of skilled master mechanics from England', that the canal company was granted loans. Among the apprentices we find there, for example, Carl Bolinder who, prior to setting up

10. F. Schütz, 'Samuel Owen', *Daedalus 1975* (Stockholm), pp.93ff.
11. Gårdlund, *Teknik och tekniker*, p.184.
12. Broling, *Anteckningar*, vol.3, p.185.

his own works in Stockholm, also worked as a foundry master at Kockum's in Malmö. There are two further well-known instances of British immigrants who set up foundries. Here we have in mind William Gibson and his foundry at Jonsered outside Göteborg, established in 1835, and James Keiller who in 1841 started the enterprise which later became Göteborg's mechanical engineering works and later the 'Götaverken'.

This account describes how British skilled workers and technicians were involved at every one of the new remelting furnaces during the later part of the eighteenth century. Even the three first engineering works, Bergsund, Kungsholmen and Motala, established early in the nineteenth century, had British experts in leading positions. The importance of these works as training centres spreading technical knowledge throughout the country is generally recognised and many of the people who were trained under the supervision of British specialists later worked at newly established foundries and engineering works; in 1850 there were twenty-two of these, in 1870 ninety-two. The following extract is from the minutes of the general annual meeting of shareholders in the Motala works in 1839: 'Whence would all these branch workshops have taken foundrymen, filers, draughtsmen, foremen, had not the Motala works existed?'[13]

4 Swedes Visiting Britain

Of considerable importance for the dissemination of technical knowledge and for bridging the existing technological gap were the study tours Swedish engineers undertook to Britain.[14] We shall be considering below a few of the better-known visits. In this context it should be pointed out that these travellers, since they were generally recipients of grants or subsidies – often from the Ironmasters' Association – were obliged to write a report, which meant that those among their contemporaries who were interested had an abundance of published material at their disposal – as we have today. Usually these travellers were not, however, sent off merely to study new techniques within the foundry industry but were as a rule to undertake more comprehensive assignments.

13. Gårdlund, *Teknik och tekniker*, p.187.
14. S. Rydberg, *Svenska studieresor till Storbritannien under frihetstiden*, Uppsala, 1951, p.204.

The experiment in the 1760s at Meyer's art foundry in Stockholm mentioned in the introduction was based on the study of reverberatory furnaces in Britain. When Meyer first attempted the casting of cylinders, he made use of knowledge gained from journeys made by his son Gerhard in Britain in 1756 and by another Swede, Reinhold Angerstein, in the mid-1750s. Angerstein visited, among other places, Coalbrookdale, where he studied wind furnaces, i.e. reverberatory furnaces. His report, however, remained unpublished.

This is also true of Bengt Andersson Quist's travel report – submitted to the Ironmasters' office, as was Angerstein's – on a journey in Britain in 1766 and 1767, where he studied both coke blast-furnaces for casting purposes and the new draught furnaces, i.e. reverberatory furnaces for melting pig iron.

Journeys undertaken by Gustaf Broling and Eric Svedenstjerna made a greater impact, however. Broling received financial support in order to study the metal industry in Britain and lived there during the period 1797–9. His travel report comprised three volumes, in all exceeding 1,000 pages, the first two appearing in 1811; the third, which contains most of the technical information, was not published until 1817 and described many technical innovations.[15] In one elaborate account, including many illustrations, Broling gives a detailed description of the design, construction and use of reverberatory furnaces, which he had been able to study at close quarters.

More important than the reverberatory furnaces, which by then had been in use for quite some time in Sweden, was his description of the entirely new cupola furnaces, which had surprised Broling:

> Among the devices which during my sojourn in London excited my attention, were small furnaces, circular or square in shape, on the outside faced with iron plates and on the inside lined with fire-proof brick, in which smaller pieces of discarded pig iron were remelted, the number of which at that time already amounted to ten or twelve in London, under the name of cupola furnaces. These furnaces can scarcely have been long in use, since I cannot recollect that I have ever heard or seen them mentioned by earlier travellers . . . Never shall I forget my astonishment when I first encountered such a foundry. I beheld eight to ten moulders busily engaged in moulding, but was unable to discover any melting furnace. To be sure I did observe a furnace half the size of a common round tile stove, but believed it to be utilised for the drying of

15. Broling, *Anteckningar*.

the moulds, the heating of the room or some such purpose. I finally asked them how they contemplated proceeding with the casting, and I was requested to return in a couple of hours. I cannot deny that, on this occasion I saw this furnace with no little wonderment. In consideration of its modest size and the feebleness of the flame emitted by carbonised coal, it in no way seemed to promise a melting capacity of sufficient magnitude, yet, at the tapping it yielded a little more than one ship's pound of molten iron.[16]

Thereafter Broling described, with illustrations, what the cupola furnace looked like and the way it worked. Broling was also able to give information on the casting of cannons in sand moulds.

Among Broling's apprentices later in Stockholm we find founders of mechanical works such as Jean Bolinder and Theofron Munktell, both of whom also made study tours to Britain. A few years after Broling, Eric Svedenstjerna undertook a journey to Britain at the expense of the Ironmasters' Association, leaving in 1802 and returning in 1803. In the following year he published his book, 'Journey through a part of England and Scotland in the years 1802 and 1803', where he tells of his encounter with reverberatory and cupola furnaces.

Svedenstjerna's experiences from the journey to England were also reflected in the journal he himself published, *Samlingar i Bergsvettenskapen* (An anthology of mining and metallurgy), which can be regarded as a kind of forerunner to the *Jernkontorets Annaler* (Annals of the Ironmasters' Association). Here he wished to make the Swedish reading public familiar with his experiences abroad, and he expressed the opinion that Swedish foundries were crude in comparison with the British. In this context there is one essay of particular interest, dating from 1807, with the title 'Additional remarks on castings of pig iron and pig iron rollers', where, by way of introduction he remarks that 'it may indeed be the lack of a more general practical knowledge in these matters, which in this country more than any ' ther cause has hindered a wider use of iron castings', and that people have as a result of certain 'incomplete experiments . . . come to the general conclusion, that certain articles of cast iron cannot be manufactured from our ore or from our pig iron'. He then goes on to describe competently and methodically the effect of various qualities of pig iron and conditions of melting on the quality, the properties and the range of applications of the finished castings. The rational classification of pig iron according to

16. Ibid., part III, p.199.

specific uses was vital, Svedenstjerna emphasised.[17]

Svedenstjerna's more comprehensive opus, 'Some considerations concerning the British iron industry' was not published until 1813. Here Svedenstjerna emphasised, *inter alia*, the fact that different types of pig iron were commonly standardised in England, which gave the purchasers, the foundries, opportunities better to determine the iron's quality and thereby to judge how best to use it; this contrasted with the anarchy which reigned in Sweden, with all the concomitant failures.

Svedenstjerna returned later to the subject of English superiority. In 1818 he wrote in the 'Annals': 'While England has decided superiority as regards moulding and casting of all heavy articles, and in particular in those artfully constructed machine parts . . . and has ores and pig iron suitable for most castings, from blast-furnaces as well as from cupola and reverberatory furnaces, so this nation can favourably compete in the field of pig iron casting of all kinds.' He further underlined the higher prices of pig iron in Sweden and 'the near impossibility of buying it sorted', a matter, which was 'most pressing for the reverberatory furnaces'.[18]

Svedenstjerna was also the first to attempt to put to practical use what he had learnt in Britain in having the first cupola furnace built at the Åkers ordnance works in 1805. It differed from the English ones in that it was heated by charcoal. Yet Svedenstjerna remarked in his report on this furnace, which was 1.80 m high, that he had never seen the inside of a cupola furnace, nor had he been able to pick up anything through reading as to the design or the construction of the shaft.[19] The experiment nevertheless turned out quite well, and soon cupola furnaces were being set up at several iron works, while at engineering works one or several cupola furnaces became the rule.

As a third example of a journey backed by the Ironmasters' Association where observations on the casting of iron were reported, we may mention Carl David af Uhr's visit to Britain in 1820, of which his account was published in 1825. He pointed to the fact that reverberatory furnaces were on the decline and were mostly used for the production of larger castings, whereas the use of cupola furnaces was spreading: 'The use of these furnaces in England is by no means a new invention or branch of industry.

17. *Samlingar i Bergsvettenskapen*, vol.6, 1807, pp.131ff.
18. *JKA*, vol.2, 1818, pp.147ff.
19. *Samlingar i Bergsvettenskapen*, vol.2, 1806, pp.160ff.

They have been in use for many years, but by degrees they have in several respects been improved upon.'[20] Britain remained an important destination for study visits during the first half of the nineteenth century. It was generally considered that much was still to be learned from there. The journey undertaken by Jean Bolinder in the company of his brother Carl has already been mentioned, as has that of Theofron Munktell. Such visits were also made by the founder of Lindbergs's engineering works, Otto Carlsund, later manager at Motala after Fraser. We could mention others. In the 'Annals' for 1838, Johan Holmgren, an apprentice at Motala, reported specifically about foundries 'by reason of the well-known need for improvements in domestic foundries'.[21] He remarked that the foundry business in England and Scotland had now become a branch of industry in its own right, separate from the blast-furnaces. This observation was reinforced in a traveller's report by Ludvig Rinman after a journey through England in 1849.[22]

5 Conclusion

The preceding account has intentionally been kept fairly uncomplicated. The task was didactically a simple one – to describe how British technological innovations (remelting furnaces in the foundry industry) were transferred to Sweden, resulting both in a reduction of costs and in improved quality – this within a branch characterised by a strong increase in demand.

Technology transfer is a vague concept. Transfer from one country to another requires the local market, the relative level of costs, and the social and institutional preconditions in the recipient country to be in tune with the new technology. For an economic change to take place, technology must interact with other factors. These are important factors, which have not, however, been considered in this investigation.[23] Instead the actual transfer has been studied concretely in a practical context, linked with skilled workers and technical experts from Britain coming over to Sweden, and

20. C. D. af Uhr, *Berättelse om de på brukssocietetens bekostnad åren 1819–1822 vid Skebo bruk verkställda puddlingsförsök jämte några upplysningar om jerntillverkningen i England*, Stockholm, 1825, pp.176ff.
21. *JKA*, 1838, p.181.
22. *JKA*, 1850, p.17.
23. L. Jörberg, 'Teknikspridning och industriell förändring i Sverige under 1800-talet', in *Festskrift til Kristof Glaman*, Odense, 1983, pp.235ff.

with Swedish visits to Britain. This survey is based on sound contemporary documentation, thanks mainly to the interest shown by the Ironmasters' Association during this period.

One final question: did Sweden in fact have anything to teach Britain in the field of iron casting? One single example has been found. The blast furnace manager Carl David af Uhr refers, in a report published in the 'Annals' in 1827, to the testimony of von Platen that the castings for the Göta canal locks, made at the Swedish Finspong works, were far superior to the castings previously imported from England, and furthermore, that this 'provided the English engineer Mr Telford with the opportunity to increase his demands for perfection in the castings of English provenance which he uses'. To be on the safe side, however, af Uhr makes the reservation 'as I have recently been told by a tolerably trustworthy gentleman'.[24]

24. *JKA*, 1827, pp.65ff.

4

Industrial Espionage and the Transfer of Technology to the Early Norwegian Glass Industry

Rolv Petter Amdam

In 1741 Norway's first glassworks, Nøstetangen, was established approximately 60 km west of Oslo, or Christiania as the capital was then called. In the early years, the glassworks produced bottles and various kinds of flint glass, and by the 1750s and 1760s Nøstetangen had developed into a leading producer of high quality crystal in Scandinavia.

Nøstetangen was run by the Norwegian Company (Det Norske Kompani), a partnership which was founded in 1739 with mainly German and Danish owners, and with the king of Denmark as co-owner and active supporter. From 1751 the majority of shares were in Norwegian hands; from 1776 to 1824 the company was under state ownership.

Until 1814 Norway was in union with Denmark; the king of Denmark was king of both countries, and Copenhagen was the capital and financial centre. In the economic field, Denmark and Norway were in many ways a common market. As regards production, the relationship was characterised by a certain division of labour, based on the two countries' respective natural advantages. Denmark was rich agriculturally, while Norway's advantage lay in the sea, in mineral deposits and in the forests. Since a good stable supply of wood for fuel was a decisive factor in the location of the glassworks, it was natural that they should be placed in Norway.

The government in Copenhagen supported the company on the grounds of national economy. It was important to develop a domestic glass industry in Denmark-Norway to avoid imports and

Abbreviations: NAD = National Archives, Denmark; NAN = National Archives, Norway; PA = Private archive.

thereby improve the balance of trade. In the middle of the 1730s the economic policy increasingly promoted industry. The task of establishing a variety of manufacturing centres to refine domestic raw materials instead of importing finished products was given a higher priority than previously.

In the eighteenth century the company, under the protection of the authorities, founded six glassworks in addition to Nøstetangen. All were located in forest districts within about 150 km from Christiania. The largest expansion took place between 1756 and 1766 when the glassworks in Hurdal, Hadeland and Biri were built. Norway was thus quickly endowed with an industry which made all kinds of glass, except plate glass, for the Danish and Norwegian markets. The company's glassworks were protected against competition through a system of privileges which applied in both Denmark and Norway. From 1760 a ban on the import of glass was introduced. This monopoly lasted until 1803.

The glassworks were run by a joint administration, and each specialised in specific products. Thereby the glass industry acquired an organisational structure which was unusual in Europe. Above all the structure reflected the idea that the development of a glass industry should be planned in order to achieve the aim of producing sufficient quantities of all types of glass for both markets.

This chapter will firstly examine how, during the period of expansion in the 1750s and 1760s, economic planning underpinned an active policy of transferring technologies from abroad to the Norwegian glassworks. We shall then investigate how the rationale behind the acquisition of new, foreign technologies altered as the glass industry in the 1780s began to free itself from the mercantilistic planned economy and became more market-oriented. Finally, we shall discuss some of the consequences the transfers of technology had for the development of the Norwegian glass industry in the eighteenth century. To a large extent we shall draw upon material from developments in the production of window panes.[1]

1. An earlier version of this article was presented at the Nordic Symposium on the History of Technology in Stavanger, 14–16 June 1988, and The International Conference on History, Technology and Industrial Archaeology of Glass in Marinha Grande, Portugal, September 1989. I am most grateful for the comments I received there. I am especially indebted to Tore J. Hanisch and Ingvild Pharo, with whom I have had the pleasure of working on a newly published book on the history of the Norwegian glass industry over 250 years. They have also actively contributed with criticisms of earlier drafts of this article. For a general outline of the history of the Norwegian glass industry, I can refer to this book (in Norwegian): R. P. Amdam, T. J. Hanisch and I. Pharo, *Vel blåst! Christiania Glasmagasin og norsk glassindustri 1739–1839*, Oslo, 1989.

1 Industrial Espionage in Britain, 1754–6

In 1754 the Norwegian Company sent a young citizen of Christiania, Morten Wærn, on an assignment of industrial espionage to Britain. The main purpose of the journey was to obtain information on the manufacture of 'crown' glass with the idea of starting production of this finer sort of window panes in Norway. Drawings of furnaces and information on glass composition which Wærn obtained were useful. However, his most important task was to recruit skilled glassmakers who were willing to move to Norway. The aim, according to his travel instructions, was to procure 'as many people as are necessary for a glasshouse' that is to say, a glasshouse master, a gatherer, a blower and a finisher. Not least it was important that they all came from the same glasshouse, so that they could work well together.[2]

Behind the decision to send Wærn was the strong desire of Caspar von Storm, who had become the company's director in 1753, to start production of window panes in Norway. At this time the company had two small glassworks, but there were in total hardly more than ten glassworkers. The majority of these seem to have come to the glassworks on their own initiative or on the initiative of intermediaries who approached the company. Christian Fillion, a Frenchman, was one of those who came to Norway in this way. He arrived from the Kosta glassworks in Sweden after a postmaster called Christiansen, from Helsingør in Denmark, had applied to the company and asked whether Fillion could get work in Norway. Earlier Fillion had also worked at German glassworks.[3]

This method, by which several of the first workers introduced themselves, had roots in a migratory culture shared by many glassworkers in Europe during the sixteenth and seventeenth centuries. Many moved – even across national borders – when they heard about new glassworks, or they moved because of crises at the works where they were employed. Denmark too, which had had a few glasshouses for short periods during the sixteenth and seventeenth centuries, had seen such migrations. For example, a company was formed in Christianshavn in 1694 to make plate glass by casting according to a new method for which the French glassworks of

2. Storm to Wærn, 5 July 1754. All letters can be found in the company's letter books in NAN, PA no.1 unless another reference is given.
3. *Kosta Glasbruk 1742–1942*, Stockholm, 1942, pp.115–120. See also Storm to Christiansen, 17 August 1753.

Rolv Petter Amdam

Saint-Gobain were renowned.[4] The enterprise, which never properly got going, was led by a Frenchman called de Mohr, who had just emigrated from France. The reason for his departure seems to have been connected with the fact that a number of glassworkers at that time travelled abroad, leaving Saint-Gobain because of operational difficulties at the works.[5]

The result of this somewhat haphazard method of acquiring workers was that the Norwegian Company had workers who could make bottles, crystal and other pieces of glass, but who did not have sufficient knowledge of window pane production. The company therefore had to take active steps to acquire such knowledge if it was going to fulfil the goal of producing what Norway and Denmark needed in the way of window panes. Consequently it was important to send Morten Wærn to England; the king's advisers in Copenhagen eagerly followed the progress of his journey.[6]

At the time the company decided to begin window glass production it was by no means obvious that the technology was to be obtained from Britain. Around 1750 there were two main methods of manufacturing window glass, and those in charge of the Norwegian Company were aware of both. In the German territories cylinder glass or broad glass was most common; in Norway this was called 'fensterglass'. The method of producing this glass was generally known throughout Europe. First, a long glass cylinder was formed by blowing and swinging molten glass over a swing hole. Both ends were cut off so that the cylinder became open. It was then opened lengthwise and rolled out and flattened on a stretching sheet.[7] The manufacture of crown glass however, was based on a completely different technology which was hardly known outside France and England.[8] The glass mass was blown into a small globe. A solid iron rod was attached to it opposite the blowpipe which was then cracked off. After intense reheating the globe was rotated at considerable speed so that centrifugal force acting on the edge of the opening where the pipe had been, caused the glass to be flung outwards, forming a flat disc or 'table' with a diameter of 35–45 inches. When this had cooled down, the result was a round table without any marks except a central 'bull's eye'

4. NAD, Sjællandske Reg. 1693–5, pp.316–18.
5. J. R. Harris, 'Saint-Gobain and Ravenhead', in B. M. Ratcliffe (ed.), *Great Britain and her World*, Manchester, 1975, pp.33–4.
6. Letters between Storm and Wærn, 1755.
7. T. C. Barker, *The Glassmakers*, London, 1977, pp.23–4, 57.
8. Ibid.; see also G. Wohlauf, *Die Spiegelmanufaktur Grünenplan im 18. Jahrhundert*, Hamburg, 1981, p.35 (crown glass in Germany).

left by the iron rod. This was cut away, and the rest was cut into window panes. Partly because the composition was better, and partly because crown glass did not come into contact with a sheet of metal, as cylinder glass did during the process of stretching, the quality of crown glass was higher than that of cylinder glass. But the price was also higher.

The director of the company, von Storm, first thought of using German technology and proposed sending a Norwegian glass blower to Pomerania or Mecklenburg to recruit workers who could produce cylinder glass. In fact, he did send someone to these areas at this time, but to recruit workers for bottle production.[9] When it came to window glass, however, he finally decided to invest in British technology rather than German, and in quality rather than low price.

There were two reasons for this: firstly, the company still had the stamp of primarily being a supplier of luxury products for a limited market in the upper social strata of the population. Nøstetangen had begun to make crystal of high quality, and the General Assembly meeting in 1753 emphasised that the company's main task was to supply Denmark and Norway with 'necessary crystal products'.[10] During this period the glassworks were strongly influenced by their role as suppliers to the royal court. For example, the king's wine cellar was the company's largest single customer in the 1750s, accounting for one-third of the sale from Norwegian glassworks during this period.[11] It therefore seems natural that the company also chose to pursue high quality in the production of window glass.

Secondly, the director, von Storm, was in general enthusiastic about what he called the British 'highly perfected factories'.[12] Britain's leading role in the industrial field had at this time begun to be recognised by the rest of Europe, and this admiration seems to have played an important role in the choice of technology for the first window glass factory in Norway.

Crown glass had come to England from Normandy with French glassworkers in the 1560s, and Britain had quickly taken the lead in crown glass production.[13] Now, however, the British authorities were concerned to prevent knowledge of the process from spreading further. Whereas a country which was intent upon building up

9. NAD, Moltke archive, Storm PM, 17 March 1753.
10. NAD, Moltke archive, General Assembly minutes, 19 June 1753.
11. NAN, PA no.1, ledger.
12. Storm to Moltke, 9 August 1755.
13. Barker, *The Glassmakers*, p.24.

a new industry made great efforts to bring technology *into* the country, it was in the interest of the country where the technology was already known to prevent it from spreading further. This was a typical feature of the relations between nations under mercantilism where the attitude that a nation's riches had to be acquired at the expense of others was prominent.

Britain had a system of laws aimed at preventing the spread of new technology. Until 1824 British skilled workers were prohibited from emigrating. If they did, they lost their citizenship, and the recruitment agents were liable to a fine of £500 and one year's imprisonment for every worker they recruited.[14] Nevertheless, studies on technology transfers from Britain to other countries have shown that these laws were not very effective.[15] Widespread industrial espionage was particularly important in the spread of technology internationally. The English historian J. R. Harris has characterised the importance of industrial espionage in the eighteenth century as follows: 'Espionage was a major means by which new technology was actually transferred or attempted to be transferred.'[16]

Even though there were opportunities to take part in the British technological miracle, Morten Wærn had no easy task. His journey was full of drama. From 1754 to 1756 he travelled in England and Scotland – and later in France. First and foremost, he visited the main centre of the English glass industry, Newcastle. He also visited glassworks in Leith, Hull, Bristol, Liverpool, London and Yarmouth. The company claimed that he visited 'all the glasshouses' in England and Scotland.[17] Morten Wærn had to proceed with great caution so that his assignment was not revealed. Letters to Morten Wærn from Norway were sent through intermediaries, such as businessmen in London, because 'several letters are opened in the post'. In the letters individuals were often referred to in code.[18]

Nevertheless, in the end Wærn was unmasked and arrested in London in the summer of 1755, accused of having obtained information about the composition of crown glass and crystal, and of having enticed workers to leave the country. After almost one

---

14. J. R. Harris, 'Industrial Espionage in the Eighteenth Century', *Industrial Archaeology Review*, vol.7, no.2, 1985, p.129; D. J. Jeremy, 'Damming the Flood: British Government Efforts to Check the Outflow of Technicians and Machinery 1780–1843', *Business History Review*, vol.51, no.1, 1977, p.3.
15. P. Mathias, *The Transformation of England*, New York, 1979, p.39.
16. Harris, 'Industrial Espionage', p.127.
17. Storm to Moltke, 9 August 1755.
18. Storm to Wærn, 10 June 1755.

month in Newgate prison he managed, after having been released on bail, to escape to France. All the drawings and information he had acquired were rescued and sent to Norway. In France he continued his journey to glassworks in Cherbourg, Nantes, Bordeaux, Le Havre, Saint-Gobain in Picardy and Strasbourg. The whole time he kept in touch with the administration in Norway, which frequently sent him instructions about how to proceed.[19]

In spite of the difficulties, Morten Wærn's expedition was successful. At the same time as he was arrested, five crown glass workers arrived in Christiania, probably from Newcastle. The five, William Fagnell, Joseph Pyne, Thomas Sims, James Swinger and Joseph Thomson, were immediately sent to Hurdal, approximately 80 km north of Christiania, where they began to construct the first crown glass factory in Norway.[20] At the same time two crystal glass workers, James Keith and William Brown, arrived from Liverpool. Morten Wærn had also been given the task of obtaining information on how English glassworks made crystal, and of recruiting skilled workers in this field.[21]

A secondary aim of Wærn's journey was to obtain insight into how the British made kelp by burning dried seaweed.[22] Kelp is soda, which was used as raw material in the production of glass. Wærn's stay in Scotland was specially designed to obtain information on the burning of kelp, and he sent a number of drawings of kelp furnaces back to Norway. But he did not manage to recruit kelp burners until he came to France in the second part of his journey. In northern France he persuaded Pierre Dufresne and his colleague Antoine to travel to Norway and teach Norwegians how to burn kelp. For several years the two Frenchmen travelled from the area around Trondheim in the north to Stavanger in the south and started up kelp production. These travels were the prelude to a flourishing business along the coast of west Norway during the later part of the eighteenth century, so that the country not only had enough kelp for its own use but also for export. In 1800, for example, kelp was sold for a total amount of 20,000rd (riksdaler) from Kristiansund, of which 78 per cent was exported.[23]

Morten Wærn's journey is not only one of the high points in the

---

19. Unfortunately only outgoing letters are preserved in NAN, PA no.1.
20. Storm to Wærn, 9 August 1755.
21. Storm to Wærn, 5 July 1754.
22. Ibid.
23. G. E. Christiansen, *De gamle privilegerte norske glassverker og Christiania Glasmagasin*, Oslo, 1939, vol.1, pp.81–8; A. O. Johnsen, *Kristiansunds Historie*, Kristiansund, 1958, vol.3, pp.658–9.

history of Norwegian industrial espionage; it also illustrates several important features of technology transfer in the eighteenth century. Firstly, it illustrates, through the five crown glass workers who came to Norway from England, how intimately technology transfer was linked with skilled workers, and that this was the most important element of technology transfer.[24]

Secondly, the journey was an expression of an active policy to bring new technology to the country, a policy which was supported by the general industrial policy of the government. The government in Copenhagen was informed about Morten Wærn's journey, and the company had frequent contact with the king's close adviser, Count Adam G. Moltke, when they worked for the release of Wærn from Newgate.[25] Strong government commitment on the part of the recipients was a general feature of the attempts to transfer technology from Britain to the other European nations during this period.[26] As regards Denmark and Norway, this encouragement and support for transfers of technology were, in addition to the use of privileges and an embargo on imports, the most important government instruments used to advance new industrial initiatives.[27] The economic policy was also characterised by a positive attitude towards immigrants. According to a 1735 manifesto for a new economic policy, it was 'advantageous to attract as many foreign workers as possible'.[28] This was followed up by using public funds to give immigrants their passage money, initial capital and resources to pay back loans taken up with their former employers. As far as the Norwegian glassworks were concerned, it was, however, the company itself which paid these travel expenses.

2   The Role of the Economic Policy

The Norwegian Company did not become a profitable enterprise, and the glassworks operated with large losses up to the end of the 1780s, that is for nearly 50 years. In the years up to 1759 incomes from sales did not cover more than 5 per cent of total costs in any

24. See J. R. Harris, 'Spies who sparked the Industrial Revolution', *New Scientist*, 22 May 1986.
25. Storm to Moltke, 9 August 1753.
26. Mathias, *The Transformation of England*, pp.25–7.
27. K. Glamann and E. Oxenbøll, *Studier i dansk merkantilisme*, Copenhagen, 1983, pp.87–9, 104–8.
28. K. Glamann, *Otto Thotts Uforgribelige Tanker*, Copenhagen, 1966, p.105.

Table 4.1 The Norwegian glass industry, expenditure and income
from sales, 1751–69 (in riksdaler)

	Expenditure	Income from sales
1751	1,512	0
1752	5,447	54
1753	6,169	37
1754	10,973	12
1755	12,249	240
1756	21,695	80
1757	21,600	600
1758	19,031	831
1759	13,434	2,274
1760	22,741	3,407
1769	61,000	21,000

Source: NAN, PA no.1, ledgers

one year. As late as in 1769 the accounts show 20,000rd in income
against 60,000rd in expenses (see Table 4.1).

The reason for this was that large warehouse stocks accumulated,
including stocks of crown glass, which the company did not
manage to sell.[29] Before the ban on imports in 1760, the customers
preferred to buy foreign glass, which was cheaper. Also the company
had only rather vague ideas about the size of the market. With
hindsight, one can see that the production capacity of the first
crown glass works in Hurdal was very large in relation to the
amount of window glass imported during the immediate period
prior to the establishment of the enterprise. During a normal week
in the early years, as much window glass was produced as Norway
imported in one year round 1750.[30] Even if the Danish market was
somewhat larger, in reality this was a rather optimistic undertaking. In the towns window glass had, it is true, begun to be
commonly used, but this was yet not the case in the countryside.

One can of course argue from today's point of view that such a
loss-making venture could not have been especially good for the
national economy. Nevertheless, as we have already indicated, it
was national economic considerations which lay behind the establishment of a Norwegian glass industry and behind the decision to

29. NAN, PA no.1, ledger.
30. The weekly production was twenty-four baskets, and the import to Norway
was approximately twenty baskets a year, NAN, PA no.1, no.35.

acquire technological knowledge to start production of crown glass. Norway was to produce glass, cost what it may, to stem the tide of imports of finished products. One is therefore left with the impression that financial considerations played a minor role in the decision to acquire new technology for the Norwegian glass industry.

Regarding the decision to bring English crown glass workers to Norway, this impression is to a large extent correct. Morten Wærn's travels cost the company nearly 5,000rd, the equivalent of wages for approximately twenty skilled workers for one year. But the profit motive even at this stage was present as a subordinate aim. Those in charge had great expectations of future profit. Storm declared as early as 1753, when he took over the management of the glassworks, that if only he could get enough skilled workers, the company 'yearly could expect over 1,000rd sheer profit'.[31] In spite of large losses, this aim was not abandoned; such optimism was caused by the economic policy of the government which fostered expectations of future profits. In particular, the prospect of being protected from competition by a ban on imports kept up their spirits. With only 6 per cent duty on imported glass in the 1750s, the Norwegian glassworks were not able to compete with foreign glass in price.[32] The company demanded an import embargo from the government and they had good grounds for expecting such a ban. This was because an import embargo had been passed in several other industrial fields. However, in order to pass such a resolution, the government stipulated that the producer should undertake to manufacture the amount necessary to meet the demand in Denmark and Norway for the product in question.[33] A number of letters which the company sent to Copenhagen at the end of the 1750s had the object of convincing the government that the glassworks were now capable of manufacturing enough glass of all types.

We can also see that the profit motive had some importance when we consider the way in which the company appraised the possibility of using coal-fired furnaces in the glassworks – something that Morten Wærn had become acquainted with in England. He concluded, however, that the price of imported coal would be so high that its use would not pay when compared with wood fuel, which was abundant in Norway. A plan to establish a glassworks

31. NAD, Moltke archive, 21, Storm PM, 11 March 1753.
32. A. Rasch, *Dansk toldpolitikk 1760–1797*, Aarhus, 1955, p.328.
33. Ibid., p.23.

based on coal in Jarlsberg, on the coast south of Christiania, was therefore rejected. For the same reasons, a proposal in 1767 to change to coal-fired furnaces at Nøstetangen was also shelved.[34] Likewise, plans to start plate glass production by casting, which Morten Wærn had also studied in France on his journey,[35] were abandoned. Such a production demanded large buildings and costly investments.[36] When plate glass production was considered most seriously in the 1780s, the directors calculated the cost of the required construction at nearly 16,000rd: 'This seems to be too much to risk'. Norway was never to embark on plate glass production.[37] Profit considerations therefore clearly influenced decisions about which technology one should not attempt to transfer.

In 1760, as mentioned, the import ban was passed. The prospect of running at a profit under the protection of an import ban provided the motivation to make larger investments and to acquire the necessary technology in order to enable the glassworks to produce as much glass as the government considered was needed in Denmark and Norway.

As we have already mentioned, the Norwegian Company operated at a loss even after the import ban was passed in 1760. Their expectations did not materialise. An apparently paradoxical situation arose when stocks of crown glass increased at the same time as Norwegian and Danish glaziers complained that they were short of window glass. Rather than buying crown glass, the consumers preferred cheaper German cylinder glass (*fensterglass*) which the company had decided not to produce because, in their opinion, the quality was so poor that it disintegrated after six to seven years.[38] The government, however, gave in to consumer pressure and permitted the import of this type of window glass during the 1760s.[39]

The company was therefore faced with the classic problem of every planned economy – producing too much merchandise the consumers did not want, and too little of the products in demand. In the end, von Storm surrendered to the wishes of the consumers, and in 1761 the company decided to establish yet another glassworks, namely Biri glassworks for the production of *fensterglass*. As

34. Christiansen, *Christiania Glasmagasin*, vol.1, p.519.
35. Storm to Wærn, 18 June 1756.
36. Harris, 'Saint-Gobain and Reavenhead', p.27.
37. NAD, Handelskomp., Delib. protokoll 1782–4/335.
38. Storm to Wærn, 18 April 1760.
39. NAD, Forestilling 1762/1, 1766/43; Christiansen, *Christiania Glasmagasin*, vol.1, pp.112–14; Rasch, *Dansk toldpolitikk*, pp.91–4.

in the case of crown glass, this decision implied that skilled glass-workers had to be imported from abroad. Production started in 1766, after Peter Holm, whom the company had sent to Pomerania and Mecklenburg, had managed to get hold of skilled workers.[40] The dispensation from the import ban as regards window panes was abolished, and the company hoped that the results would improve.

The government had, by its policy of creating expectations about producing in a protected market, encouraged both an expansion of the Norwegian glass industry and an active attitude towards the transfer of technology for this purpose. In this way the ideas behind the planned economy played a positive role in the establishment of a Norwegian glass industry. From the 1770s onwards, however, the negative aspects of this policy became more obvious.[41]

3 Adjusting to the Market and the Transfer of Technology

At the beginning of the 1770s the glassworks experienced a major crisis accompanied by shutdowns and running down of stocks. Underlying the crisis was the owners' emerging reluctance to cover losses as long as the glassworks were run according to the authorities' directives. The reason why we can argue that the development of the Norwegian glass industry involved an element of planned economy is twofold. On the one hand it was a process of gradual, planned expansion to meet demand as estimated by the government and the directors. On the other hand, planning is also evident in the severe restrictions imposed upon the company; for example, they were obliged to produce and stock at all times in the largest towns of Denmark and Norway all the 600 different products which were presented in a sales catalogue of 1763.[42] This was irrational, since it meant that they were left with sizeable stocks of scarcely salable products but not allowed to lower prices to speed up sales.

An increasing number of owners argued that these regulations were an obstacle to profitable management, and since they did not manage to obtain any relaxation of the rules, they sold out to the government in 1776. Government resistance to the former owners'

40. Christiansen, *Christiania Glasmagasin*, vol.1, p.116; vol.2, pp.215–20.
41. This is thoroughly discussed in Amdam, Hanisch and Pharo, *Vel blåst!*, chap. 1.
42. NAN, PA no.1, Weyses katalog.

request was based on the fear that such changes would lead to the glassworks no longer being able to manufacture the types of glass required for the Norwegian and Danish markets. National economic considerations thus stood opposed to business economics, and the former proved the stronger.[43]

Paradoxically, the inflexible regulations were repealed during the course of the first ten to fifteen years after the government took over. The state management of the glassworks during the 1780s began to organise the running of the works with the aim of returning a profit, and both the obligation to manufacture and stock hardly salable products, and the fixed prices were abolished. With the government in command, the glassworks entered a process of liberalisation which culminated in the abolition of the system of privilege and the lifting of the embargo on imports in 1803.

During this period production was organised to meet demand, and liberated from a policy of planning. From now on production growth followed the business cycle (see figure 4.1). This was accompanied by changes in technology policy. While decisions concerning technology acquisitions by the Norwegian glassworks had earlier been tied to the question of how to fulfil the plans that had been set up, technology was now increasingly acquired in order to enable the glassworks to meet demand in periods of prosperity.

In 1781 the state directors of the glassworks decided to implement the old plans to use coal at a new bottle glassworks on the coast just south of Christiania – Schimmelmann's glassworks. Skilled workers were recruited during a trip to Hesse.[44] At this time the glassworks had experienced a period of quite astonishing increases in sales. In 1781 the glassworks sold to the value of 113,000rd compared to 28,000rd in 1776 and approximately 20,000rd in 1769. Growth in production had lagged behind, but its value had even so increased from approximately 60,000rd in 1769 and 35,000rd in 1776 to 87,000rd in 1781 (see figure 4.1). Complaints poured in from dealers who did not get enough glass. This applied especially to bottles, which were the main product in addition to window glass, and which showed the biggest sales increase. In 1781 bottles comprised 60 per cent of total sales. In the 1790s bottles accounted for 40 per cent of turnover, window glass about 40 per cent, and flint glass about 20 per cent.[45]

43. See Amdam, Hanisch and Pharo, *Vel blåst!*, chap. 1.
44. NAN, Finansarkivene XXI,11,3, Fabrikkdir. to Statsbalansedir., 7 July 1784.
45. NAD, Finanskollegiet, 3, Schimmelmann papers.

Figure 4.1 Norwegian glassworks: production, turnover and imports, all types of glass (in Riksdaler)

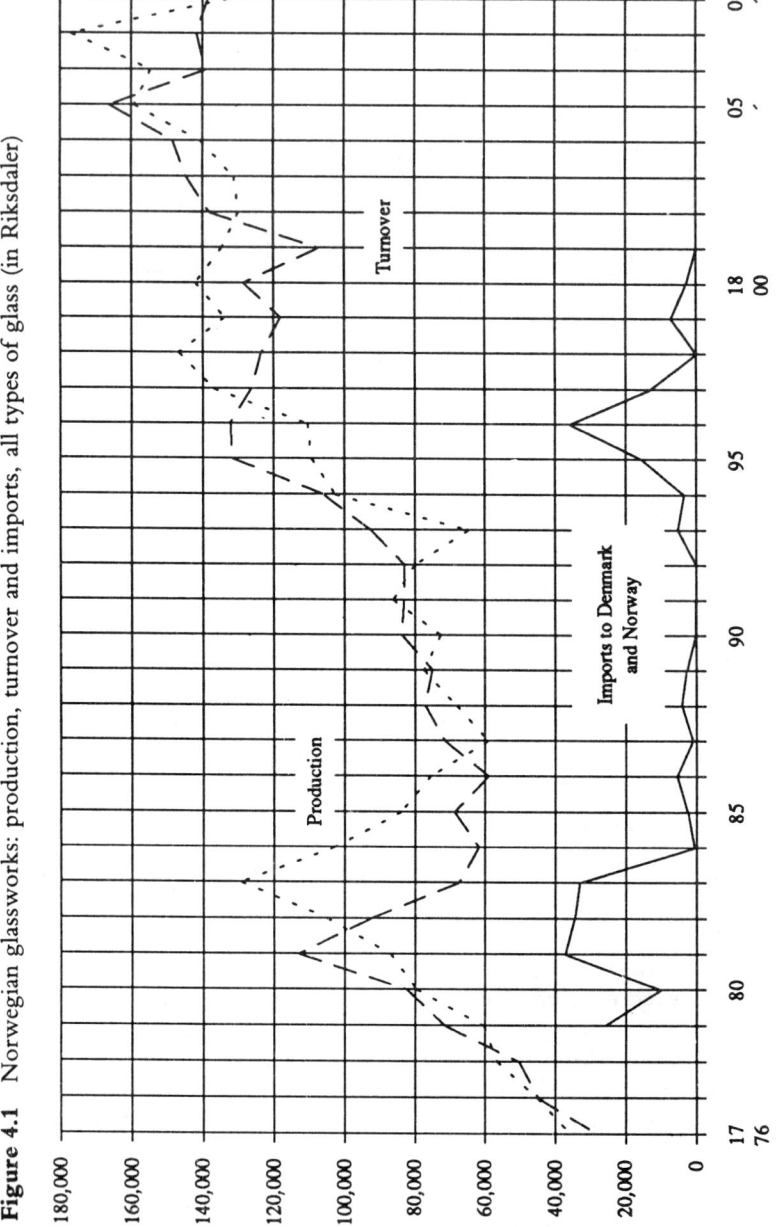

Note: Imports after 1800 are not known.

During the sellers' market round 1780 it was not just a question of producing enough glass, but of delivering it quickly enough to Copenhagen, which was the largest market. Norway's natural advantage as a glass manufacturer was an ample supply of wood for fuel. Consequently the glassworks were situated near forests. As glass could best be transported in wintertime on snow-covered tracks and lakes which were frozen over, their location was not very suitable for delivering large quantities of glass quickly. This was the background for the launching of a plan by the management of the glassworks to move glass production nearer the coast.[46] It was in this connection that the use of coal-fired furnaces was of great potential interest; rapid transport between the glassworks and the market was more important than the high price of imported coal. Norway had no coal deposits. This was a decisive factor in the management's wish to recruit glassworkers who were able to produce bottles using coal as fuel.

In 1784 the new coal-fired glasshouse was ready, and German glassmakers were brought to Hurum both to build the furnace and to start up production. But then business conditions deteriorated, and the sale of bottles was reduced from 72,000rd in 1781 to 32,000rd in 1784. Transport distance now hardly mattered, and the attempt to run the works on coal was abandoned after a short time because it was too expensive.

Attempts to apply this technology were repeated at the same works in 1795 after a new upward trend. From 1792 to 1795 turnover for all glassworks increased from 83,000 to 131,000rd. The sales of window glass in particular grew, and a new group of British glassmakers were brought to Norway to produce crown glass with coal as fuel. But again the glassworks suffered a drop in sales at the same time as the British workers came to Norway. The cost factor again became more important than the need for quick delivery to the market, so that this attempt to apply new technology was abandoned after a short time also.[47] The next attempt to run a glassworks with coal in Norway took place when British interests established the Høvik glassworks just outside Christiania in the 1850s.

As mentioned above, the upswing in the 1790s was above all linked with the growing market for window glass. The decrease in sales which the glassworks experienced in the 1780s was primarily linked to bottle production, while sales of window glass, however,

46. C. Anker PM, 31 January 1778.
47. Christiansen, *Christiania Glasmagasin*, vol.1, pp.272–4; vol.3, pp.238–40.

increased a little: crown glass from 12,000rd in 1782 to 16,000rd in 1788, cylinder glass (*fensterglass*) from 8,000 to 12,000. In 1788 this corresponded to 1,000 baskets – or 90,000 lb – of crown glass, and about the same amount – 1,200 chests – of *fensterglass*.[48] The glaziers in Copenhagen complained that they did not get enough glass, and that they were not supplied with the type of glass their customers wanted.[49] This was the background for the management's decision to start production of a new type of cylinder glass, called 'tafelglass'. The method of producing *tafelglass* closely resembled that of the simpler *fensterglass*, but in quality it came close to crown glass. In price it was about halfway between the two. To start this production, technological know-how had to be acquired from abroad. The management arranged the recruitment of *tafelglass* workers from Bohemia, and with the appointment of one Frants Landgraf as glasshouse master a new glassworks was ready in 1794 near Lillehammer.[50]

Whereas coal-based crown glass production did not survive the recession from 1796 onwards, the transfer of *tafelglass* technology from Bohemia proved more viable. This transfer represented an important addition to the Norwegian glass industry when trading conditions again improved from 1803. In 1806/7 window panes accounted for 45 per cent of total sales from Norwegian glassworks, and reached a figure of 250,000–300,000 lb. Of this amount, *tafelglass* made up 40 per cent in terms of value, crown glass 37 per cent, and *fensterglass* 23 per cent.[51]

The years round 1800 were a golden age for the traditional Norwegian glass industry. The establishment of an industry capable of supplying more or less the entire demand for glass in Denmark and Norway had successfully been accomplished. Only during the boom years round 1780 and in the mid-1790s did registered imports to Denmark and Norway exceed 5 per cent of the Norwegian production. During these periods, however, it reached as high as 35 per cent (figure 4.1). For a couple of years in the early 1780s and in 1796, imports of window panes equalled between 70 and 80 per cent of sales from Norwegian glassworks; otherwise it rarely rose above 5 per cent.

In 1824 the government withdrew from its engagement in the administration of the glassworks, and the glassworks were sold to

48. NAD, Finanskollegiet, 3, Schimmelmann papers.
49. A. J. Barclay, *Monopoliers skadelige Følger* . . ., Copenhagen, 1788.
50. Amdam, Hanisch and Pharo, *Vel blåst!*, chap. 1.
51. NAN, Finanskassedir., journal.

private interests. When Norway broke away from Denmark in 1814, the Norwegian glassworks lost Denmark as a domestic market. Moreover, Denmark soon established her own glassworks, with Holmegaard as the principal one. This contraction of the market, in conjunction with a general recession in Norway, led to severe sales difficulties for Norwegian window glass. The problems were made worse since imported window glass now began to gain a firm foothold in the Norwegian market. A customs duty of 20 per cent was not enough to stem the tide of foreign glass.

Up until 1835 the annual import of window glass to Norway was around 10,000 lb, equal to less than 10 per cent of Norwegian production. In 1835 imports increased to 42,000 lb, and in 1845 to 177,000 lb. At the same time the stocks of the Norwegian glassworks increased, and in about 1845 imports overtook domestic production.[52]

The success of foreign glass was due to the backwardness of the Norwegian window glassworks in comparison to technological development on the Continent: the Norwegian *fenster-* and *tafelglass* was based on the cylinder glass method, which, on the Continent, had been significantly improved. The most important improvement was in the cutting and stretching of the cylinder – a method which gave better glass quality than *fenster-* and *tafelglass*, and equalled that of crown glass. Furthermore it only cost about 75 per cent of the cost of crown glass.[53]

In Norway the old *fenster-* and crown glass were not abandoned, even when the import of the new glass began to be a serious competitor. However, it was not only in Norway that the new glass became a serious challenge. In Britain too crown glass was favoured for a long time for tax reasons, and the result was that especially Belgian glass increased its share of the market at the expense of the British glassworks in the 1830s.[54]

The problems caused by the new wave of imports at the end of the 1830s led the directors of the Hurdal glassworks to decide to introduce this new technology to Norway as well. In this instance the clear trends of a looming crisis for the old method of window glass production provide the explanation of why the new technology was brought to Norway. We know little about how this transfer took place, except that the Hurdal glassworks dispatched the first crates of this new glass, or 'stretched glass' as it was called

52. NAN, Central Bureau of Statistics, various cases, 24 and 55.
53. NAN, PA no.580,46, C. Berg to Suhr 1 December 1840.
54. Barker, *The Glassmakers*, pp.56–9, 117–19.

at first in Norway, in 1840. Initially the firm seems to have had some problems in making glass of high quality. These manufacturing problems prompted the owner, Petter Tanberg, to travel to Germany to recruit skilled workers in the mid-1840s. So as not to arouse suspicion, he acted as a clock trader on his visits to the glassworks. When he returned to Norway, he not only had with him several glassmakers, but also several wall clocks, which the glassworks subsequently sold off.[55] This journey seems to have played a crucial role in solving the problems of the glassworks; in the early 1850s the decline of Norwegian window glass production was halted.[56]

4 Foreign Workers in Norway

The building up of the Norwegian glassworks in the eighteenth century was based on an active policy of technology transfer through the use of foreign skilled glassworkers. Four years after the arrival of the five British crown glass workers in 1755, Kay Brandt was sent on a new journey of industrial espionage to Mecklenburg and Pomerania to recruit German workers.[57] As we have seen, these journeys were followed by several others.

The transfer of technology through immigrant workers was in most cases a success. The director, von Storm, maintained as early as 1759 that the company had succeeded in developing the manufacture of crown glass 'to a state of perfection which cannot be rivalled anywhere else in Europe'.[58]

The building up of the glass industry took place, then, with the help of foreign workers. The main method of expanding the workforce, especially when introducing new technologies, was to recruit additional workers from outside. In 1801 the Norwegian glassworks employed 160 men. Of these, approximately 110 were skilled production workers, of whom fifteen were of British origin, working in crown glass production. Just under sixty were of German descent, and they worked mainly with the production of flint glass, bottles and cylinder glass. Approximately forty were Norwegians who had been trained by the foreign glassmakers.[59]

55. Christiansen, *Christiania Glasmagasin*, vol.2, p.198.
56. Amdam, Hanisch and Pharo, *Vel blåst!*, chap. 3.
57. NAN, PA no.1, 53, travelling instructions, 23 October 1759.
58. Storm to Moltke, 11 August 1759.
59. NAN, national census, 1801.

One of the main questions is why foreign craftsmen moved to Norway. Certainly they were enticed by high wages. Those who went to recruit foreign workers were specifically instructed to offer high wages. When Morten Wærn visited England in 1755, the director of the Norwegian Company wrote to him saying that he could certainly offer the workers 26 English shillings per week. 'But if there is one who is indeed a great master . . . then it matters not if the pay is to be 30–4 shillings per week', he added.[60]

The company paid high wages in spite of the fact that it ran at a loss throughout almost the whole of the eighteenth century. Thus the Norwegian glass industry is at variance with the findings of the British historian Peter Mathias. He refers to several attempts at technology transfers to unprofitable enterprises in the eighteenth century and argues that 'unprofitability is doubtless one of the most common explanations for the failure to diffuse an innovation'.[61] In the case of Norway, strong government involvement was not an obstacle to successful transfers; on the contrary, it was a prerequisite for successful adaptation as it enabled the glassworks to offer high wages. Large wage expenditures were compensated for by high prices, which during the eighteenth century were 10–50 per cent higher than for foreign glass, depending on the type of glass. Because the glassworks were protected through the import ban, it was possible to charge a higher price and at the same time dominate the market.

Of the five crown glass workers who came to Norway from Britain in 1755, Thomas Sims and James Swinger seem to have returned after a short time. William Fagnell, Joseph Pyne and Joseph Thomsen, however, remained at Norwegian glassworks until they died. In the same year, 1755, a crystal glassmaker called James Keith and an assistant arrived from Liverpool. They were also recruited by Morten Wærn. It was James Keith who above all put his mark on the design of Norwegian crystal during its heyday in the 1760s. One year after his arrival his wife came to Norway. The others were either followed by their wives, or married in Norway. The traditional crafts were through several generations passed on from fathers to sons.[62] Most of the German glassworkers likewise settled in Norway.

It was vital for the development of the glass industry that the majority of the foreign workers remained in Norway. High wages

60. Storm to Wærn, 10 May 1755.
61. Mathias, *The Transformation of England*, p.39.
62. Christiansen, *Christiania Glasmagasin*.

were the main incentive to stay more than a few years. A master in the 1760s earned as a rule 24rd a month, and a skilled glassworker about 16rd. This was twice as much as the masters and skilled workers were paid at the copper works and at the country's only salt works. The smith at the glassworks, who was Norwegian, earned 9rd.[63]

However, high wages were probably not the only factor. The government's positive attitude towards immigration was matched by an equally encouraging immigration policy on the part of the directors of the glassworks. While the general flow of knowledge from Britain in this period was carried both by entrepreneurs and skilled workers,[64] it was only the latter who came to the Norwegian glassworks. In general the administration of both the individual glassworks and of the whole glass industry was managed by Norwegians – possibly with a Danish background – without any technical knowledge of glassmaking. Caspar von Storm, the driving force in the 1750s and 1760s, was a typical example of this. He had a military background and had no experience in the running of glassworks prior to his appointment as director. In his application he wrote about why he was qualified for the job: 'I can most humbly assure you that I know the country, have honest intentions and posess the maturity required by the position.'[65]

Storm recognised the vital role of foreign skilled workers, perhaps because of his own lack of technical knowledge, and he actively sought to resolve the conflicts which developed both between immigrant workers and the management, and amongst groups of immigrants. He attached importance to putting together working teams according to nationality, because 'of the prevailing hatred between nations'.[66] It did occur that German and English workers came to blows, and at times the friction between them led to sabotage; in one instances workers damaged the molten glass by putting water in the furnace.[67] Storm also put great emphasis on ensuring that the newcomers' religious customs were respected, and he saw to it that priests from the Reformed and Roman Catholic Churches were available. Furthermore Storm criticised the management in some of the glassworks because they disregarded the British workers' demand for no work on Sundays.[68]

63. Amdam, Hanisch and Pharo, *Vel blåst*, appendix 1.1.
64. Mathias, *The Transformation of England*, p.30.
65. NAD, Moltke archive, 21, Storm to Moltke 15 April 1753.
66. Storm to Wærn, 30 August 1755.
67. Storm to Schierraad, 15 September 1755.
68. E. Mæhlum, 'Glassverk og glasspustere', in *Norsk Kulturhistorie*, Oslo, 1938, vol.3.

The glassworks increasingly came to be surrounded by groups of cottages where the workers lived. The works were quite clearly separated from the neighbouring peasant community and had, for instance, their own schools. For generations the traditions and customs of workers from Britain, Bohemia, Pomerania and other German areas were kept alive, partly because new groups of workers kept coming. Consequently we can speak of a gradual assimilation into Norwegian society. Initially the local population was hardly ever directly employed in glass production, but was connected to the works through deliveries of firewood and the transport of glass. Thereby the peasants were absorbed into activities which gave them a considerable extra income. This probably helped to soften the state of tension which tends to arise when different cultures meet. The answer to why the first large contingents of foreign glassworkers stayed on in Norway may be connected with the respect they were shown upon arrival and the opportunities they were given to maintain their cultural traditions over a longer period within the glassworks community. This is an area of research in Norwegian history which has so far been hardly explored.

5

Borrowing and Adaptation of British Technology by the Swedish Iron Industry in the Early Nineteenth Century

Rolf Adamson

1 General Observations about Transfer of British Technological Know-how to the Swedish Iron Industry, 1760–1840

During a considerable part of the eighteenth century and at least the first half of the nineteenth century, very important transfers of technological know-how took place from the industrial pioneer country, Britain, to the Continent. Three technological fields have often been emphasised: energy (the steam–engine), textile machinery and iron production methods. In this chapter a limited part of the third field, the borrowing and adaptation of certain British production methods by Swedish industrialists, will be discussed.[1]

Three forms of technology transfer within the iron industry can be identified: one in which entirely new products and processes

1. This article has a long history. Some components were included in an unprinted licentiate dissertation, *Järnexport över Göteborg under 1800-talets första hälft, avsättningsmekanismerna och marknadsförhållandena* (1956). During the end of the 1960s and most of the 1970s efforts were made in spare time to extend the treatment. Some results were summarised at a conference for Scandinavian economic historians at Utstein Kloster, Norway, *Överföring och anpassning av brittiskt tekniskt kunnande till svensk järnhantering i början av 1800-talet*, mimeo (1980). Parts of sections 5(b), 6(a) and 6(b) in that paper have been published in Swedish, 'Lancashiresmidets införande vid Bäckefors bruk', in *Dalsland, Svenska turist- föreningens årsskrift*, NACKA, 1981, pp.107–14. During the early 1980s a broader source material was scrutinised concerning marketing conditions in Britain, exporting merchants in Swedish ports, and wages and work organisation at Swedish iron factories. Because of other pressing duties, I have not yet been able to write about those more comprehensive subjects. Thus I here give a slightly revised version of the paper from 1980, which also means that few recent works of interest have been included.

were introduced; a second in which existing Swedish processes were modified from within by the use of foreign technologies; and a third in which related products were developed. The first category of transfer corresponds quite closely to ordinary ideas about technology diffusion to the Continent. Production processes which were earlier nearly unknown in Sweden were introduced there: from about 1760, a small number of new foundries and mechanical workshops were built. Unlike the established Swedish bar iron works, which were situated at waterfalls in the countryside, these new enterprises were established in towns. The intricate regulations in force for the older iron works, concerning raw material procurement and production privileges, were not extended to the new types. Britons took part in important functions at the latter as owners, technical supervisors and skilled workers.[2]

Somewhat later, after an interval of a few decades, two prominent Swedish technicians learned, in Sheffield, the details of Huntsman's famous crucible steel production technique. With this knowledge they themselves were able to transfer the process to Stockholm: after their study tours were ended they needed no further help from Britain.[3]

This first category of technological borrowing was relatively less important during the period of concern here. Therefore it will not be treated separately but will be touched upon in section 6 (c) in a comparison relating to preconditions and realisation for the second category of technological transfer, namely the acquisition within the production of Sweden's most important export goods, bar iron.

Within this second form of transfer, borrowing from British technology became especially interesting in Sweden during the period 1810–40. At that time a series of experiments were carried out with the objective of improving quality, and hence the international reputation and competitiveness of Swedish bar iron. Even so, its competitiveness was weakened as the quality of puddled and rolled British bar iron gradually improved.

Preconditions for transferring British technology to Sweden were much less favourable than in the cases of Belgium or Germany. There was no coal in Sweden to be used for puddling,

2. B. Boëthius and Å. Kromnow, *Jernkontorets historia*, part III(1), *Jernkontoret och tekniken före götstålsprocesserna*, by Bertil Boëthius, Stockholm, 1955, pp.80–2, 226, 351–3.

3. Ibid., pp.76–80, 356–62.

although there were very large forests which gave sufficient quantities of charcoal. This laid the basis for continued competition by the so-called German method of wrought iron production which had dominated in Sweden since the seventeenth century. But although Swedish iron factories were traditional and in many respects uniformly organised, for instance in book keeping, there were some circumstances favouring the possibility of transforming important parts of the industry.

During the eighteenth and nineteenth centuries several spokesmen for Swedish iron manufacturers kept a close eye on technological progress in Britain. The Ironmasters' Association (IMA, in Swedish Jernkontoret), an effective national organisation representing every single iron producer, had built up instruments for that. To a certain extent it was thanks to the IMA that a continuous though slow productivity increase took place. Both in pig iron production and in the subsequent process of making bar iron the consumption of charcoal diminished within the scope of the German method. The publication of the review *Jernkontorets annaler* was begun in 1817, and there the employees of the IMA reported each year on their work for members in both important and small questions affecting quality of output and profitability for the owners.

That Swedish technological borrowings within the second category from Britain differed markedly from those of several other industrialising countries depended on more than the different industrial structures mentioned above. Marketing conditions also played an important part.

The third, again less significant category of Swedish technological borrowings from Britain concerned by-products from bar iron manufacture. During the 1830s two British workers were engaged to teach Swedes to produce high quality blister steel and good files. These efforts will be mentioned in section 6 (c) in another comparison with the second category.

2 General Observations about the Transfer of British Technological Know-how to Swedish Bar Iron Production

In some European countries the public authorities played an important part when British technology was introduced.[4] This was

4. The transfer of British technological knowledge to the European continent has been treated especially by W. O. Henderson in a series of articles collected in books.

not the case in Sweden. The public Board of Mines and the lower administrative level had it as their principal task to manage the comprehensive mercantilistic licence and regulation system. In a general way, many talked about liberal moves but few of them had a consistent view of the iron manufacturing process as a whole. Defence of the regulatory system was difficult, and the defeatism which gradually spread among the public authorities made reform steps easier. Through it violations were made possible of laws and regulations, which, if observed, could have obstructed technological initiatives.

Within Sweden new ways of thinking had two differing origins. At the beginning (1815–25) the IMA was most important. Although its efforts continued on a rather large scale, from the middle of the 1820s a few manufacturers working on their own account played an increasing part. The characteristic aims for the IMA were above all technical, where costs but not marketing aspects were fairly thoroughly considered. On the private side, however, technical experiments were subordinated to a clear view about which new products could be exported successfully.

The IMA was a democratic organisation with an extensive publication programme, and therefore its arguments and decisions can be followed. They have been well analysed in a detailed monograph about the IMA on which I draw in the first part of section 3.[5] Of course the efforts by private ironmasters are not documented publicly in the same way. However, a very large and informative body of unprinted source material is preserved (section 5 (a)). It has not been much used by economic-historical research for the kind of questions asked below.

However, it is not possible to analyse the whole course of events. Instead only some clearly demarcated parts of it will be taken up. Only a general outline will be given about those marketing conditions in Britain that formed a starting-point for the private manufacturers' plans (section 4). Following this a chronology will be given for private experiments for technological innovation (section 5).

Generally speaking, most information is given in his *Britain and Industrial Europe 1750–1850: Studies in British Influence on the Industrial Revolution in Western Europe*, 2nd edn, Leicester, 1965. For the more special aspects relevant here, the following is as valuable: 'The State and the Rise of the Coal and Iron Industries 1740–1870', part I of his book *The State and the Industrial Revolution in Prussia 1740–1870*, Liverpool, 1958, pp. 1–75.

5. Boëthius and Kromnow, *Jernkontorets historia*, part III(1), is an indispensable work. Gustaf Ekman's unprinted letters give some information. About them see section 5(a).

Thereafter follows a more systematic treatment of some of its aspects (section 6), in particular:

1. public and private use of new technical know-how
2. instruments for technology transfers
3. technological change and realistic commercial goals

For the last of those points comparisons will be made with what have been called the first and third categories of borrowings. Developments will not be followed further than to the late part of the 1830s, which means that the completion of the reform movement during the 1840s and the early 1850s will be omitted. By then Swedish technological contacts with Britain were less important.

3 The IMA and British Technology

During the eighteenth century British iron manufacturers gradually learned to replace charcoal as fuel with coal (in the form of coke). Early in the century coke could be used profitably in the reduction of iron ore to pig iron for use in one single, not especially important product, thin-walled castings. Soon after the middle of the century, however, coke pig iron became cheaper than charcoal pig iron and began to compete very well. Progress was next made around 1760, this time in the converting of pig iron to wrought iron (bar iron) through the roundabout potting process. In 1783–4 Henry Cort patented a puddling and rolling process for the same part of iron production, but it was not really competitive before 1792, after some improvements made by Richard Crawshay. After the mid-1790s it rapidly took over nearly the whole, strongly expanding British conversion of pig iron to bar iron. By then British ironmasters were producing a much larger part of the nation's consumption than earlier and began to export. They became very serious competitors to the Swedes and Russians, not only in Great Britain but also on the Continent.[6]

6. The impressive standard work is C. K. Hyde, *Technological Change and the British Iron Industry 1700–1870*, Princeton, 1977, pp.1–116. For readers accustomed to Swedish terms the following information must be given covering British conditions before the potting process and Swedish experiments during the whole period. Based on Hyde, *Technological change*, p.10, I use 'refining pig iron in the finery' for 'smälta tackjärn i smälthärd' and 'reheating blooms in the chafery' for 'välla och räcka smältstycken i räckhärd'. Both steps together will be called 'converting pig iron to bar iron at the forge'.

99

The transformation of British pig iron conversion was mentioned by Swedish observers as early as 1789 and 1793. The first serious evaluation of the puddling process ordered by the IMA was done by its leading metallurgist, E. T. Svedenstierna. In 1802–3 he travelled through Britain and visited many factories, publishing a popular travel book as early as 1804. The full, much more ambitious analysis of what he had seen was not printed until 1813, and even then the author did not give a comprehensive account of the puddling process.[7] Svedenstierna was convinced, however, that puddling could not be imitated in Sweden, but that rolling could; on the whole, he argued, the Swedish programme for the future should be to improve quality on the basis of domestic preconditions.

Apart from the IMA's great decade of experiment (1815–25), this opinion remained dominant.[8] Puddling became a marginal phenomenon in Sweden. On the other hand, rolling had had traditions there since Polhem's technical contributions at the beginning of the eighteenth century, even if these had been less demanding both of the surface of the rollers and of their suspension devices. He had only rolled out bar iron to thinner dimensions, not – like Cort – heavy blooms to bar iron.[9] A rolling process like the British one was developed in Sweden step by step up to 1850.[10]

Only during the decade 1815–25 were direct borrowings of modern British iron technology seriously considered, and Svedenstierna's general recommendations questioned. At the beginning of the 1810s an individual iron producer who needed good material for making sheet iron had tried puddling for his own use. His experiments were observed by Svedenstierna's successor, C. D. af Uhr and were carried out on a small scale. In 1817 af Uhr proposed that the economic viability of the method ought to be examined on a larger scale at the expense of the IMA. Decisions to this effect were made.

The practical adaptations to Swedish conditions were much more difficult than had been anticipated. It was even questionable if the firewood that was used instead of coke would be able to

---

7. Boëthius and Kromnow, *Jernkontorets historia*, part III(1), pp.100, 108–9, 121–2. Svedenstierna's first work is available in English, translated from the German edition (of 1811), with a new, informative introduction by M. W. Flinn, as E. T. Svedenstierna, *Svedenstierna's Tour of Great Britain 1802–3: The Travel Diary of an Industrial Spy*, Newton Abbot, 1973.
8. Boëthius and Kromnow, *Jernkontorets historia*, part III(1), pp.110, 193–4.
9. Ibid., pp.366–8.
10. Ibid., pp.493–6.

generate the strong welding heat that was needed for the product to be good.[11]

In summer 1820 af Uhr travelled to Britain at the expense of the IMA to study puddling thoroughly. During the journey his optimism that the process might well be used in Sweden was strengthened. Later he tried to engage a skilled British puddler, but in 1822 his application was turned down as a result of British prohibitions on the emigration of engineers. In spite of all efforts, the puddling experiments up to and including 1822 gave negative results: unsatisfactory blooms were produced.[12]

C. D. af Uhr did not question the disappointing outcome, but a group of younger, progressive technicians within the IMA led by Emanuel Rothoff and Per Lagerhjelm, were not as easily discouraged. They insisted on separate rolling experiments and got their way. If puddled blooms were not suitable, it was important to use blooms reheated with charcoal. The rolling experiments of 1824 gave an unsatisfactory bar iron, however. With the report from af Uhr printed in 1825, large-scale efforts by the IMA to transfer and adapt *modern* British technology were ended, even if puddling was not altogether passed over and even if critics within the organisation still sometimes pleaded for big rolling-mills.[13]

The experiments had shown the importance of a series of factors which seemed insignificant to those who were not well informed. Fire-brick able to withstand strong heat, and furnaces which could make the most of the possibilities of different fuels were needed. In sum, technicians who had shared Svedenstierna's reform ideas could notice that improvements had to be made throughout the process from pig iron to bar iron, i.e. both in the refining and in reheating with hammering.[14]

As late as 1825–35 the IMA still made the largest total contribution to technological reform within the Swedish iron industry. But it was divided into several approaches, partly competing with, partly complementing each other. One of these ways was the transfer of some British methods not yet mentioned here.

On his journey in Britain af Uhr had noticed that in a few places bar iron was still produced with charcoal as fuel – like the 'German'

---

11. Ibid., pp.371–2.
12. Ibid., pp.373–81. About legislation and reality: Henderson, *Britain and Industrial Europe*, pp.5–7, and in much greater detail, D. J. Jeremy, 'Damming the Flood: British Government Efforts to Check the Outflow of Technicians and Machinery, 1780–1843', *Business History Review*, vol.51, no.1, 1977.
13. Boëthius and Kromnow, *Jernkontorets historia*, part III(1), pp.378–87.
14. Ibid., p.388.

method in Sweden. He thought this a very specialised product, unimportant to the market as a whole. It fetched prices as much as fifty per cent above those for common Swedish bar iron and had a still wider price margin over British puddle iron.[15] Per Lagerhjelm was one of the leaders of the opponents within the IMA, and also henceforth a clear advocate of rolling. In 1826 he reported in the review *Svea* about experiences from a visit to Great Britain the year before. Among other information, he gave a detailed description of the production of material for sheet iron at Llanelly in South Wales, where charcoal was used in the finery.[16] Neither af Uhr nor Lagerhjelm seems to have considered that the British charcoal process could give Sweden valuable lessons. However, one of the younger employees of the IMA, Gustaf Ekman, would soon suggest that it should be tested.

Ekman was enrolled as assistant to one of the organisation's two forging directors (*översmedsmästare*). From spring 1828 to autumn 1829 he was free from his duties there (with full salary) for a study tour which initially took him through Britain. He saw the charcoal refining mentioned by Lagerhjelm at Llanelly, visited Sheffield and then went to Lancashire 'to learn about the only complete charcoal iron production' still existing in Britain. According to Ekman there were only four charcoal furnaces working, three of them at Ulverston in Lancashire and one in Scotland. Most of their pig iron was refined with charcoal. Ekman specifically mentioned that the finery was built to save charcoal.[17]

Unlike af Uhr, Ekman compared the work process to what in Sweden was called 'Walloon' forging. In both cases refining was done in one hearth and reheating in another, whereas with the German method both stages were performed in the same hearth. Walloon forging was used only for a limited part of Swedish bar iron, namely a specialised product made of ore from the Dannemora mines in Upland and called steel iron, because it was the raw material for crucible steel production in Sheffield.

In the spring of 1830 the IMA instructed Ekman to try Walloon forging with other than the usual ores. The decision had been delayed because IMA leaders had feared opposition from its Upland members and therefore tried to advance slowly. His experiments were carried out at Dormsjö. Although it had not been

---

15. Ibid., pp.191, 380.
16. P. Lagerhjelm, 'Om svensk och engelsk jernberedning', *Svea*, vol.9, 1826, pp.70–4.
17. Ekman's official report for 1829, *Jernkontorets annaler* (to be abridged *JKA*), vol.14, 1830, pp.149, 152–4.

formally included in his task, he used some observations from Britain.[18] An imitation of the British way of working in the finery was tried in the autumn when Ekman was experimenting at Sö-derfors. As an official of the IMA, his terms were then and later on that individual factory owners engaged him as a consultant, paid the outlays for raw material and gave him adequate remuneration. In his summarising assessment of the experiments at Söderfors, Ekman was cautiously optimistic. He thought that with the British method he could 'establish the quality of the iron produced much more reliably than with our old methods'.[19]

During his continued service at the IMA Ekman concentrated his efforts on thoroughly mastering the British forging method. It was transferred by him personally during longer stays at the two factories at Färna and Furudal, and besides was spread more freely to some other areas. But Ekman was not satisfied. Production was still much too slow. His observations during his journey in Britain in 1829 had not been enough to make him a confident user of the method. A further journey, paid for by the IMA, was made in the autumn of 1831, but it did not remove the defects. For some years workers at British charcoal factories had emigrated in different directions, and this made it hard for him to enter the factories. It was only with difficulty that, during his short stay, he 'was able to obtain casual information to assist with removing those several inconveniences' he had had to fight against during his experiments. After the journey he was very busy for three months around the turn of the year 1831–2 at Lesjöfors – a factory of which his father was part-owner – before at last he thought that the work process in the finery was satisfactory.[20]

In the literature about Ekman's efforts, his own testimony of his insufficient observations in Britain has been neglected. Nor has the long list of unresolved problems which he mentioned in his official report of 1833 been noticed.

After many experiments over several years in order to adapt the new processes he gave it as his view that there were a number of points in the production outside the finery that had to be improved before the intended results were reached:

18. G. Ekman to G. H. Ekman, 28 February, 28 March, 4, 11, 16, 25 April, 9 and 22 May 1830, vol.E I:68, Ekman & Co.'s archive, (to be abridged EA), Göteborgs Landsarkiv.
19. Ekman's official report for 1830, *JKA*, vol.15, 1831, p.105. A more detailed account about the experiments at Söderfors is in *JKA*, vol.14, 1830, pp.333–76.
20. Ekman's official report for 1831, *JKA*, vol.16, 1832, pp.173–5.

However carefully the work may be done, it is always through the quality of the pig iron that those properties will be created that strictly speaking will characterise the manufactured bar iron. Therefore I think it is totally futile to transfer British Walloon forging or any other new method for making an improved product if you are not willing or able to prepare good pig iron.[21]

In addition, he pointed out that the Swedish blowing machines (used both in pig iron and bar iron production) must be improved, as well as the hammer which forged the refined metal to blooms. The usual complaints about the chaferies which could not give enough heat before the final hammering ended the long list, which began with the problem of teaching Swedish workers new production methods.[22]

As those two aspects of Ekman's official reports about his occupation during 1831 and 1833 have been overlooked by researchers, it is natural that they have been surprised by the pessimistic tone in his often cited conclusions, formulated after a series of practical forging experiments by different methods – ending in 1835 and reported in 1836.[23]

He asserted that the British forging method – like puddling – gave iron of excellent value, but only for special purposes. Production was not cheaper than by the German method, now rendered more efficient than earlier. Therefore it did 'not deserve more general consideration'.[24] This judgement marks the end of the ongoing discussion within the IMA about the British forging

21. Ekman's offical report for 1833, *JKA*, vol.18, 1834, p.65.
22. Ibid., pp.69–78. That a sufficient temperature should not have been reached during the reheating, is a major point with several researchers: J. A. Leffler, 'Lancashiresmidets införande i Sverige', in *En bergsbok, några studier över svensk bergshantering tillägnade Carl Sahlin 15/12 1921*, Stockholm, 1921, p.123; G. Ekman (II), *Gustaf Ekman, svenska järnhanteringens nydanare för 100 år sedan*, Stockholm, 1944, pp.12, 36; Boëthius and Kromnow, *Jernkontorets historia*, part III(1), pp.482–4. During the 1830s the term 'det engelska smidet' – literally translated as 'the English forging method' – was normally used in Sweden. As it actually had roots in both Wales and Lancashire, I have used the geographically neutral 'British method'. In Sweden the terms 'Lancashire iron' and 'Lancashire method' were prevalent from the 1840s. The very trifling role this type of forging had in Britain during the nineteenth century is evident from Hyde, *Technological Change*, p.207. Information about the Ulverston region and its special iron industry is given in J. D. Marshall, *Furness and the Industrial Revolution: An Economic History of Furness (1711–1900) and the Town of Barrow (1757–1897) with an Epilogue*, Barrow-in-Furness, 1958, pp.17–38.
23. Surprise is shown by Boëthius and Kromnow, *Jernkontorets historia*, part III(1), p.483.
24. Ekman's official report for 1835, *JKA*, vol.20, 1836, p.350.

method. It seems logical if the whole sequence of reports from Ekman is included. However, it can be enriched considerably, when Ekman's private endeavours as a business consultant are studied.

On the IMA's first board of directors in the middle of the eighteenth century there were both factory owners and big merchants. During a great part of that century commercial questions were regarded as particularly important.[25] But gradually technical interests got the upper hand. However, this did not mean that sales aspects were entirely forgotten. In the major programme for changing the organisation presented in 1825, an ambitious follow-up of foreign customers was advocated. But only some components of this plan were carried through.[26]

No iron merchant any longer acted for or advised the IMA. Discussions about the future of Swedish iron were thus mainly pursued from the supply side: if Swedes managed to produce better iron than earlier at unchanged costs, no one seemed to question its future sale. For example, the reformers did not reflect on how Swedes could sell puddle iron (possibly rolled) in foreign countries in competition with British ironmasters. When special qualities of or uses for Swedish iron were mentioned, they often concerned what has been called steel iron. Yet ideas about the market for such a product were very vague.[27]

4 Demand in Britain during the 1820s for Swedish Quality Iron: Some Starting-points

Iron merchants and a few commercially interested factory owners had more realistic opinions. They did not speak about a general transformation of Swedish iron forging but concentrated their efforts on delivering good steel iron. As a background, an outline is given of the development in the Sheffield region – up to 1840 the only market of importance both within and outside Britain.

The year 1815 had been a record one for British exports, among other things of two kinds of finished iron and steel products:

25. Boëthius and Kromnow, *Jernkontorets historia*, part I, *Grundläggningstiden*, Stockholm, 1947, pp.213, 230.

26. Boëthius and Kromnow, *Jernkontorets historia*, part III(1), pp.193–4.

27. In his official report for 1828 the forging director August Åkerman (Ekman's chief) thought that nearly half the Swedish export could be needed for such use. Eight years later his estimate was only $\frac{3}{8}$ as high. See *JKA*, vol.13, 1829, p.102, and vol.21, 1837, pp.28–9.

'hardware' (many different products, mostly from the Birmingham region) and 'cutlery' (edge-tools, a speciality for Sheffield and its area). After that figures decreased, except during a recovery in 1818. In Sheffield 'distress was universal' in 1820. By then, however, the turning-point was near. The price of British pig iron, which gives an indication of general economic activity, and of bar iron reached a trough at the beginning of 1822. That year sales of hardware and cutlery were distinctly improved. In the factory districts people thought times were good. Whereas steel producers in Sheffield normally were rather small firms, in 1823 what has been called the first steel factory of the town opened. The real rush followed with the very strong general price increase during the last quarter of 1824 and the first half of 1825.[28]

During the years around 1820 on average 3,500 (long) tons of steel iron were shipped from Upland factories via Stockholm to Hull, as were about 2,000 tons of Russian bar iron, marked 'Old Sable', for similar but somewhat less qualified uses. These suppliers could not enlarge their production to any high degree. In 1823 iron export from Göteborg to Britain was at its lowest for many decades. Hardly any of this insignificant quantity went to Hull. During the industrial expansion in 1824 iron wholesalers in Hull began to ask for Swedish bar iron stamps, of higher quality than those made by the usual German method. In July 1825 customs tariffs on iron imported to England were lowered substantially – on products shipped in British bottoms from £6 10s to £1 10s per ton. The reduction was equivalent to about 40 per cent of the usual bar iron price fob in Swedish port. Swedish iron exports to Great Britain were considerably larger than during the preceding years. The Göteborg figure for 1825 was five times higher than that for 1823. In this strongly increased quantity were included whole yearly productions of such iron bar marks (stamps) that were supposed to be suitable for new, less demanding steel uses.[29]

28. A. D. Gayer, W. W. Rostow and A. J. Schwartz, *The Growth and Fluctuation of the British Economy 1790–1850*, Oxford, 1953, vol.1, pp.126–8, 151–2, 177, 192–4; G. I. H. Lloyd, *The Cutlery Trades: An Historical Essay in the Economics of Small-scale Production*, London, 1913, pp.182, 340–1. As my picture here is rather general, I have not tried to incorporate findings from two new and comprehensive works, K. C. Barraclough, *Steelmaking before Bessemer*, London, 1984, vol.1, *Blister Steel: The Birth of an Industry*, and vol.2, *Crucible Steel: The Growth of Technology*, as well as G. Tweedale, *Sheffield Steel and America: A Century of Commercial and Technological Interdependence 1830–1930*, Cambridge, 1987.

29. Customs and Excise, Ledgers of Imports, Customs 4/10–20 (1815–25), Customs 5/11–16 (1822–7), PRO, London; R. Adamson, *De svenska järnbrukens storleksutveckling och avsättningsinriktning 1796–1860*, Göteborg, 1963, p.136; my own

The very high prices did not last even the whole year 1825. The year after was marked by stagnation within many branches of British trade. For Sheffield the last part of the 1820s has been characterised as 'almost continuous depression'. This did not exclude increasing production and exports. In 1832 the export volume of finished steel products was nearly double that of a decade earlier. This indicates a considerably increasing demand for base material, i.e. steel iron, from at least the middle of the 1820s.[30] That British iron buyers were not able to meet their needs is evident from the development of prices on Upland steel iron marks. Their contract prices nearly doubled between 1826 and 1827. This meant that much more than the whole customs decrease went to the producers without rising costs for buyers. These extremely satisfactory conditions for Upland factory owners continued for six years.[31]

No outright substitution for Upland iron was tried. The most advanced efforts aimed at making 'middle marks' in Sweden, marks on a somewhat lower level, i.e. for similar uses as Russian iron. From the beginning these experiments were not tied to borrowings of British technology, but later they were. In section 5 two subsequent achievements with this aim will be followed. Both were principally made in western Sweden. One of them was for nearly ten years led by a British-Swedish iron merchant.

Less elaborate attempts to meet the increasing need for steel iron were also made, as will be evident from section 5 (b). Simpler steel iron (called 'common marks') was imported to England. Requirements were rather vague but may be summarised in the following way:

1. The iron had to be prepared more carefully than usual, which meant that it took more time to produce. The forging method was of no concern as long as the bars were fairly even. Thus it

---

study of unprinted accounts (*tolagsräkenskaper*) from the town of Göteborg; firm circular from Cowie & Brändström (Hull), 22 July 1825; C. F. Waern & Co.'s archive (to be abridged WA); D. Carnegie & Co. to T. Hudson (Newcastle), 8 November 1824, 24 January, 30 May and 4 July 1825, and in direct continuation to Beckinton, Wilson & Co. (Hull), 19 September, 3 October and 14 December 1825; vol.8, D. Carnegie & Co.'s archive (to be abridged CA), Göteborgs Landsarkiv.
    30. Gayer, Rostow and Schwartz, *Growth and Fluctuation*, vol.1, pp.192–4, 229; Lloyd, *Cutlery Trades*, pp.340–1.
    31. A. Attman, 'Vallonjärnets avsättning på världsmarknaden 1800–1914' in J. Norrby, M. Nisser and W. Ekman (eds), *Forsmark och vallonjärnet*, Stockholm, 1987, pp.192–7.

was not acceptable that well prepared parts lay close to those full of slag or that carbon content varied much.

2. The British buyers gave clear instructions about the length of the bars. They wanted them much shorter than ordinary west Swedish iron.

3. Only some prescribed dimensions could be used. During the period in question 2–3 x $\frac{5}{8}$ inches were the usual ones.

Chemical analyses of the composition of the iron were not yet made. The problem was that the first requirement could seldom be fulfilled for any extended period without important changes in production methods. When after some time customers complained or refused to continue buying, the second and third requirements were important obstacles to selling the rest of the production as ordinary bar iron in Swedish ports, where buyers insisted upon a variety of dimensions and longer bars.[32]

5 Private Achievements in Steel Iron Manufacture in Western Sweden, 1825–35

(a) Source Material

The possibility of closely following the two efforts towards a west Swedish steel iron production is given by two unusually well preserved archives from family firms. C. F. Waern (later C. F. Waern & Co.) and Ekman & Co. were merchant houses in Göteborg whose senior partners had owner interests in iron factories. From 1823 C. F. Waern had owned Bäckefors iron factory in Dalsland, and G. H. Ekman, father of the metallurgist Gustaf Ekman mentioned above, was part-owner of Lesjöfors in eastern Värmland.

The richness of the sources is due to the fact that these principals always kept strategic decisions for themselves and had to be very carefully informed about all those business operations they did not supervise in person. C. F. Waern resided during a major part of the year at his mansion, Baldersnäs in mid-Dalsland. There he received reports from his representative in Göteborg and from the iron production at Bäckefors. Gustaf Henrik (in Göteborg) and Gustaf Ekman corresponded regularly both during the latter's journeys in

---

32. D. Carnegie & Co. to Beckinton, Wilson & Co., 19 June, 3, 10, 24 July, 21 August and 11 September 1826, vol.8, CA.

the service of the IMA and later, when he normally lived at Lesjöfors. In all these cases the letters are very informative. Still more information is given in the archives of a third Göteborg merchant house, D. Carnegie & Co., and of several iron factories.[33]

(b) Brändström's and Waern's Achievements

Two firms had been importing west Swedish iron to Hull since 1824. One of them was Beckinton, Wilson & Co., which carried on a regular trade without any technical initiatives. Therefore it will not be treated here. The other was Cowie & Brändström and its sister firm John Cowie & Co.[34] The most important of the partners was John Peter Brändström, born in Hull in 1794 but brought up in Gävle. He was a son of the merchant and Swedish consul in Hull, Simon Brändström, and grandson of a famous merchant and shipowner in Gävle named Peter Brändström.[35]

From the start Brändström worked on the assumption that it was raw material that was crucial for the production of good steel iron. Obviously he needed to try those ores that were most similar to those in Upland. In March–April 1824 he was in Söderhamn, occupied in finding out which ores were used for different iron factories in Gävle's hinterland.[36] During 1825 he displayed interest in ores from eastern Värmland, from the Långban ore-field. A post of pig iron made from such ore was sent to Bäckefors. It was divided into several parcels. Each of these was worked in a special way and was given a special mark on the bars before it was sent to England in the autumn of 1825. Experiments were made, although

---

33. Since my investigation was carried through, WA has been reorganised, but as far as I understand, this does not affect my reference system here, although it could have been done more exactly today.

34. John Cowie and J. P. Brändström were partners of both firms. Moreover Lancelot Haslope was a sleeping partner of John Cowie & Co.

35. G. Anrep, *Svenska slägtboken*, Stockholm, 1872, vol.1:2, pp.121–3; P. Elfstrand, 'Peter Brändström', *Svenskt biografiskt lexikon*, Stockholm, 1926, vol.6, pp.601–4; 'Enskilta antekningar rörande svenska exporthandeln på England', vol.8, 'Handel och sjöfart', Riksarkivet, Stockholm. No less a person than Gustaf Ekman appreciated the partners' understanding of steel iron in the following way: 'I could never imagine that Cowie & Brändström had so much knowledge about iron sales and of different types of iron as I have now found. They have deep insight about Swedish and British iron and steel industry and have given me much information that even manufacturers in Sheffield (i.e. masters or owners) *could not* give me.' G. Ekman (from Hull) to G. H. Ekman, 1 July 1833, vol.E I:74, EA.

36. John Brolin & Son (Söderhamn) to I. G. Clason II, 20 March and 10 April 1824, vol.5, Furudals bruksarkiv, Uppsala University Library.

unsuccessfully, to persuade the owners of Lesjöfors to forge dimensions suitable for steel iron from similar raw material.[37] Furthermore Brändström asked Waern to look for any furnace owner in the Långban who might be willing to sell his share. At the beginning of 1826 a chance emerged. John Cowie & Co. was offered half of Gåsborn furnace, where ore from Långban was smelted to pig iron. As it was not possible without special arrangements for foreigners to acquire Swedish real estate, the agreement was made in the name of Waern. The aim was that converting pig iron from Gåsborn to steel iron should be done at Bäckefors.[38]

In the spring of 1826 Brändström and Waern tried to buy mine shares in a distinct part of the Långban field. Thus prices increased very sharply. The manager of Lesjöfors, Sten Helling, acted as an agent. Very cautiously G. H. Ekman took part in this rush for mine shares, but he did not want to form a partnership. Brändström's motive for acquiring some special shares was that the ore there was poor in iron but instead rich in manganese.[39]

In summer of the same year Brändström reported to Waern about the experience in Sheffield of the special delivery from Bäckefors in the autumn of 1825. There were quality differences between parcels. In sum production had not been successful. Both in purity and compactness in fractures, they were still far from their aim. Uneven and badly worked iron like that consignment would be inappropriate for many requirements and very difficult to sell. Brändström wondered whether he had assessed the properties of the ore correctly.[40]

For the post just to be made at Bäckefors Brändström recommended *small* blooms. These should be worked through so accurately that the finished bar iron should be free from slag and also free from obvious surface unevenness. Waern answered that the suggested methods implied such heavy expenses that they could probably not be used on a large scale, but Brändström replied in August 1826 that they must not be deterred by high costs during

---

37. Cowie & Brändström to C. F. Waern, 6 August 1825, WA; D. Carnegie & Co. to Cowie & Brändström, 11 July and 14 December 1825, vol.8, CA.

38. Cowie & Brändström to C. F. Waern, 7 July 1825; O. Ek (Gåsborn) to C. F. Waern, 18 July 1825; Schön & Co. (Stockholm) to C. F. Waern, 22 May, 5 June and 17 July 1828; WA; S. Helling to Ekman & Co., 4 April 1826, vol.E I:60, EA.

39. J. P. Brändström to C. F. Waern, 23 May 1826; Cowie & Brändström to C. F. Waern, 26 June and 22 August 1826, WA; Ekman & Co. to S. Helling 10, 29 April, 15, 18 May and 10 June 1826, vol.B I:44; S. Helling to Ekman & Co., 8 May 1826, vol.E I:60, EA.

40. Cowie & Brändström to C. F. Waern, 26 June 1826, WA.

experiments. He himself would pay for them.[41]

From the year 1826 Brändström also had contacts for three to four years with at least two other iron factories. Helped by Waern he bought from Uddeholm a special kind of bar iron, produced from Långban ore. Evidently he had no direct influence over the production there. But through this operation he got into touch with Emanuel Rothoff, one of the opposition leaders of the IMA. Through other firms he bought steel iron from Liljendal. Single posts were purchased from still other factories.[42]

In August 1826 Brändström noted for the first time that he was to visit British factories which still produced with charcoal. In October he informed Waern that he had got new ideas on his visits. Now he thought he was able to improve the production process at Bäckefors.[43]

A considerable effort was made at the beginning of 1827. The assistant at the IMA, S. J. Cleophas, was at Bäckefors in February and March and led experiments to produce an acceptable steel iron from especially carefully smelted Långban pig iron. He complained about the unwillingness of the workers to learn anything new. Chiefly he examined different arrangements and variants of the German method. As usual at that time of the year, Waern was in Göteborg to balance the books of the merchant firm, but Brändström participated all the more actively. Nothing indicates that any important experience from Britain was used.[44]

In summer the forge at Bäckefors was reconstructed, evidently to allow the installation of better bellows. Through it a higher heat and a more thorough working of the iron could be achieved. When the performance of the year 1827 was summarised, Brändström was moderately optimistic. He did not doubt that good iron could be made in two variants of the German method. However, several preconditions had to be met: the charcoal must be clean, the pig iron heated beforehand, slag not included in the blooms, bars forged in strong heat, and just for that reason they had to be shorter than usual.[45]

41. Cowie & Brändström to C. F. Waern, 26 June and 29 August 1826, WA.
42. D. Carnegie & Co. to Cowie & Brändström, 11 July and 22 August 1825; to John Cowie & Co., 9 April 1827 and 3 March 1828, vols8–9, CA; J. P. Brändström to C. F. Waern, 14 July 1827; John Cowie & Co. to C. F. Waern, 19 September 1827, WA; bar iron account in general ledgers, 1826–9, for Uddeholm Co. Värmlandsarkiv, Karlstad.
43. J. P. Brändström to C. F. Waern, 29 August 1826; Cowie & Brändström to C. F. Waern, 11 October 1826, WA.
44. S. J. Cleophas to C. F. Waern, 10 February and 10 March 1827, WA.
45. M. Fjellman (manager, Bäckefors) to C. F. Waern, for example 5 July 1827;

If Bäckefors from 1825 had given Brändström the best possibilities to experiment on his own conditions he was still in 1828 willing to try a new idea. F. W. Grill, who had just bought out his joint heirs from the Godegård iron factory in Östergötland, was allowed to purchase pig iron and receive a description of the forging method at Bäckefors. Grill was soon convinced that he had found the right way – a Walloon method – and made no secret of it. From fervour 'for public weal and . . . the honour of our nation' he would not any longer see the excellent Swedish iron 'bungled unchallenged through the miserable German method'. He was prepared to place his findings at public disposal. However, Grill's expectations were premature. His efforts gave no lasting result, even if his steel iron was hardly given a fair trial.[46]

During the spring and early summer of 1828 Brändström concentrated more and more on using the British charcoal method. In May he asked Waern that one ton of pig iron from Gåsborn should be sent as soon as possible to Hull for conversion to steel. In June he declared to Waern that it would be a less radical change in the iron production than Grill's idea, but still quite satisfactory for reaching a strong heat, to use a British small furnace for reheating, a hollow-fire. Towards the end of August he urgently asked for the Swedish pig iron, which he then had an opportunity to get refined through a business contact. It is an example of the slow transport of the period that not until 25 September could Waern's office in Göteborg inform its chief that three lumps of pig iron which were not allowed for export had been taken care of 'privatum' by a shipmaster sailing for England.[47]

During the year 1829 Cowie & Brändström at last came to a decision after its so far broad and diversified efforts. The firm would no longer buy steel iron from Liljendal, and that year's delivery from Uddeholm was the last one. Certainly Grill's Walloon forging should be tested, but it was not to be an obstacle to the important new achievement by Brändström and Waern.[48]

In a letter of 1 July from the firm to Waern the steel iron that had

---

J. P. Brändström to C. F. Waern, 26 December 1827, WA.

46. Cowie & Brändström to C. F. Waern, 13 May and 10 June 1828; F. W. Grill to C. F. Waern, 4 October 1828, 17 May, 10 July, 13 August 1829, 12 June 1831 and 5 August 1832, WA; G. Ekman to G. H. Ekman, 28 February 1830, vol.E I:68; G. H. Ekman to G. Ekman, 8 May 1830, vol.B I:46, EA.

47. Cowie & Brändström to C. F. Waern, 13 May, 10 June and 28 August 1828; C. E. Billman (Göteborg) to C. F. Waern, 25 September 1828, WA.

48. D. Carnegie & Co. to John Cowie & Co., 16 March 1829, vol.9, CA; Cowie & Brändström to C. F. Waern, 2 September 1829, WA.

been sent from Bäckefors was rejected in very explicit terms. Waern was requested to discontinue production for British use. On the other hand the three pig iron lumps which were sent over in 1828 and refined in an English charcoal factory had given a very fine result: an altogether pure and even iron. To be sure, more pig iron than usual had been used, but much less charcoal. Within a fortnight the firm would report its future steps to a breakthrough with the help of Waern. The inclination was to engage British workers.[49] At the beginning of August it was confirmed that such smiths were to accompany John Peter Brändström to Bäckefors to work according to their own method. The letter ends with a postscript: 'Be careful to keep the proposed work as secret as possible.' The British workers were somewhat delayed because they needed a lot of cast iron models. They arrived at Göteborg on 1 October, saw that the castings were passed on, and came to Bäckefors on the 12th.[50]

About this great summit of achievement, after nearly five years of effort, there is information from two persons. The manager at Bäckefors, Almquist, wrote once or twice a week to Waern, who lived fifteen miles from there at Baldersnäs but was never present. In his characteristic handwriting Brändström jotted down some short, not fully dated, letters to the same addressee.

The work was carried on strenuously. Brändström had been given full authority to·arrange it in the best way. Workers got additional pay for evening work. On 28 October the manager thought that the preparations, also for the new British hearth, were nearly ready so that production ought to begin. But problems with the bellows arose. They could not give enough air to attain the intended high temperature. Changes must be made. On 19 November Almquist reported that good iron was coming from the British finery, although the bellows were still not satisfactory to the smiths. Brändström lamented bitterly about the unpardonable ignorance of the bellows-maker which had caused the work with the bellows to become much more tiresome than expected.[51]

As November turned to December the problems were overcome. The parties concerned regarded what had happened as a

49. Cowie & Brändström to C. F. Waern, 1 July 1829, WA.
50. Cowie & Brändström to C. F. Waern, 5 August and 2 September 1829; C. E. Billman to C. F. Waern, 1 and 8 October 1829; A. T. Almquist to C. F. Waern, 12 October 1829, WA.
51. A. T. Almquist to C. F. Waern, 28 October, 1, 10, 19, 24, 28 and 30 November 1829; J. P. Brändström to C. F. Waern, Thursday morning, Sunday morning, Thursday evening, WA.

technological breakthrough. Brändström sent a messenger to Waern with the news. Waern in his turn notified a person who had helped them earlier with pig iron production at Gåsborn and was the public officer most concerned in the region, namely Franz von Schéele. After congratulations, Schéele promised in his reply on 5 December 1829 that what had been confided to him had been hidden in his memory for freemasons' secrets, and that he had immediately destroyed Waern's letter. The process ought to be kept as secret as possible and its diffusion through workers and salaried employees ought to be prevented.[52]

However, mastering the purely technical problems was not the only decisive point for the future of British forging at Bäckefors. The last experiments, from October 1829 to the beginning of 1830, resulted in heavy incidental expenditure, amounting to nearly 3,000rd banco – roughly £250 – evidently to be paid by the Brändström firm. It was much more alarming that running production costs were high. An early estimate, from Christmas Day 1829, by Brändström and Waern together showed that one ton of the new steel iron was nearly £2 15s more expensive to produce than by the normal method. A somewhat later calculation by Waern was about as high, whereas Brändström believed in a lower metal loss during the refining and in better charcoal economy, and therefore thought £1 10s extra cost more accurate. The heaviest item was the high wages of the British workers. The abnormally large consumption of pig iron could be nearly as expensive. The extra costs were together much higher than those extra high prices given for west Swedish steel iron during the latter half of the 1820s. If production was to continue, the costs must be reduced, or the prices in Britain for this much improved iron had to be forced up substantially.[53]

However, the possibilities for such changes seemed good, and the new activity at Bäckefors was made permanent. On 19 July 1830 John Cowie & Co. entered into a contract with smiths from Monmouthshire about steel iron production in Dalsland. Their leader, Samuel Houlder Sr., was one of the two among them who had taken part in the experiments the year before. At the beginning of October they arrived at Bäckefors, where Cowie already was. Their speciality was work in the finery, the refining, for which

52. F. von Schéele to C. F. Waern, 5 December 1829, WA.
53. Memorandum respecting the British forging at Bäckefors in 1829, costs higher than usual; cost account for the British forging, account with J. P. Brändström, wholesale dealer, WA.

the pay was fixed at £1 per ton, against £1 3s 6d during the previous period. Excluded from this sum was the remuneration paid to the foreman, Houlder. He was to supervise the refining and also to assist at the reheating – whether it was performed in British hollow-fire or in a Swedish Walloon hearth. His weekly wages were £1 15s.[54]

The plan was to render production more effective through procuring more appropriate machinery and tools, coordinating available resources better and reducing wages. The first two points will be treated briefly here, whereas that about the wages of the British workers will be taken up in section 6(b).

For the 1829 experiments Brändström brought some British castings. With the larger 1830 group of smiths came thirty-two pieces of castings, two of which were hammers and walls for the hollow-fire. Evidently the bellows at Bäckefors had not functioned well. When Cowie & Brändström received a clear plan of the works, they made inquiries in Great Britain about alternative solutions and ordered a blowing-machine which arrived in the autumn of 1831 together with two new British workers. An English mechanic followed to set it up. In November it was reported that it had already been out of order once but had been repaired.[55]

What seems to have been the last initiative to acquire direct British mechanical knowledge concerned a heavy hammer. It was taken by Cowie & Brändström in the beginning of 1832. Probably it was this idea that was carried through the following year when C. F. Waern visited England and ordered 'a lot of castings to the work equipment, among which was a bar iron hammer with a pig iron shaft of about $3\frac{1}{2}$ tons weight'. Because of many difficulties during transport it could not be installed until the autumn of 1834.[56]

After the beginning of 1832, however, the technical initiative seemed to move to the Swedish side. From there the question was brought up of using hot blast – surely a Scottish invention – to save

54. The contract, which has the title 'Memorandum', is written in a clumsy English with several repetitions, WA.
55. C. E. Billman to C. F. Waern, 9 September and 17 November 1830; Cowie & Brändström to C. F. Waern, 21 June 1831; A. W. Melin (Göteborg) to C. F. Waern, 19 September 1831. In summer 1831 Gustaf Ekman had visited Bäckefors, among other things to check the older blowing-machine. A. T. Almquist to C. F. Waern, 29 June and 31 October 1831, WA.
56. Cowie & Brändström to C. F. Waern, 2 May 1832 and 26 October 1833; C. F. Waern (from London) to A. W. Melin, 7 June 1833; A. T. Almquist to C. F. Waern, 24 September 1834, WA.

fuel. Later when Waern learnt that it had been installed at Lesjöfors by Gustaf Ekman, the latter was asked to procure similar apparatus for Bäckefors. A series of adaptations concerning the best organisation of the work process were carried through.[57] Coke was considerably more expensive but seemed at the beginning to give a much better product during reheating in the hollow-fire. Some years later, however, this kind of reheating had not any advantage over the use of charcoal.[58]

There is no doubt that Bäckefors steel iron was quickly and successfully introduced in Sheffield by Cowie & Brändström. No crises occurred when its quality could have been seriously questioned during the next few years. The higher price against ordinary Swedish bar iron, which was reached at the latest in 1831, more than covered the extra high production costs. Not even a breach of contract by the British firm, which definitively ended this connection, led to a serious setback.[59]

(c) The Ekman Achievement

Part of section 3, 'The IMA and British Technology', treated Gustaf Ekman's first two journeys to Britain in 1828–9 and 1831 and his attempts during his service to introduce the British forging method at some factories. This public activity has been well documented in scientific literature.

However, his other, private work has scarcely been noticed. Even if it is not possible to draw a sharp line between public and private, I shall here try to isolate some traits in his private business, including technical aspects of his task of introducing the British method to the family factory at Lesjöfors.[60] From Ekman's differ-

---

57. Cowie & Brändström to C. F. Waern, 8 May 1832 as an answer to a letter from Waern; L. Uggla to C. F. Waern, 20 September 1834; G. Ekman to C. F. Waern, 17 November 1834; A. T. Almquist to C. F. Waern, 23 May and 1 August 1831, WA.
58. A. T. Almquist to C. F. Waern, 16 November 1830, 29 June, 14 August, 23 October and 21 November 1831. In 1832 the cost for coke was half of the total cost for reheating in 'hollow-fire' and corresponded nearly exactly to the whole expenditure for reheating in a Swedish hearth. The hollow-fire had also a much greater loss of pig iron and an important sum for fire-brick and clay. C. F. Waern's calculation, WA.
59. C. E. Billman to C. F. Waern, 24 June 1830; C. F. Waern to Beckinton, Wilson & Co., 13 May 1835, WA; G. Ekman to G. H. Ekman, 30 April 1832, vol.E I:72, EA.
60. Besides, the letters well document his work at the end of 1832 and beginning of 1833 in introducing the method at the neighbouring factory Liljendal and give

ent activities and from Brändström's and Waern's experience, the problems connected with technology transfer and the possibility of benefiting personally will be discussed more systematically in section 6(a). The earliest glimpse of private action came during Ekman's first assistant period at IMA (1827 to the first half of 1828). In August 1827 a sample of steel iron, hammered according to British measure, was forged at Lesjöfors. It is uncertain whether it was ever used for that purpose. The composition of a test delivery, designed for Sheffield, was discussed at the end of September. After enquiries, a Scottish acquaintance reported shortly before Christmas that the best way to try a small post of iron for production of cast steel would be to send it to a steel works near Glasgow called Old Monkland Steel Company. At the end of February 1828 pig iron suitable for the Walloon method was blown, and at the beginning of March forging followed. In october, when Gustaf Ekman had just arrived in Britain, Ekman & Co.'s first iron post for export in many years was shipped to Leith in Scotland. Beside the firm's own delivery of about three tons, a batch of 1-ton test iron from Ph. Åkerman was included.[61]

Led by his older colleague in the IMA, August Åkerman, Ekman took part in experimental forgings, among other factories, at Lesjöfors, owned by his own family, and Forsbacka, the estate of the Åkerman family. It is not surprising that the family factories used the IMA services in their own interest. They must have paid for them, as all other customers did. What is important about the organisation is that from the very beginning Gustaf Ekman planned a reliable scrutiny of the properties of the new production.[62]

The sample delivery was small. Only few preparations had been made for its production. The information about the outcome seems to have been obtained verbally by Ekman and did not stimulate direct continuation. As late as May 1830, when he was experimenting with Walloon forging at Dormsjö (above, p.102), he wrote to his father that this was of no interest to Lesjöfors.[63] From

many glimpses of similar activities at other places such as Färna, Furudal and Dormsjö.

61. S. Helling to Ekman & Co., 10 August 1827, 27 February and 6 March 1828; G. Ekman to G. H. Ekman, 29 September 1827; J. R. Bald (Edinburgh) to Ekman & Co., 13 December 1827; Ph. Åkerman to Ekman & Co., 29 September 1828, volsE I:62, 64; Ekman & Co. to [J.] R. Bald, 30 September 1828, vol.B I:45, EA.

62. G. Ekman (II), *Svenska järnhanteringens nydanare*, p.16; A. Edestam, *De dalsländska järnbruken*, Vänersborg, 1977, p.107.

63. G. Ekman (from Garpenberg) to G. H. Ekman, 9 May 1830, vol.E I:68, EA.

April 1830 to the end of 1831 he held a deputyship at one of the IMA forging director posts, but was very hesitant about his future. In a long series of letters over several years, he discussed with his father his prospects inside and outside the IMA.[64] In the latter case a purchase of an iron factory in Värmland could possibly give him a proper position.[65]

His ambiguous situation *vis-à-vis* Lesjöfors was not talked over explicitly in the letters but can easily be read between the lines. When the trusted Sten Helling left his position as manager there, he was succeeded by persons who, though not technicians, supervised the work to such a great extent that it would not be necessary permanently to engage Gustaf Ekman himself also as a special forging manager.

Another aspect was that his father surely led the enterprise, but did not own more than half the factory. The other half was held by his father's brother-in-law and earlier merchant partner, Prytz, who in the Ekman letters was thought to be erratic. It was only when Prytz became insolvent (in 1833) that the problem was solved. After some parleying, G. H. Ekman bought his share.[66]

Gustaf Ekman does not seem to have reached a clear judgement about his future until 1836, when at last he decided definitely to leave the IMA. But he had stood outside that organisation in 1832 and most of 1833. Roughly speaking, from the beginning of 1831 he gradually moved towards private activity based on Lesjöfors, but with connections in other quarters.

In section 5(b) the extended experiments by Brändström and Waern to reach a middle mark of steel iron have been fairly thoroughly discussed. Nor could Gustaf Ekman achieve his targets without a long process of adaptation. He had begun his private endeavours in 1827–8, had then in Britain, in 1828–9, studied the method he wanted to introduce, and had during 1830 practised it at some Swedish factories.

In 1831 he was ready to introduce it, both at some additional factories owned by others and at Lesjöfors. In May when he was at Färna (above, p.103) the proposition emerged that this factory,

64. I. e. G. Ekman to G. H. Ekman, 10 April, 6, 30 May, 14, 19 December 1831, 3 January, 16 April, 7, 13, 19 and 28 May 1832, volsE I:70, 72, EA.

65. G. H. Ekman to S. Helling, 13 January, 24 February, 6 and 18 March 1830; to G. Ekman, 17 February 1830, volsB I:45–6; G. Ekman to G. H. Ekman, 30 January 1830, and 17 January 1832, volsE I:68, 72, EA.

66. G. Ekman to G. H. Ekman, 6 May 1831, 11, 28 March and 11 June 1833, vol.E I:70; G. H. Ekman to G. Ekman, 21 May 1831, 9 February, 28 May and 8 June 1833, volsB I:46–7, EA.

together with Lesjöfors, should engage a British worker. In order not to be forestalled by other producers who might be interested in making steel iron Ekman was to go to Britain to study the production there once more. The latter justification was the only one to be mentioned in public. He almost succeeded in recruiting a skilled worker from Lancashire 'when the manager found out somewhat too early and I had to withdraw secretly'.[67]

Another point of departure for his subsequent private job was that he heard in Sheffield that steel firms, faced by still higher prices on Upland marks, had with good results used iron made from east Värmland ore. Mostly it was products from Liljendal but to some extent also from Lesjöfors, which iron now ought to be seriously introduced on the English market.[68]

The first blooms according to the British method were made at Lesjöfors in the middle of November 1831, since a new hearth had been built there. Time and again Ekman was worried by the problem of getting the workers to imitate the new way of working:

> The smiths of Lesjöfors are among the very worst I ever saw. I must admit that they are very willing, understand the better quality of the new iron when they manage the work and generally let themselves be shown the way, though with insignificant effect . . . If the smiths do not change their drinking habits, there will be no other way to have order at the factory and accuracy in the work than to kick them.[69]

In the middle of December he thought that the forging had succeeded fairly well, so that without exception good iron should be made from the beginning of 1832. On Christmas Eve he wrote that forging had been improved step by step, and that the result ought to be an iron as good as that made at Bäckefors. In mid-January he was obliged to report that the pig iron had not been of sufficiently high quality. One month later he wrote: 'Many, many times the conditions of the smiths have taxed my patience, not to say that I was brought to despair of the success of the change of forging method.' During the next few weeks, however, he thought that the

---

67. G. Ekman to G. H. Ekman, 28 February, 6 May and 25 September 1831, vol.E I:70, EA.

68. G. Ekman (from Sheffield) to G. H. Ekman, 4 September 1831, vol.E I:70, EA.

69. G. Ekman to G. H. Ekman, 15 and 22 November 1831 (citation), 1 December (citation), vol.E I:70. Skilled or careful smiths visiting Lesjöfors from other factories seemed to improve the performance of their own people. G. Ekman to G. H. Ekman, 14, 20 February and 19 March 1832, vol.E I:72, EA.

problems were solved, since the reheating under the hammer had been improved. His rather optimistic IMA report for 1831, sent from Lesjöfors on 25 February 1832 and discussed above (pp. 103–4) dates from just that period.[70]

When during the later part of March 1832 he was again to go to Färna, he left behind 'distinct memoranda for the running of the forging during my absence'. Late in April a few days' visit gave him the possibility of attending the forging at Bäckefors and he 'won much valuable experience which I have already used successfully'.[71]

One month later he said that he understood more for each week that he had had to extend his stay at Lesjöfors to teach the smiths the method carefully: 'Just now I am beginning to master my art. The saving of charcoal will be much more considerable than I ever thought, the forging as good or better than in the old way, but the metal loss will not diminish, possibly increase somewhat.'[72] In mid-July he thought he had not yet reached his aim, although the work was markedly better.[73]

During the last quarter of 1832 the work in the chafery, carried on by new smiths on new contracts, began to succeed. Then he concentrated his efforts on the one hand at the furnace belonging to Lesjöfors, on the other on introducing the British method also at the neighbouring factory, Liljendal. The first aim meant important improvements: the building of a roast kiln with storehouse and hauling installation for ore, and the repair of blowing-machine, water wheel and water grove. At the turn of the year 1832–3, more than one year after the beginning of the very intense efforts, he judged that pig iron production was still imperfect in many respects. At the beginning of February he thought that the roast kiln was in order.[74]

Finally in mid-March 1833 he gave an altogether satisfactory report from Lesjöfors which he did not need to retract:

At my return here yesterday I found that the forge and the furnace had managed exceedingly well during my absence, but I was especially

70. G. Ekman to G. H. Ekman, 24 December 1831, 17 January, 14 February (citation), 20 and 26 February 1832, vol. E I:72, EA.
71. G. Ekman to G. H. Ekman, 19 March (citation), 30 April (citation), and 7 May 1832, vol. E I:72, EA.
72. G. Ekman to G. H. Ekman, 28 May 1832, vol. E I:72, EA.
73. G. Ekman to G. H. Ekman, 17 July 1832, vol. E I:72, EA.
74. G. Ekman to G. H. Ekman, 28 August, 6 November and 22 December 1832, 1 January and 1 February, 1833, volsE I:72, 74, EA.

pleased to see the important improvement of the bar iron since the pig iron blown this year had been used. Now I think I have mastered also the pig iron production. The bar iron that is made now is really so superior that it must win an advantageous sale.[75]

At the end of 1833 it was reported that hot blast – which had been noticed in Sweden nearly immediately after its introduction in Britain – would reduce the charcoal consumption of the furnace considerably. After one more year it was used there as well as in the conversion to bar iron. Beside fuel-saving, an increased weekly production was registered.[76]

The key point here is that it took Sweden's most skilful practical metallurgist – with three years' knowledge of the method and with access to other experts – one and a half years to introduce and adapt the British method at his own factory. A series of changes in the iron manufacture had to be made before the export product was satisfactory. Then still more important improvements during another period of one and a half years established the iron from Lesjöfors and Liljendal as middle marks beside Bäckefors. Efforts at Färna were less sustained and did not give the same result.[77]

The second iron sample, sent out too early, at the beginning of the shipping season of 1832, was deemed by one customer as bad. During 1833, however, Edward Spence, one of the well-known steel iron merchants in Hull, began to sell Ekman & Co.'s iron from Lesjöfors. The contract prices of the Upland marks fell in 1833 by 10 per cent, and were to fall still more in 1835. New middle marks had begun to establish themselves beside Russian iron.[78]

75. G. Ekman to G. H. Ekman, 21 March 1833, vol.E I:74, EA.
76. G. Ekman to G. H. Ekman, 8 October 1833, 22 July, 7, 26 September, 21 October and 7 December 1834, volsE I:74–5, EA; Boëthius and Kromnow, *Jernkontorets historia*, part III(1), pp.445–7.
77. The contact between the new holder of the entailed estate Färna, C. J. von Hermansson, and Gustaf Ekman was arranged in 1830 by Sten Helling, who in November 1829 left Lesjöfors for Färna. During the years 1831–3 Ekman stayed for several periods at the factory to teach the British method for part of the main production. Despite these efforts he thought that he had less success there than at other factories. S. Helling to Ekman & Co., 12 March 1830; G. Ekman to G. H. Ekman, 6, 30 May 1831, 6, 26 March, 28 August 1832, 11 March and 26 November 1833, volsE I:68, 70, 72, 74, EA. The factory owner did not find the economic result very encouraging. C. J. von Hermansson to S. Helling, 1 January and 16 April 1835, vol.B 7, Färna bruksarkiv, Uppsala Landsarkiv.
78. G. Ekman to E. James (Bristol), 13 March 1832 and 18 February 1833; to E. Spence, 18 February 1833, volsB I:46–7, EA. Attman, 'Vallonjärnets avsättning', pp.192–7. On his third British journey Gustaf Ekman wrote about the development of the prices: 'Ugglas [owner of two factories in Upland] is going to the manufacturing towns, as the Dannemora iron has lost in demand, because as a consequence

121

In February 1834 Gustaf Ekman wrote his official report to the IMA concerning 1833 (in this case only October–December): 'It is to be delivered by the end of this month, and as my real measures to be reported for these three months are not worthy of mention, I have to produce some pages by discussing projects.'[79] Actually he used this report to discuss ideas about what was needed to introduce a new forging method based on lessons from his own arduous task at Lesjöfors and Liljendal.

6 More Systematic Observations about Technology Transfer

In the main, the long sections 3 and 5 have had a chronological outline. This is chiefly because the course of events has not been sufficiently known. Below, however, I shall try in a more systematic way to point out some aspects of special interest.

(a) Public and Private Use of New Technological Knowledge

It is evident that over the years the IMA made important efforts for technological development and adaptation. Yet it was scarcely by chance that it was the private sector efforts, not those of the IMA, that succeeded. After 1825 the organisation could not follow any uniform policy. The individual employee earned a basic salary and had to be at the disposal of the members. Either he could be occupied with several small experiments, widely differing from each other, or he could work for, and become loyal to, someone or a few members who wished to employ him more extensively. Ekman was troubled about this loyalty aspect.[80]

The survey in section 5 has shown that stubborn perseverance was necessary. On one occasion Gustaf Ekman worded the central dilemma for the IMA employees thus:

---

of this much other iron has been used. It is fortunate that steel consumption really has increased. Many articles formerly made with iron are now made with steel to withstand wearing. This brings it about that the iron marks manufactured by me ought to be popular, because, for the uses just mentioned, I think they are better than Dannemora iron.' G. Ekman (from London) to G. H. Ekman, 30 May 1833, vol.E I:74, EA.

79. G. Ekman to G. H. Ekman, 25 February 1834, vol.E I:75, EA.

80. It is observable here and there but is seldom explicitly discussed in a series of letters to his father.

The postponement in my utilisation by the IMA which has occurred is, as far as I understand, rather fortunate. Now I have time to try my own strength better in order to judge if really anything about forging can be accomplished by an employee who has to share his toil between many places. From my own experience I realise that such a change in forging method must be followed up very watchfully over a long period, and if I were to stay at one single task for half or a whole year, a lot of factory owners would soon grumble.[81]

IMA had inadequate means for letting the new technique become known uniformly to the majority of its members. Its review was of course one way. But the content was difficult for many readers, and it was no proper substitute for thorough practical demonstrations, for which individual employees had no time. The problems were more than usually troublesome at the introduction of the British method, when the marketing situation demanded quick results.

The basic attitude of those factory owners who tried technology transfer privately was that as much as possible of the fruits ought to go to themselves.[82] In the first instance concealment, in the second locked factories were the most obvious defence measures. Thirdly, intending imitators were to be discouraged by warnings about the very heavy expenses.

The first defence line was very weak. In spite of both Brändström's and Schéele's discretion, the knowledge soon spread that something important had happened at Bäckefors. In spring 1830, when G. H. Ekman asked Waern for information for an acquaintance who had heard about the experiments, Waern sent a memorandum concerning them.[83] Ekman's activities in 1830 aroused so much attention and such great hopes that on several occasions he was markedly irritated and asked to be left alone.[84]

Visits from far-off and unknown persons who asked for permission to come could be prevented.[85] On the other hand a pushing

81. G. Ekman to G. H. Ekman, 3 January 1832, vol.E I:72, EA.
82. From the beginning F. W. Grill at Godegård (p.112) held an altogether different opinion, which, however, changed when he had suffered setbacks. Some years later his application for 'privilegium exclusivum' was sent to the authorities. F. W. Grill to C. F. Waern, 18 November 1832, WA.
83. Lovisa Fersen to Ekman & Co., 29 April 1830, vol.E I:68; G. H. Ekman, 8 May 1830 to G. Ekman, letting him read the description, vol.B I:46, EA.
84. G. Ekman to G. H. Ekman, 20 June 1830 and 20 February 1831, volsE I:68, 70, EA.
85. G. Strömbom (Nissafors) to C. F. Waern, 23 October, 6 and 20 November 1830 (citation), WA: 'Dear Mr Ironmaster, Your first esteemed letter gave me the sure presumption that you had made sacrifices, which would lead also to the general

neighbour, the managing director at Upperud, Henry St Cyr, could appear uninvited at Bäckefors and find an opportunity to see Brändström's preparatory work in November 1829 and imitate parts of it.[86] When in spring 1830 Baron Mannerheim, the influential leader of the IMA, asked for admittance to the Bäckefors forge for its employee Ekman, Waern granted it without requesting Brändström's consent.[87] Factory owners who had consulted Ekman during his IMA service were not aware that the same person, as a representative of a private firm, could be less communicative when they wanted their staff to visit Lesjöfors.[88]

A central contradiction within industrial ethics was of course that the technology adapters had acquainted themselves without any obvious scruples with production methods in Britain. Only in unimportant respects were they pure inventors who could have applied for patents for their contributions.[89]

However, the long and expensive process of adaptation and introduction from which the technology adapters had obtained their experience, without which others could have achieved little, justified the view that in the longer run they should have some compensation. Their combination of technical and commercial knowledge was exceptional.

When the British method was tried out at Färna and introduced at Liljendal by Gustaf Ekman, their steel iron, thanks to his knowledge about the market in Hull and Sheffield, was exported by his father's firm, Ekman & Co. in Göteborg. That was part of the agreement when the factory owners made use of the private technician Gustaf Ekman's skill. The deliberations between him and the Myhrman family at Liljendal on the introduction of a new forging method resulted also in a payment to him partly based on the price gap down to ordinary iron that the improved mark fetched. The risk was, of course, that such a royalty would seem less and less well-founded as the years passed.[90]

---

benefit of the Swedish iron industry. But when you later associated yourself with foreigners on such binding conditions that the method must not be known to any Swede, I understand my mistake.'

86. J. P. Brändström to C. F. Waern, 'November, Thursday morning' 1829, WA.

87. G. H. Ekman to G. Ekman, 8 May 1830, vol.B I:46, EA.

88. G. Ekman to G. H. Ekman, 17 January and 28 August 1832, vol.E I:72, EA.

89. The important exception is his reheating fire (vällugn) from 1843, when Gustaf Ekman neither patented nor let the results go to Lesjöfors. One possible explanation is that he did not feel himself compensated enough earlier within the family business.

90. 'They thought that I ought to be compensated for my supervision of the

For the Waern family a similarly favourable situation came towards the middle of the 1840s when a new generation had grown up well at home in Hull and Sheffield, and given the firm a high commercial, and at the same time some technical, competence. Thus profit from their own steel iron was increased by commissions for exporting other factory owners' similar produce.

(b) Means for Technology Transfer

The foregoing survey involves a range of channels:

1. learning from books
2. machine imports
3. British organisational improvement and investments
4. engagement of skilled British workers
5. Swedes' study tours in Britain

These will be developed in more detail below.

1. Of course the IMA technicians studied international literature and foreign journals. They knew their own authorities such as J. J. Berzelius, the most prominent Swedish chemist of the period and a man of distinction in the development of his branch of science. But not only pure but also applied research was difficult to use. An attempt to rely only on reading was a feature of af Uhr's earlier puddling experiments. When they failed they had to be supported by a study tour to Great Britain (p.101). Perhaps book-learning may have given the stimulus in one important case. It is evident that J. P. Brändström had read Lagerhjelm's long article in *Svea* (1826) where British charcoal factories were mentioned, the same year as he himself first brought them up for discussion (p.111). However, it is not possible to establish whether he had known them before.[91]

2. The survey in section 5(b) has shown that for several years British castings were of importance for Brändström and Waern.

production. Would it not be most convenient that I got a fixed part of the higher price this mark could obtain over common iron? Then my interest would be to increase production and provide the most profitable sale for it. In the long run perhaps they will grow tired of this expense to me but the same will probably apply to all arrangements.' G. Ekman to G. H. Ekman, 7 May 1832, vol.E I:72, EA.

91. John Cowie & Co. to C. F. Waern, 24 May 1827, J. P. Brändström to C. F. Waern, 14 July 1827, WA.

Ekman was an able constructor, but sometimes he had to engage other persons to have his intentions carried through. He seems to have used Swedish engineers. Comments about them are few. Sometimes it was difficult to procure good castings.[92] No indications have been found that British material would have been necessary for the hot blast. Of course, Sweden's long iron industry traditions made its needs from foreign countries small.

3. On the Continent British manufacturers were now and then active in introducing coke furnaces, the puddling process and rolling.[93] But in Sweden mercantilist regulations and lack of coke hindered new departures in bar iron forging. The only example of British reorganisers within this branch shows traits differing from those usual in other countries.

It would be most correct to call J. P. Brändström British-Swedish. He started as a merchant. The later chain of events showed that for him industrial development was a means, not a goal in itself. He did not try to start new ironworks in Sweden, but only took part in a limited and rather unsuccessful acquisition of real property in eastern Värmland. Thus, even if he deviated from the usual behaviour in technology transfers, Brändström's multifarious and enthusiastic efforts in 1825–30, which have not been well known hitherto, make him one of the pioneers during the introduction of the British method in Sweden beside Gustaf Ekman and C. F. Waern.

In the light of these facts, his contribution during the 1830s is remarkably small. He does not seem to have returned to Bäckefors or to have met Waern other than during the latter's journey to Britain in 1833, when he demonstrated iron factories to the Swede. One reason for his small activity seems to be that the firms in which he was a partner had been lacking in working capital since 1831. It seems that he had not yet in 1832 paid those extra costs from the 1829–30 experiments he had promised. Possibly he did so in the spring of 1833, when Waern mentioned in general terms that an agreement had been made. Though Göteborg merchants several times complained that Cowie & Brändström did not even answer letters, Waern carried on the connection through 1834.[94] Obvi-

---

92. G. Ekman to G. H. Ekman, 9 December 1832, vol.E I:72, EA.
93. For an example, Henderson, *Britain and Industrial Europe*, pp.118–24, about the many-sided John Cockerill's large ironworks at Seraing in Belgium.
94. A. T. Almquist's report 22 April 1832; C. F. Waern (from Grimsby) to A. W. Melin, 16 May 1833; A. W. Melin to C. F. Waern, 5 July 1831, 21 August, 13 October 1834 and 7 March 1836; C. E. Billman to C. F. Waern, 11 August and

ously, having a heavy stock of unsold steel iron bought from Bäckefors, the Hull merchants broke the contract about continuing purchase. Some years later Brändström was partner in a new firm, Brändström & Thompson, about which opinions were hesitant.[95] In the year 1837 half of the Gåsborn furnace was sold. The loss then seems to have been borne only by the earlier partners, John Cowie and Lancelot Haslope.[96] Probably at the same time Brändström emigrated to the United States. Presumably at the beginning of 1837 Waern answered a farewell letter from Brändström which is not preserved but evidently was bitter:

Concerning the final success for the new forging method you surely first urged me to introduce in this country, I am willing to admit that it – along with all kinds of iron manufacturing – seems promising. But it was a time, and you must remember it well, when after three years of experiments the whole Bäckefors had produced only losses, and besides owed more than 40,000 in unpaid advances. The hope for ultimate profit appeared [?] to both of us; maybe the financial risk was greatest for me, and I think you left all this [?] just when it began to take a fortunate turn.[97]

The letter indicates some kind of partnership during the initial phase. Then and later Waern made his great contribution through continuously running an economic risk for technology renewal. However, his personal participation in experiments was markedly slight.[98]

4. Even if it was forbidden before 1825, many British workers were engaged on the Continent. The general problems for the new

---

24 November 1831, WA; G. Ekman to G. H. Ekman, 13 September 1833 and 20 May 1834; Cowie & Brändström to Ekman & Co., 1 October 1834, volsE I:74–5, EA.
95. Beckinton, Wilson & Co. to C. F. Waern, 13 May 1835; C. F. Waern to A. W. Melin, 7 February and 17 November 1836; A. W. Melin to C. F. Waern, 21 November 1836, WA.
96. L. Haslope to C. F. Waern, 13 May 1831, 25 August 1836 and 17 April 1837; C. F. Waern to J. Cowie (Liverpool), April 1837, WA.
97. C. F. Waern to J. P. Brändström, probably the beginning of 1837, WA.
98. In 1827 when Cleophas experimented with forging (p.111), he was disappointed at Waern's absence. The manager, Almquist, usually rather subservient, made the same point explicitly, when the British hammer was to be installed (p.115) in the same way as in England, which only Waern had observed. Nor did he attend Brändström's experiments in 1829. A. T. Almquist to C. F. Waern, 24 September 1834, WA.

employers in using this way of recruitment are well known. It was rarely the best British workers who emigrated. Beside lack of skill, they were also accused of lack of ability to adapt themselves, which showed itself in abuse of alcohol. For their exclusive skill they were paid very high wages with the intention that they would gradually teach native workers, which would make them redundant. If they refused to do so, at least they ought to be stopped from passing information to real or presumptive competitors. Bäckefors gives some illustrations.

It has been mentioned (p.115) that the refiners got lower pay in 1830 than in 1829. When two more smiths came in 1831 nothing was changed. Next spring Waern gave them all notice of termination of their contracts because they were too expensive. Their foreman, Houlder, informed Ekman who, behind Waern's back, went to Bäckefors and hired the two most skilled ones. Immediately thereafter Waern changed his instructions, and all British smiths had their earlier terms extended.[99]

In the long run it was impossible to reach a solution acceptable to both sides. Beside those adult workers who lived at the factory in December 1831, there were two years later several sons between fourteen and seventeen years old who needed employment. The manager suggested that the adjacent Öxnäs factory should be bought. A son of the foreman working in East India wanted to come to Sweden.

During his British visit in the spring of 1833 Waern had found out that there the pay for refining was 17s per ton. In the autumn he himself offered 15s instead of the previously paid 20s. The result was a postponement. In November 1834 two smiths left for a Norwegian iron factory, and in 1835 two others were employed in IMA forging experiments. In summer 1835 it was evident that two British smiths were willing to stay for the reduced rate. Now for the first time Swedish workers would be instructed in earnest and paid higher wages. The rest of the British smiths left for other areas although Houlder returned and ten years later made a distinguished contribution as teacher of Swedish smiths. Concerning technology and sales (section 6(a)) it was mainly factories among Waern's own customers that were helped.[100]

Even if he reflected a great deal, Ekman did not formulate any

99. G. Ekman to G. H. Ekman, 16 and 30 April 1832, vol.E I:72, EA.
100. A. T. Almquist to C. F. Waern, 7, 17 November 1833, 10 and 16, November 1834; C. F. Waern to A. W. Melin, 11 June 1835; J. G. Möllenhoff (manager, Bäckefors) to C. F. Waern, 25 November 1835, WA.

consistent view of profitability for the factory owner of the British workers employed. When he had failed to engage some of Waern's smiths, he gave it as his opinion: 'But to my surprise, and I think also to my advantage, this agreement was cancelled.'[101] As a representative for factory owners he saw the benefit of using the British in their better working capacity. On the eve of efforts planned in the spring of 1831 he wrote from Färna:

> During my absence the forging had given very fine iron but had been carried on too slowly. I know that the British make twice the quantity. Certainly their pig iron is more suitable, but anyhow I am uncertain about how much workers' skill may achieve, just as I clearly perceive several shortcomings in our workers which can be corrected only if they have an example to follow. In view of this, Count Hermansson brought up the question if you [Ekman Sr.] and he could not agree upon engaging a British worker.[102]

However, his failure to carry out this plan in Lancashire (above, p.103) seems to have resulted from his own moderate effort. On the other hand, some years later he was interested in finding an English steel burner and file maker for a shorter period in Sweden.

Ekman combined study trips to Britain, assistance from a foreign worker and the several opportunities he had over the years to observe British smiths working at Bäckefors (or at least their equipment). The time he himself especially emphasised was April 1832, but also January 1831; probably his fourth and second visits there in a year and eight months.[103]

5. More than others, Ekman appears to be the representative for those experienced travellers who were able to learn from study tours. However, his private correspondence shows that he was serious in his official report of 1831 where he belittled the value of his own travels:

> In my opinion the introduction of the method I am now trying is the safest way to give our iron an improved quality and an increased

---

101. G. Ekman to G. H. Ekman, 30 April 1832, vol.E I:72, EA.
102. G. Ekman to G. H. Ekman, 6 May 1831, vol.E I:70, EA.
103. A visit in May 1830 which Waern had accepted was cancelled by the IMA, because then the forging had ended. Visits were actually made in August 1830, January 1831, June 1831 and April 1832. A. T. Almquist to C. F. Waern, 21 August 1830 and 29 June 1831, WA; G. Ekman to G. H. Ekman, 17 January 1831 and 30 April 1832, volsE I:70, 72, EA.

revenue, at least to those who are the first among factory owners. For the introduction of the method there are two alternatives: one is to engage workers as Waern did, the other that you yourself take cognisance of the method. Then the surest way ought to be to use both. In addition it has now been an ambition of mine to have my own method, and I confess that during the few hours I saw the method in England I could not acquire sure knowledge of it. I had not time enough and could not afford to stay long enough to find it out perfectly. If, contrary to expectations, I should not be able to recruit any skilled or cheap worker, then I count on myself.[104]

Of course Ekman's remark that 'both/and' would give better results than 'either/or' is not very informative, as enough consideration is not shown for different outlays of time and money. Yet in practice in the two private Swedish efforts there was something of a 'both/and'. The introducers had wide understanding of metallurgic circumstances in Britain and were at the same time obliged to learn from, or use, skilled British workers. That their central efforts were different in character is clear.

(c) Technological Change and Realistic Commercial Aims

Production from the Swedish iron industry was chiefly intended for export. Attempts to change without considering that fact were bound to meet serious obstacles. So if Swedish iron masters had been able to make good rolled puddled iron they had met discrimination against it in favour of hammered iron in the United States, Sweden's most important buyer.[105]

The technological acquisition which was to be used successfully in Sweden was on its way out in Britain. No strong established interests there had any reason to defend its few producers. Even if it took some trouble, the method could be adapted to the earlier production process in Sweden. Its output, steel iron, was a priori clearly coveted in Great Britain just at that time. Demand increased strongly, at the same time as conditions were lacking for the domestic manufacture of steel iron. When commercial interests in Sweden settled choice and adaptation of technology, a marketable product was made.

104. G. Ekman to G. H. Ekman, 23 May 1831, vol.E I:70, EA.
105. A detailed analysis of the very important export to the United States is given in R. Adamson, 'Swedish Iron Exports to the United States, 1783–1860', *Scandinavian Economic History Review*, vol.17, 1969, pp.58–114.

For those few west Swedish iron factories (Bäckefors, Lesjöfors, Liljendal and, to a certain degree, Upperud) which before the mid-1830s were capable of climbing over the high technical threshold, the British method in the finery, together with a series of improvements in the preceding and following processes, resulted in a great commercial success.[106] The much higher price for middle marks against ordinary bar iron indicates that a really good product was made. But the way towards it was so difficult and called for such tight control at all points that only few firms managed it.

Earlier scientific interpretations have maintained that few factories definitively accepted the British method because the technical chain was not complete at its end. It was not possible to reach a sufficient reheating (welding heat) before the final hammering of the iron.[107] But the survey above of the big private efforts shows that to produce a middle-mark steel iron a long series of technical adaptations was necessary. Therefore lack of good reheating could scarcely have been a decisive obstacle for the further diffusion in the 1830s of the British method. Many factories introduced it tentatively for small parts of their production and then soon ceased. Normally this might have signified that in the beginning of the 1830s they had not skill or endurance enough to solve the different technical problems. Nor were they able to wait for a market introduction of an untried production eventually to give a good profit. But had the technical threshold been lower, the price of the middle marks would have been pressed down.

In the course of the 1830s a broader circle of ironmasters carried through at least part of those measures necessary for steel iron manufacture. They began to pay attention to exactly and consistently used mixtures of ores. Hot blasts were applied at furnaces, better blowing-machines were built and different fuel-saving hearth constructions and processes were taken up within the forging. These successive improvements must be held as technical preconditions, when the important invention for the later phase of the forging, Gustaf Ekman's welding-fire in 1843, could very quickly be used by many west Swedish ironmasters. Just then unusually favourable market conditions, much better than fifteen–twenty years earlier, were at hand. The railway mania in

106. About Upperud for example D. Carnegie & Co. to T. Hudson (Newcastle), 8 November 1824; to Beckinton, Wilson & Co., 12 March, 9 May, 14 November and 19 December 1831, vols8–9, CA; J. P. Brändström to C. F. Waern, 1829 'November, Thursday morning' 1829; A. T. Almquist to C. F. Waern, August 1830 and 8 October 1831, WA.

107. Boëthius and Kromnow, *Jernkontorets historia*, part III(1), p.484.

Great Britain in the mid-1840s led the iron-importing merchants to scramble for Swedish steel iron, especially of common marks.[108]

As a contrast, some other cases will briefly be mentioned, when technically sufficient knowledge was not combined with commercial understanding or possibilities. When production of crucible steel (category 1, p.96) was started twice during the eighteenth century, a problem was that the Swedish market was too small. From the scientific literature we find no serious sales efforts even in Sweden, still less outside.

Another case (category 3, p.97) deals with the production of steel in Sweden. During the first decades of the enlarged steel iron production, Swedish ironmasters sometimes had to use only the best-worked middle parts of the blooms for export. Even though the end parts had also involved extra costs, they had to be used in a less remunerative way. They formed raw material at the home factory for blister steel. In several connections steel iron producers tried to make steel from this raw material much better than the usual one and sell it for a higher price. The engaging in the middle of the 1840s of a skilled English steel burner was intended to further this aim.

Here technical skill was not enough in selling the finished product. During this period most Swedish steel was shipped to markets where high quality was not valued, to East India as well as the United States. To sell fine and expensive Swedish steel to Sheffield – instead of steel iron to Hull or Sheffield – was impossible for another reason. Many masters in Sheffield themselves partly burned iron to steel, partly made finished products such as files. Some profit was made in each link. The integration of consecutive processes in the Sheffield region prevented a Swedish intrusion in this most important steel centre of the world.[109]

7 Implications

Eli F. Heckscher once argued that the intense efforts of Sweden's ironmasters, metallurgists and factory workers during the first half of the nineteenth century possibly 'made up the most glorious page

108. On the railway mania Gayer, Rostow and Schwartz, *Growth and Fluctuation*, vol.1, pp.304–5, 309, 315–18.

109. A. T. Almquist to C. F. Waern, 19 June, 1830; C. F. Waern to Cowie & Brändström, 12 May 1834; Wilson, Hudson & Co. to C. F. Waern, 19 April 1838, WA; G. Ekman to E. Spence (Hull), 1 June 1834, vol.B I:48; E. Spence to Ekman & Co., 17 June 1834, vol.E I:76, EA.

in the history of the Swedish iron industry and even in Sweden's whole economic history'.[110]
The first phase of this process, marked by technology transfer and adaptation, has been studied above in detail. In somewhat less elevated words, the investigation shows that merchants also played an important part in saving the Swedish iron industry from very threatening competition.

In a broader perspective, the introduction of the obsolete British forging method in Sweden returned to its native country a material of vital significance, i.e. steel iron. Through this, an important possibility for expanding the use of high quality steel, one of the basic products in the industrial revolution, was created for the thirty-year period 1830–60 before the Bessemer process.

---

110. E. F. Heckscher, *Svenskt arbete och liv, från medeltiden till nutiden*, Stockholm, 1957, p.254.

6
Artisan Travel and Technology Transfer to Denmark, 1750–1900

Poul Strømstad

Being situated at the European cultural periphery, it is not surprising that Denmark has received important impulses from other countries through the ages. New knowledge about foreign technical developments has always been of great importance for the production of Danish goods. Medieval metal forging techniques, brick and tile firing, modern steam power and, today, electronics are marked examples of technology transfers to Denmark.

Since foreigners in older times reached Denmark only in limited numbers, Danes themselves had to go abroad if they wanted to keep abreast with European developments. From medieval times we have information about numerous Danish students studying at foreign universities. Most of them returned to Denmark after some years, some continued their studies, and others ended up as teachers at universities or schools abroad. From the turn of the fifteenth century many young learned people and nobles were sent on educational visits abroad. The journeys could last for a number of years, and such visits without doubt greatly influenced Danish cultural development, as they gave the otherwise somewhat provincial Danes insights into foreign circumstances and scientific knowledge, expanded their knowledge of languages and generally improved their education.

Of far more widespread importance, and absolutely decisive for production to meet daily needs, however, was the enormous exchange of technical knowledge and skills within the craft trades. In a period when only a small minority were able to read, and the opportunity to extend one's knowledge by help of books was minimal, the only way to learn a trade was through apprenticeship in a workshop. Here the apprentice was introduced to the trade by a master or journeyman, and if he passed the apprentice examination after a learning period of four to five years, he was made a journeyman. Even during the sixteenth century this was not, however, regarded as sufficient; the journeyman was expected subsequently to seek to improve his education by moving from place to place, working for longer or shorter periods in workshops abroad. Owing to the dominant position of the German crafts, the operations of the Danish crafts were to a large extent influenced by German traditions and customs, the so-called *zunft*, which regulated operations in every detail. The craft societies' rules (*skråerne*) included regulations for apprenticeship, year of travel, the conduct and relationships between artisans, and much else.

Artisanal wanderings were to a large extent influenced by the business cycles and functioned as the employers' regulator for work. During good times, with high employment, there were relatively few wandering artisans on the country lanes; however, when there was a slump and marked unemployment, the journeymen had to seek areas where labour demand was greater. Many Danish journeymen were content with wandering in Denmark, but work experience from abroad was regarded as a better qualification. Thus many wandered abroad, especially to the German-speaking areas. As a consequence of this extensive artisanal brotherhood, many foreign journeymen came to Denmark. Most of them spent a limited period there, but some settled in Denmark. The wandering artisans contributed to the diffusion of professional knowledge between fellow craftsmen, and to the international character of the trades.

This international technology transfer was of very great importance with regard both to traditional working methods and to innovations. In consequence of the frequent exchange of knowledge it was possible to follow developments within the different crafts. A very considerable number of journeymen wandered along Europe's country lanes in search of new workplaces. In the nineteenth century the average number per day is estimated as not less than 200,000, and even today professional as well as everyday language are marked by this strong cultural influence.

1 Mercantilistic Industrial Attempts

The consumption of Denmark's population was to a high degree based on self-supply. All homes sought as far as possible to produce what they needed. The most important exports were cereals and livestock, but almost all other produce went to the home market. The system was slow and difficult to change, but from early times the king had sought to promote domestic production of some key products. Attempts were made by offering subsidies, loans and tax relief to attract foreign artisans, artists and experts, but the outcome of these attempts was usually meagre. The products they produced, could not, because of the absence of economic and distributional preconditions, be sold at a profit in large quantities, and these often somewhat random experiments often came to little.

With the spread of mercantilistic ideas from about the middle of the seventeenth century some more direct attempts were developed to establish domestic production of goods which otherwise would have been imported from abroad. Under Frederik IV a commercial college (*kommerce-kollegiet*) was founded to promote the establishment of factories and encourage interest in industrial production. This experiment, too, was unsuccessful, and the commercial college was dissolved after only a few years in operation. A serious slump in the 1730s forced the government to try to fight the crisis, and in 1735 one of the deputies in the finance collegiate, Count Otto Thott, presented a proposal based on central mercantilistic ideas: the encouragement of domestic production, as far as possible based on Danish raw materials; the encouragement of exports; the limitation of imports; subsidies and prizes to the producers; and the instruction that both the state and the individual citizen were to adopt the most economical conduct. Teaching was to be improved, useful sciences such as experimental mechanics and physics to be encouraged, and important branches such as textiles, iron and glass to be developed. It was critically important that money was made available for production and commerce, as money was for 'commerce the same as blood in a human body'. Otto Thott's text thus contains a kind of programme for the development of Denmark's industry, and again a commercial college was established, which, this time, proved longer-lasting. In the period 1730–46 approximately sixty manufacturers from Britain, Holland and Germany was called in, and each year 30,000–40,000rd (1 riksdaler is worth approximately 2 Danish Krone) were set aside for this purpose.[1]

Among those called in we find for example the cloth manufacturer Pierre Gandil, who came from Magdeburg with his family and eighteen workers, of whom four were children. From the state he initially received 3,220rd to purchase tools, establish a cloth manufactory and pay for his travel and other expenses – a considerable sum, but one of the smaller amounts invested in this field.

---

1. Kristof Glamann, *Otto Thotts uforgribelige Tanker om Kommerciens Tilstand*, Copenhagen, 1966, pp.106, 114.

Others received far larger sums of money, but the profit accruing to the state was usually meagre. Frequently the foreigners would disappear abroad, leaving behind debts and shipwrecked projects, which had cost Denmark dearly and to no avail. It was rare that well-qualified people with enough capital could be persuaded to leave their homes and settle in an unknown foreign country. The system produced a number of charlatans and swindlers, who lived well at the expense of the state and fled when contracts had to be fulfilled. Even those who arrived with the best intentions and with the best qualifications would find it difficult to make a break-through. Opposition from the guilds to the grant of work permits without guild membership, difficulties in getting the necessary raw material and qualified workers, lack of capital, sluggish turnover and fierce competition were all factors which might beat the best among them. It was, of course, impossible to foresee what the newcomer was capable of, and many were tempted by premiums and privileges. The eighteenth century was the golden age of project-making.[2]

To try to ensure the domestic manufacture of complex military weapons and equipment, attempts were made to improve casting techniques. Bronze cannons were expensive, and the metal had to be imported. Therefore it was sought to make them with Norwegian iron, and in 1751 a French cannon maker, Jandin de Peyrembert, was called in. He had pledged that he was able to forge iron cannons which were both less expensive and lighter than bronze cannons. He arrived with a master caster, bore-master and workers, and was given a loan in order to set up a smithy in the old Frederiksværk foundry. Sole right of manufacture was granted him and his offspring for a period of fifty years. Several buildings were erected, among them a boring-mill. The project did not, however, bring him, or the Danish state, much joy. When attempts were made to fire Peyrembert's cannons they tore apart or exploded, and he was duly sacked in 1756, albeit 'in grace', and left Denmark.[3] The experiment cost the Danish state dearly, but luckily some of Peyrembert's workers stayed on in Denmark. The boring-mill mentioned above was enshrined in a mist of secrecy, but operation was carried on by the bore-master Lorentz Juncker, originally from Strasbourg, and he, together with the Swedish-born casting-master Henrik Hornhaver, manufactured the main part of Denmark's

2. Bro Jørgensen, *Industriens Historie i Danmark*, Copenhagen, 1943, vol. 2, p.72.
3. Egon Eriksen, *Dansk artilleri i Napoleonstiden*, Copenhagen, 1988, pp. 26ff., 91, and *Krudtværket på Frederiksværk 1785–1985*, Copenhagen, 1958, pp.31ff.

Figure 6.1   Cannon works

Source: C. A. Lorinzen, *Frederiksværk*, 1773; 2nd edition, 1986; p.523

armament production for many years. Their sons continued the production, and during the period 1756 to 1819 deliveries to the Danish army and navy were based on these two families. Although Peyrembert cost the Danish state large sums of money, his short-lived activities nevertheless resulted in the development of a Danish cannon foundry, which for more than half a century delivered up-to-date armaments.

2   Textile Production

Textile production underwent frequent technical developments which it was important to get to know. The ban on exports of machinery from Britain, and great unwillingness to let foreigners get to know about innovations and improvements, made it difficult to procure certain information about new production methods. In 1750 a Briton resident in Denmark, the dyer John Smith, smuggled, at the risk of his life, a model of a calender to Copenhagen.[4] In recognition he received 200rd and the sole right to employ the machine, which was built according to the model. To prevent others getting to know the construction of the calender, the workers, who were engaged in building it, were never to see it complete, a measure which presumably must have been impossible to implement. Further precautions included the demand that the workers took an oath that they would not divulge the secrets.

4. C. Nyrop, *Niels Lunde Reiersen*, Copenhagen, 1896, p.185.

Many attempts were subsequently made to end this secrecy, but it took a long time before attitudes to this problem became more open. When a mechanic had managed to find out a construction detail or had himself made an innovation, he was of course not interested in divulging it to one and all. In 1765 a special fund was established, the Fonden ad usus publicos (fund for public utility), whose aim was to subsidise travel for artists and craftsmen.[5] Those who received support were obliged to report on the usefulness of the travel, and if they received help to buy or build machinery, others were to be allowed to see and possibly copy it. The Travel Fund, which was founded in 1795, and the Industrial Fund of 1797, which was under the Commerce College, stipulated similar conditions.

It was sought, through loans, subsidies and other forms of support to supply the textile industry with new initiatives and methods. To promote cotton production the authorities accepted the Swedish manufacturer Charles Axel Nordberg's offer to come to Copenhagen and set up a so-called 'Manchester' factory. Nordberg had experience partly from his time as a factory owner in Stockholm, partly from a fifteen-year stay in Britain, and he also had knowledge of finishing, a textile technique little known in Denmark.[6] With support from the Commerce College, that is the Danish state, the Royal Privileged Cotton Manufactory was built, started operating in 1780 and within a few years became Denmark's largest manufacturing concern. At one point there were about 125 weavers and 800 workers employed. Nordberg had intimate knowledge of British textile machinery and had brought home with him drawings to enable him to build the machines. He constructed new spinning machines based on his knowledge of Hargreave's jenny, Arkwright's waterframe and Crompton's mule. Thus, only ten to fifteen years after the invention of these machines Nordberg was able to introduce them in Denmark. But the constructor's insight and intentions were one thing; the qualifications of the workforce were another matter. The different machine parts were constructed by skilled workers from a wide range of craft backgrounds, which caused problems and long delays. One problem was inadequate precision, which was of decisive importance in the spinning and weaving processes, and it was difficult to explain to the mechanics, who had never seen the machine, how

5. Axel Nielsen *Industries Historie i Danmark*, 1944, vol.3.1, pp.183ff.
6. Bro Jørgensen, *Industriens Historie*, pp.174ff. See also Trine Parmer's chapter in this book for Nordberg's activities.

the specific parts were to be constructed. In many instances Nordberg had to travel to Britain to collect supplementary information, and he also had to seek help from a British mechanic living in Denmark, John Smith.

The Danish state spent considerable amounts of money on the 'Manchester' factory in Copenhagen, with a view to improving knowledge about cotton manufacture.[7] At the factory Danish producers were able to gain insight into the production processes, to see the machines in operation and to assess the finished products. In addition, workers received and learned how to operate the new machinery.

In spite of the advantages due to the machines and large-scale operation, the products were expensive and sales too small for the production to be profitable. In 1795 the factory was sold at auction, and in 1798 Nordberg became the manager of a machine engineering works, which delivered textile machines to several factories in the provinces and also produced agricultural machinery.

It usually took some time before new machines and production methods made a breakthrough in Denmark. Investment in machinery often depended on public subsidies, and the problem of getting the workers to accept the labour-saving machinery further dissuaded the manufacturers from switching to more modern production methods. The one-man-operated flying-shuttle loom, which was invented by Kay as early as 1730, arrived in Denmark only in the 1790s, and then only in a limited number. This was to a large extent due to the cloth apprentices' refusal to work these looms – just as in Britain – and negotiations about new improved wage rates did not succeed.[8]

The introduction of the Jaquard loom in Denmark probably dates from the end of the 1820s, and in 1830 we know that a Danish carpenter made Jaquard looms modelled on looms used in Germany and France.[9] He had, among other things, reduced the height of the loom, so that the attached punch-hole machine could be used in rooms of ordinary size. At the same time another Danish carpenter applied to the Reiersen Foundation for 600rd to construct a so-called 'punch-machine' for making the holes in the cardboard which was used in pattern weaving. In his application he maintained that no other person than himself built Jaquard looms in Denmark, and that he would be able to supply cards for all kinds of

7. Ibid., p.178.
8. Ibid., p.213.
9. Nielsen, *Industriens Historie*, vol.3.2, p.126.

patterns for shawls, carpets, tablecloths, blankets etc. with his punch-machine.[10]

3 Artisans' Travels

The traditional wanderings of artisans to different workshops to see the improvements which were taking place within their crafts continued, if not unaffected by the new technical improvements, then to a high degree in accordance with established tradition. This practice frequently generated technology transfers of considerable importance to the Danish economy, which is illustrated by the examples below.

C. C. Hornung was born in 1801 in the provincial town of Skelskør, in Zealand.[11] From the age of seven he helped his father, who was a hatter, in his workshop. When he was fourteen years old, he was able to make a good quality hat though he had not learnt to read or write. He did not care much for hat making and was allowed to volunteer for work with a carpenter, and later with a wood turner. When his elder brother returned home from his wanderings he had to set out, as it was too expensive to keep both brothers at home. He travelled in Germany as a hat maker, but worked as an apprentice carpenter, which was illegal, in different towns. By chance he came across a grand piano made by one of the best musical instrument makers at the time, Ritmüller in Göttingen. It was very elaborate, with inbuilt drum, bell, cymbal and bassoon. Hornung became very enthusiastic and set off to Göttingen to seek employment at Ritmüller's, but was rejected, as he was no instrument maker. For some time he made a living by making furniture, but succeeded, in Kassel in getting employment as instrument maker, despite being a 'mere' carpenter. He was employed to make a piano, which he had never previously attempted. He would have liked to see the other journeymen's work and study their technique, but did not dare to do so in their presence. Only when they had left the workshop did he steal in, and he succeeded in making copies of some of the pianos. When he left the workshop he told the master about this, but the master replied that it did not matter. He had himself done likewise.

After three years in Germany he returned home and immediately

10. R. A. Reiersen Fond (National Archives, Copenhagen), Sager til Deliberationsprot, 1 October 1830, no.88.
11. C. C. Hornung, *Et Haandværkerliv i forrige Aarhundrede*, Odense, 1904.

started to build a fortepiano in his parents' sitting room. The Kassel piano was the model, and he rapidly sold it. In order to formalise his trade position he subsequently built his master's qualifying piece as carpenter and instrument maker. In Copenhagen he got the opportunity to see one of Marschall's pianos, which was considered among the best in Europe. He made a drawing of it and subsequently made six or seven pianos each year. His main employment, however, was furniture making, where he employed about six men. He nevertheless kept abreast with developments in piano making and built pianos modelled on the Briton Collard's designs, which were better than Marschall's.

Hornung's business grew. He now made a couple of pianos every week, and in 1842 moved to Copenhagen, where he rapidly became the leading instrument maker. It was not only replicas Hornung made. In 1842 he got the patent rights for ten years for a piano where the strings were fixed in a cast iron frame placed across the sound board, without touching it. Makers had begun to strengthen or brace pianos with iron, but it had not been possible to make the box sufficiently strong. At concerts the instruments therefore had to be tuned between numbers, but on Hornung's pianos it was possible to play throughout a whole evening's concert without setting the tuner to work. The iron frame was an important improvement, and Hornung found himself at the opposite end of the same situation which he had been in as a youth: the leading instrument maker in Berlin bought one of his pianos to use as a model. In 1844 he stayed for longer periods in London and Paris, where he hired instruments in order to examine their construction.[12] He made drawings of some of them, but built his pianos somewhat differently – not as mere copies. In 1846 he got the sole rights for five years on his upright fortepiano, which was of less height than the older instruments and therefore easier to fit into ordinary living rooms. His factory became the largest in Copenhagen, and Hornung became royal supplier. He took the instrument maker H. P. Møller into his firm, and during the years 1842–9 approximately 1,200 pianos were built by the firm Hornung & Møller, which employed approximately sixty men. The factory was still growing and was, until the middle of the twentieth century, one of the leading firms. Hornung had worked very hard to reach the position he held in about 1850. He now wanted to retire, and by selling the firm could ensure for himself and his family a quiet existence.[13] In

12. O. J. Rawert, *Kongeriget Danmarks industrielle Forhold*, Copenhagen, 1850, p.233.
13. Ibid., p.233.

1850 he made his last two pianos for the world exhibition in London in 1851. They were highly esteemed, and the same year Hornung retired to his home town. He died in 1873 in Copenhagen. Nearly 100 years later, in 1972, the firm Hornung & Møller closed down. During its 150-year lifespan it had delivered more than 55,000 pianos.[14]

Jørgen Balthasar Dalhoff was born in 1800 and was apprenticed to a goldsmith in Copenhagen in 1815. Dalhoff sought to acquire insights into all fields of metal production in addition to that of the goldsmith, e.g. metal casting, chasing and engraving techniques, as well as drawing and modelling. On the completion of his apprenticeship in 1820, he went abroad for three years, subsidised by the Royal Danish Academy of Fine Arts.[15]

He worked for some time in a larger metal works in Berlin, where he was informed that journeymen were not to move between the different workshops within the works; the work methods were to be kept secret. However, prior to this instruction Dalhoff had already been in the machine shop, where the inventory was particularly secret, making drawings of some of the machines. He sent detailed descriptions of tools and machines to his brother, who was a brazier in Copenhagen, and models and drawings to the Royal Danish Porcelain Factory in Copenhagen.

Following a six-month stay in Vienna, where Dalhoff worked with a leading engraver, he went on to Rome. Employed at the works of the Danish sculptor Thorvaldsen, he was taught bronze casting by the famous Hopfgarten. To receive instruction in bronze casting was one of the conditions for financial support from the Academy, as Denmark was lacking in skilled people in this field. In Rome Dalhof drew and modelled objects and motifs from Herculaneum and Pompeii, and this experience, combined with later drawing lessons from Professor Hetsch in Copenhagen, greatly influenced his use of classical style and ornaments in the decoration of the numerous art objects he made during his long working life.

On his return to Copenhagen Dalhof was made master goldsmith, without submitting a master's qualifying piece. His work testified to his skill, and in 1833 he became goldsmith to the king. In the following years Dalhof applied what he had learnt and seen during his travels; he built, among other things, a machine for pressing larger silver works, presumably a development of a ma-

---

14. Knud Meister, *Ulige børn leger bedst*, Copenhagen, 1988, p.50.
15. The following draws on N. Dalhoff's work: *Jørgen Balthasar Dalhoff. Et Liv i Arbejde*, Copenhagen, 1915.

chine he had seen in the machine shop in Berlin. In his memoirs and letters he relates how difficult it had been to get the opportunity to observe firms' trade secrets. During a stay in Britain in 1837 he visited the factory of the famous file manufacturer Hall in Sheffield. Although he had a letter of introduction recommending him to the owner, he did not get the opportunity to see what he wanted. The day following the visit he wrote:

> Yesterday [the owner's] brother would not show me how open brass work is made, which is here widely used in larger pieces and so fine that they are used for Venetian blinds, but said that they did not produce this. I now asked him if it were not paper which had been painted, and he assured me to the contrary. We were almost ready to start betting on it, but it ended with him promising to show them to us. It was not his intention, however, to show us how they were made, because we arrived there at a time when no work is done; scarcely had we entered, however, before I secretly put half a pound in the hand of an old worker, and more thereafter, and I was able to see everything in detail, while the others conversed in English with the guide; I repeated this conduct in a plate work shop.

In Birmingham Dalhof was also accompanied by a guide

> When we were outside again, we gave him half a pound, and he managed to get us access to places which were especially forbidden. When we later returned to Quest's [the Quest factory], [the owners] became apprehensive that . . . [we] might get too involved with the other firms, and therefore they preferred to guide us around themselves, and thus we got access to almost everywhere we wanted, especially after we gave the guide who was to look after us £1 10s. In addition to some workshops we have also visited thirty factories in Birmingham and Sheffield, and surely not one in a hundred persons has seen as much as myself; often I have myself been able to attempt working, and I have samples of thimbles, buttons, sewing needles, pins etc. at every stage in production until finished. But it has cost large sums of money; we have for 11 days spent £1 10s pounds a day.

While in Germany Dalhof writes in 1840:

> I saw a number of interesting things in Nuremberg. I was not permitted, among other things, to enter a foundry, where candle plates, etc., were cast in clay moulds, but when I saw an open window at a distance, I fell into conversation with a moulder on the inside, and by means of some guilders, the outcome was that I passed him my handkerchief, and it

came back out again wrapped around two clay moulds, and he willingly showed me how to use them.[16]

Back home in Denmark, Dalhof kept up to date in all the fields of metal working, and was assigned numerous decorative jobs. In 1837 he started bronze casting, and proved his ability, based on his apprenticeship in Rome, by casting several statues, including that of Christian IV in Roskilde Cathedral, four colossi for the façade of Christiansborg Castle in Copenhagen, and the largest, and most difficult, the goddess of Victory with her four-in-hand carriage on the roof of the Thorvaldsen Museum. Furthermore, Dalhof worked with terracotta firing, galvanoplastic, electroplating and German silver methods, and he made seven statues in zinc casting for the façade of the Church of Ansgar. That Dalhof kept abreast with the most recent developments is evident from the fact that he made objects in galvanoplastic in 1840, only three years after the method was invented by M. H. Jacobi, and the same year it was made public.[17]

In his youth Dalhof's ambition was to become a sculptor, but he realised that his talents were within the applied arts and crafts, and he developed into one of the leading artisans educated during the first half of the nineteenth century. He brought to Danish craft production, through tireless experimenting and constant search for new methods and materials, valuable experience and initiatives. He attached great importance to the education of artisans. He published a collection of drawings, 'Models for goldsmiths', taught for almost forty years at the Royal Academy of Fine Arts, and in 1843 was a founder member of the Danish Industrial Association in Copenhagen.

Both Hornung and Dalhoff worked in areas outside the range of their original education. A characteristic feature of early Danish industrial development is, presumably, that enterprising people tentatively applied themselves in novel areas, often ending up as specialists in techniques far removed from their professional starting-point. One of Dalhof's employees, for example, was originally a chimney-sweep, another had started work as a stocking weaver. The latter, Thomas Chr. Thomsen became Denmark's first plasterer, a profession which so far had been dominated by Italians, and he later worked in bronze casting.[18] Thomsen made

16. Ibid., pp.54–249.
17. Knud Holm, *Metalskulpturer i forrige århundrede. Natonalmuseets Arbejdsmark*, Copenhagen, 1988, p.172.
18. Ibid., p.174.

the 'Lion of Isted', the statue of King Frederik VI in the Frederiks-
berg garden in Copenhagen and 'Moses' in front of Copenhagen
Cathedral, to mention some of his independent, larger works. His
works show that Dalhof's workshop had produced a professional
sculptor of the highest ability.

4 A Technician in Search of Knowledge.

Anthon Frederik Tscherning spent long periods travelling to a
number of distant destinations. The reasons for his journeys were,
however, very different from those of the artisans.[19] Tscherning
did not travel as a working craftsman, but as an officer in the
artillery. Already as a small boy he gained insights into the casting
of cannons and other heavy armaments at the Frederiksværk
foundry, where his father was employed as director. At twenty
years old, in 1817, Tscherning was in France with the Danish
occupation forces, and was given the opportunity to study artillery
in Paris and Metz. He later worked as a teacher at the Danish
military academy, and was subsequently promoted to the rank of
captain. However, Tscherning rapidly found himself in conflict not
only with the officer corps but also with Frederik VI personally
because of his critical views on the current management of the
army and of the country in general. Unperturbed by the king's
repeated warnings, Tscherning continued publicising his criticisms
in the papers and was consequently discharged from his duties at
the Academy in 1833. He was not expelled from the army, however, but sent abroad on a compulsory 'study tour'. He was not
told how long the journey would last and was only gradually
informed where he was heading. In this way Tscherning spent five
years being ordered around in Europe by the commanders of the
artillery corps.

Already during his first journey to France in 1817 he had visited
several factories and sent descriptions and drawings of machines to
his father at the Frederiksværn. He wrote from Paris, on 13 August
1817, that His Royal Highness the Prince (later King Christian
VIII) had given him the task of examining

the art, which is here known as lithography, that is the printing of
scripts or drawings by means of a stone; he wanted to know whether

19. See Anthon Frederik Tscherning's papers (National Archives, Copenhagen),
1876, vol.1, p.103.

this art of reproducing drawings would be practicable at the military office. I have been allowed to make several drawings on the stone and had them printed, but as they make a great secret of both how the stone is prepared and how to prepare the ink and the pencil, I have had to spend much time and money (from my own pocket) to enable me to give a correct and detailed description of whether this art could be of use for the army.[20]

Tscherning sent home numerous letters during his compulsory stay abroad from 1833 to 1838, and kept diaries in which detailed reports of his experience and activities were written down. Here he gave vent to his political views, which, under the influence of contemporary revolutionary European currents, became increasingly radical. Further, he wrote in detail about the people he met and the local circumstances where he stayed, but a main part of the diaries deals with all kinds of industrial techniques. He was unbelievably active and travelled long distances to see specific factories. Organisation, equipment, machines and products are described in detail and often illustrated with sketches. Tscherning attended trial firings with cannons and visited several factories: a saltpetre refinery, a porcelain factory, a lead rolling-mill, rocket bases, foundries, gun factories and so on. He took notes and made drawings everywhere. In December 1833 he visited the Frankfurt gasworks of Knoblauch & Schiele and noted:

They also produce oil gas, find resin the cheapest, which they melt at low heat and later dissolve in ether oil from the gas production. The retorts are made of iron, cylindrical, each retort consisting of two parts: the actual retort and the rear part. The retort is in itself about 2 ft long with an inside diameter of 9 inches, with edges at each end, the one at the front for the lid, the hindmost for the rear part.[21]

The description is several pages long, contains sketches of the retort and the gear mechanism, and has attached to it an advertisement from the firm with a drawing of the gasworks. It must be borne in mind that Denmark's first gasworks was opened in 1853, that is twenty years after Tscherning had seen the gas installation in Frankfurt.

This is only one example illustrating Tscherning's insatiable thirst for knowledge. During what for him must have felt like five

20. Ibid., p.110.
21. Rigsarkivet (National Archives, Copenhagen), Tscherning's diaries, 6464/6, 27 December 1833.

very long years, he investigated the macadamised roads in Paris, was at the Villette canal to see 'one of the so-called English flying boats', visited paper mills, iron foundries, arms workshops, telegraph stations, instrument collections, explosives manufacturers, railways and numerous factories. As a Danish officer equipped with letters of recommendation from the highest authorities Tscherning stood a good chance of being admitted where others were refused.

His knowledge of languages, deportment and general education made it easier for him to gain entry, and only rarely did he experience any difficulties in getting information about techniques and manufacturing methods.

It is difficult to assess the effects Tscherning's investigations had on Danish development; his personal gains, however, are beyond doubt. His extensive knowledge was important when he later became manager of a colliery in France, and his experience invaluable in his subsequent work as a politician, war minister and participant in several industrial undertakings.

5 Developments in Printing Techniques

In the early nineteenth century many improvements and new inventions took place in printing. In 1810 Friedrich König invented the machine-operated book printing press, which in 1814 was introduced for printing *The Times*. The new printing methods were applied in Denmark after a slight delay. In 1829 the printer J. D. Qvist in Copenhagen bought 'a specimen of the printing press invented in Philadelphia (in 1817) called the Columbia press'. He made the purchase in Brunswick, and it was the first press of its kind in Denmark.[22]

It was rapidly used as a model for imitation by the mechanic Schiøtt, and one was installed in 1830 in Peter Nikolaj Jørgensen's book printing press in Copenhagen. This was described as a machine 'which, in the accurate and solid construction of its various parts, [surpasses] the German presses of this construction'.[23]

Schiøtt was the leading Danish mechanic at the time, and delivered machines to a number of factories.[24] However, like so many people engaged in industry he went bankrupt in 1833 and

22. Rigsarkivet (National Archives, Copenhagen), Det Reiersenske Fond, Deliberationsprot, 1830, no. 70.
23. Ibid., 1831, no. 7.
24. Nielsen, *Industriens Historie*, vol. 3.1, p.390.

escaped to America, leaving behind him debts exceeding 12,000rd. Frequent bankruptcies were to a high degree caused by the problems of getting the machines to function perfectly. Machines purchased from abroad were often defective and had to be repaired after a short time. This caused interruptions in production and consequently loss for the owner. Machines built in Denmark, on the other hand, could often not be delivered on time, because essential machine parts had to be replaced, and frequently the running-in period lasted longer than expected. In addition, the workforce was unaccustomed to the precision and alertness required by the new machines. Many of the entrepreneurs who established engineering works and iron foundries were practical, technically gifted men, but only rarely did they possess insights in construction calculations.[25] Consequently the machines were often unsuitable, broke down rapidly and required time-consuming repairs. When buying machinery from abroad, it was common to get foreign workers to go with them, and during the running-in period these men taught Danish workers how to operate them. However, in the case of Danish machines, operation depended on Danish workers who did not possess similar thorough knowledge of the machines.[26]

The first high-speed printing press arrived in Denmark in 1825. In 1836 there were still only five, but in the 1840s the number increased rapidly.[27] The Danish machine builder Hilarius Lind had in 1841–2 worked as a draughtsman, machine fitter and commercial traveller for the Viennees firm of Helberg & Müller, which was a branch of the engineering firm of König & Bauer in Württemberg – the firm which delivered the above-mentioned press for printing *The Times*. When Lind returned to Denmark in 1844 he brought with him complete drawings for a 'Wiener-Schnell-Presse' and built several presses of this type.[28]

Meanwhile J. H. Hüttemeier, a master locksmith in Copenhagen, had made the first Danish high-speed printing press in 1836. With him worked the smith J. G. A. Eickhoff who learned its construction and subsequently became the leading manufacturer of printing presses.

Eickhoff is a good example of the itinerant journeyman who settled in Denmark, married, and by industry and skill created a

25. Ibid., p.391
26. Ibid., p.392
27. *Skandinavisk Bogtrykkertidende*, 1875, col.24.
28. Ibid., 1874, cols75ff., 102ff.

Figure 6.2   High-speed printing press built at Eickhoff's machine
works in Copenhagen

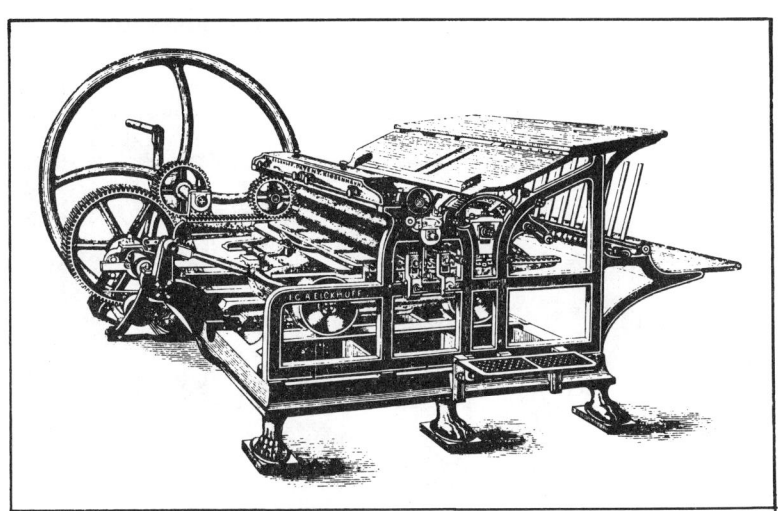

Source: *Opfindelsernes Bog*, 1st edn, vol. 1, 1877.

solid firm. He was born in 1809 in Mecklenburg-Schwerin, and in
1824 was apprenticed to a smith in Lauenberg, which at that time
was part of Denmark.[29] On the completion of his apprenticeship in
1828 Eickhoff moved around Germany for some months, and then
came to Denmark. He worked in different places but came, in 1833,
to Frederiksværk, which under the direction of the mechanic von
Wurden was Denmark's most important machine factory. Here
Eickhoff worked until 1836, followed by a six and a half years'
employment at Hüttemeier's. In 1848 he started his own firm and
became, on the closing of Hilarius Lind's works mentioned above,
Denmark's main supplier of printing machines. At first he delivered a couple of presses a year; in 1861 he had completed a total
of fifty, reaching no.100 in 1869 and no.200 in 1874. Production
expanded constantly, and in the 1870s the firm delivered one press
each week, and exported more than half of the machines.

29. Ibid., 1875, cols102ff.

Figure 6.3 Emil Chr. Hansen's yeast-apparatus constructed in collaboration with S. A. van der Kühle.

Source: Kristoff Glamann, *Louis Pasteur et Carlsberg*, 1988.

6 Bavarian Beer Comes to Denmark

The brewing of beer in Denmark had been carried out by traditional methods reaching back several centuries, until J. C. Jacobsen succeeded, as the first Dane, in brewing Bavarian beer. Up until 1846 'white beer', or household beer, was the most important beer in Denmark. The quality varied considerably, but usually it was not very good. Jacobsen wanted to improve the quality and, according to family tradition, he experimented with different kinds of brewing methods, using his mother's laundry pan in the old brewing house in Brolæggerstredet in Copenhagen. Results were slow in coming, and Jacobsen therefore set out for Germany to make a thorough study of brewing methods. The brewer Gabriel Sedlmayr in Munich, welcomed him with kindness and openness. Sedlmayr lived by the principle that a person's experience and knowledge should be disseminated and passed on to others, and J. C. Jacobsen received all the information he wanted. On a later visit he was given some of Sedlmayr's special yeast to use in his experiments. Tradition has it that Jacobsen brought home the yeast in a tin can which, during his long and strenuous journey, he carried inside his top hat. Whenever the stage coach made a stop at various inns Jacobsen had to pour cold water on the yeast to keep it fresh.

In 1846 Jacobsen sold his first brew of Bavarian beer, and since then the 'lager' has been increasingly successful. Demand for the new bottom-fermented, or partly fermented, beer increased rapidly, and Jacobsen continually expanded his production.

Jacobsen's acquisition of the Sedlmayr yeast and his persistent brewing experiments caused a radical change in Danish drinking customs and in his own position; he became Denmark's leading brewer and never forgot Gabriel Sedlmayr's principle that all information about brewing methods was to be accessible to one and all. Undoubtedly, this was in his mind when he established the Carlsberg brewery laboratory and formulated the 'Golden Words', which still form the basis for Carlsberg's activities. This was reflected, among other things, in Carlsberg not keeping secret Emil Christian Hansen's epoch-making breakthrough in yeast cultivation, a method which Carlsberg on the contrary publicised, even though they thereby surrendered important knowledge to their competitors.

7 New Ways of Travel

The Reiersenske Fund, which was, as earlier mentioned, established in 1795, made it possible for many Danish artisans and technicians to travel abroad throughout the nineteenth century. From 1850 subsidies, in total exceeding 10,000rd per year, were distributed as travelling scholarships, and the foundation has continued such activities into the twentieth century. The majority of the travellers remained wandering journeymen who worked their way from place to place.[30] When unemployed they lived on the

30. Poul Strømstad, *Håndværkervandringer i lavstiden*, Haderslev, 1955, pp.36ff.

benefit (*Geskænk*) which the guild system guaranteed them. During the first half of the nineteenth century Danish law prescribed that journeymen must report to the police on arrival in the towns they passed through on their wanderings, and the extant registers record the number of journeymen passing through the various towns. In 1823, 1,500 passed through Horsens, approximately 4,000 through Kolding, 1,600 through Roskilde, and 800–900 came to Ringsted each year.[31] Close on half of the wanderers were foreigners, and of these 80 per cent were German. At the outbreak of the 1848–50 war between Denmark and Prussia almost 1,000 German artisans were expelled from Copenhagen, and in 1880 12 per cent of artisans in Copenhagen had been born abroad. It was quite common that of every four journeymen within a profession one was a foreigner and in some professions they formed the majority as, for instance, in tailoring, where 60 per cent were foreigners in 1897.

The growing importance of trade unions altered the character of artisan wandering. Because union members received financial support during periods of unemployment, although on a modest scale, it meant that they no longer needed to wander to other places to find a job. They could stay in their home towns awaiting better times.

Many went abroad, however, but now the journey was made by train or ship and was usually better organised; artisans often secured for themselves letters of recommendation, for instance, which opened doors to firms which otherwise would remain closed to foreigners. A lithographer who was in Paris in 1880 described the difficulties he experienced in obtaining detailed information about production methods: the French answered willingly, but 'if you got too close to them' they said it was a secret of the firm. They were cautious about giving foreigners permission to visit the workshops and it was therefore imperative to have a letter of introduction. If not, one had to pay for their secrets.[32]

The tower clock manufacturer Julius Bertram-Larsen got, by the assistance of the Danish consulate in Paris, the opportunity to see a pneumatic mechanism which worked four clocks in a tower, while the master smith Th. Henze, who wanted to see German screw and nut factories, experienced great difficulties in getting entry. When

31. John Logue, 'Svendevandringer og internationalisme i fagbevægelsens barndom', *Arbejderhistorie*, vol.20, Copenhagen, 1983.
32. Rigsarkivet (National Archives, Copenhagen), Det Reiersenske Fond, Rejseberetning, 1880.

he later travelled in England it was considerably easier, because he brought with him letters of recommendation.[33]

In 1882 the Travel Stipend Association was established, with the aim of subsidising artisans and manufacturers travelling abroad. The Association organised group travel for its members; in 1885 a visit was arranged to the industrial exhibition in Antwerp, and 207 people participated in a trip to England and Scotland in 1887.[34] During the latter journey visits were made to a number of factories where the managers showed the visitors around and explained everything. One of the factories had even forwarded an illustrated description of the works prior to their arrival. The Danes were not, however, always met with the same kindness, but the printed report emphasises that admission was granted to factories normally closed to foreigners.

8 Conclusion

The examples presented here show that both the Danish state and private individuals persistently tried to keep up with foreign technical development by procuring information about innovations and improved methods from abroad. The government's official policy was to recruit foreign manufacturers and experts, and to give financial support to Danish craftsmen and industrialists who wanted to perfect their knowledge in foreign countries. Furthermore, private funds and organisations granted travel stipends.

The possibilities of being admitted to workshops and factories varied; firms did not willingly unveil their production secrets to people who might become competitors. Certainly the foreigners must have missed much information, but since this is a matter of trade secrets, it is of course difficult to assess how much was withheld. In many cases it was possible to gain whatever information was wanted, but the fear of industrial espionage kept many firms closed to foreign visitors.

Innovations in craft manufacture, from the Middle Ages onward, were brought to Denmark mainly by wandering journeymen, in particular from the German-speaking areas of Europe. As industrialisation increased artisans and mechanics increasingly looked to England, which during a long period in the eighteenth and nineteenth centuries constituted the preferred destination

33. Ibid., 1881.
34. N. Davidsen, *Fællesrejsen til England og Skotland 1887*, Kolding, 1887.

for a growing number of Danish craftsmen and technicians.

This paper has focused on Denmark as a recipient of new technology, and it remains to mention that Denmark only to a very limited extent acted as a starting-point for technology transfer during the period we have been considering.

7
Transfer Patterns of Technology: Theory and Evidence

Fritz Hodne

1 Introduction

Studies of technological innovation fall under four headings: invention, innovation, diffusion and long-term impact. The two first aspects deal with origins and causation, the latter with consequences. Diffusion studies analyse both aspects. They often ask why adoption of commercially viable innovations proceed smoothly in some cases, in others only with lengthy time lags, in some cases not at all. The present paper limits itself to the diffusion aspect. Data have been assembled from the Norwegian economy, 1840–1914, with references to Denmark and Sweden. The aim is to uncover patterns, if any, in the diffusion of new technology in Scandinavia during this period, to distinguish factors that determined timing, to identify the initiators, discover sources of capital and indicate the channels through which the novelties were propagated. The lessons have bearing on policies for technology transfer in general.

In the absence of a settled taxonomy, let alone an accepted transfer theory, the paper, in order to establish a framework for the observations, will first summarise two theories dealing with diffusion, the Vernon theory and the Hayami–Ruttan theory. Insights contributed by the Stanford School will also be acknowledged.[1] The procedure suggests that while the theoretical strands provide less than a synthesis, at present they appear to offer the best tools by which to assess evidence as to how technology transfers occur.

---

1. Amongst contributions from Stanford University, see Paul David, *Technological Choice, Innovation and Economic Growth*, Cambridge, 1975; Nathan Rosenberg, *Perspectives on Technology*, Cambridge, 1976; Nathan Rosenberg, *Inside the Black Box: Technology and Economics*, Cambridge, 1982.

Fritz Hodne

Raymond Vernon's product cycle theory identifies successive stages of standardisation observable in a product's life cycle. These in turn, are linked with a sequence of international trading in the product. At first production and consumption of the product is limited to high-income nations. In step with increasing standardisation and price reductions imitators in other industrialised countries start to compete; they are followed by Third World manufacturers. While consumption finally becomes universal, production gets confined to Third World countries, whose cost advantages will outweigh differences in levels of expertise.[2] Figure 7.1 shows the trade pattern that is expected to unfold over time.

The limitations of the Vernon theory, even in predicting likely trade patterns, are obvious; its focus is restricted to changes in the commodity composition of trade. It would, however, be compatible with the new idea that information is not a free market good, available to all, instantaneously and at no cost. From the point of view of development, the Vernon theory links levels and composition of commodity trade in a consistent way with levels of overall economic development.[3]

The Hayami–Ruttan theory, though based mainly upon data from agricultural development, also seeks to identify patterns of technological diffusion.[4] Their theory distinguishes between three phases of international technology transfer: (1) *material* transfer; (2) *pattern* transfer; (3) *capacity* transfer.

1. The first phase is characterised by simple import of new materials, such as new seeds, animals, plants, machines and techniques

2. For the Vernon theory, see Raymond Vernon, 'International Investment and International Trade in the Product Cycle', *The Quarterly Journal of Economics*, vol. 80, 1966, pp. 190–207, and Raymond Vernon (ed.), *The Technology Factor in International Trade*, New York, 1970, articles by R. Vernon and William H. Gruber, 'The Technology Factor in a World Trade Matrix', pp. 233–72, and G. C. Hufbauer, 'The Impact of National Characteristics and Technology on the Commodity Composition of Trade in Manufactured Goods', pp. 145–233; also P. Kelly and M. Kranzberg (eds), *Technological Innovation: A Critical Review of Current Knowledge*, San Francisco, 1978; L. Nabseth and G. F. Ray, *The Diffusion of New Industrial Processes*, Cambridge, 1974.

3. Nathan Rosenberg, 'Economic Development and the Transfer of Technology: Some Historical Perspectives', *Technology and Culture*, 1970, pp. 550–75, and 'Factors Affecting the Diffusion of Technology', *Explorations in Economic History*, 1972, pp. 3–35.

4. Yujiro Hayami and Vernon W. Ruttan, *Agricultural Development: An International Perspective*, Baltimore, 1971. Also in Rondo Cameron, *The International Diffusion of Technology and Economic Development in the Modern Epoch*, Sixth International Economic History Congress, Copenhagen, 1974, pp. 83–93.

Figure 7.1 The balance of trade for a new product through the successive phases of the product cycle.

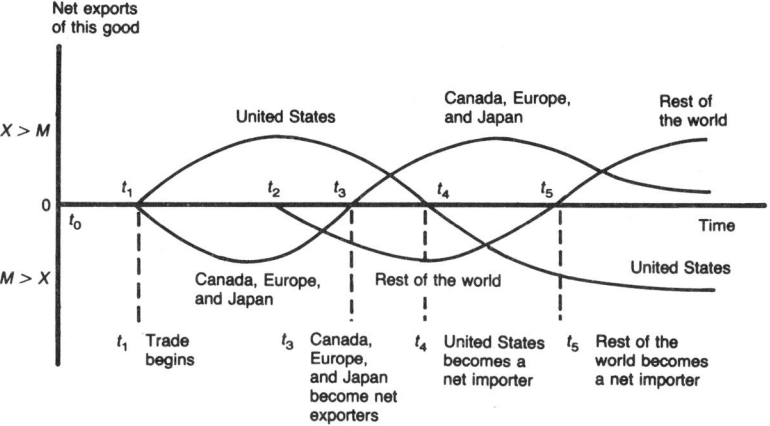

Note: At time t_0, time measured along the horizontal axis, the innovation and the first commercial production of a new good take place in the high-income country, for example the USA, where income levels and consumption patterns justify such an effort. The invention may, of course, have its origin in an outside country. At time t_1, American firms begin exporting the new product to other industrialised countries. By permitting economies of scale, the widened markets hasten the process of standardisation. After a time lag, during which the product reaches a mature stage, the imitator countries develop sufficient technical competence to make the product, aided in some cases by skills acquired at a subsidiary of the mother firm. The transfer erodes the advantage originally possessed by the pioneer country. At point t_3 the imitating countries have reached a stage of sophistication at which they become net exporters. At point t_4 the USA, having lost its comparative advantage, has become a net importer. After this reversal of roles between the industrialised countries, at point t_5, the rest of the world has also closed the technological gap in this ageing product technology and achieved a net exporter role.

Source: Charles P. Kindleberger and Peter R. Lindert, *International Economics*, 6th edn, Homewood, Ill., 1978, p. 77.

159

associated with these materials. Local adaptation is not conducted in any systematic way. The naturalisation of animals and plants tends to occur primarily as a result of trial and error by local gentlemen farmers.

2. In the second phase technology transfer occurs primarily through the import of certain designs, blueprints, formulas, books etc. New plants, animals and equipment are subject to orderly tests and knowledge of them is spread through public institutions for agricultural research and experimentation. Domestic production of the machines imported in the previous phase is initiated.

3. In the third phase the transfer of technology is predominantly made through the transfer of scientific knowledge. This enables the adopting country to produce its local versions of the prototype technology over a wider range, whether biological or mechanical, by virtue of a newly achieved technological maturity. In each case the successful transfer depends critically on personal contact and association between local firms and the leading international centres.

If we compare the two diffusion theories, we observe both similarities and differences. The Hayami–Ruttan theory is cast on a higher level of aggregation; its scope and time span are broader. The Vernon theory is limited to the era of mass production, beginning in the 1870s. Yet they complement each other. Both theories link commodity flows and technology imports to successive levels of development in the importer country. Both would be compatible with the prediction that sustained economic growth, whether reflected in exports or imports, will fail in the absence of internal structural changes. Imports in other words, provide a proxy for the initial levels and subsequent advances made by a country during the march to technological maturity. Instead of studying the changing composition of a country's imports, the paper by Lennart Jörberg included in this book argues that the patent statistics might be used instead to identify patterns in the growth of technical competence.

2 Material Transfer of Technology

(a) Consumer Goods

The general point to be made at the outset is that the consumer goods industries generally include familiar products that meet existing needs: food, drink, shoes and clothing, which, other things being equal, lead to the expectation that in these areas at least, the introduction and diffusion of new products and processes should be governed by the demand side of the economy. In the eighteenth century that meant state and crown. In the free trade era of the nineteenth century, demand implied a market, in which price and quantity rather than quality serve as yardsticks for decisions. The government would play a passive role. Demand existed before-hand, so that in some cases, but not in all, the new products simply replaced old ones. The period also saw entirely new products, such as margarine or beet sugar. But how, where, when, by whom? Some examples will provide answers.

The novelty of the German, or rather Bavarian, beer from the mid-1820s touched off a wave of town breweries in the mid-1840s, among them Schreiner's (1849) and Dahl's (1859) in Trondheim, and in Christiania Schou's (1842) and Frydenlund's (1959). Entre-preneurship and capital came from native merchant families, oc-casionally of Jewish origins, with contacts and training at the Munich pioneering breweries. The new products caught on be-cause of superior quality and price in a growing urban market. Hence the crucial, perhaps decisive, importance of the demand side for the timing of the transfer. Frydenlund even went on to develop export sales of 'undercooled' beer in the 1860s, thus offering an example of the Vernon trading cycle.[5]

Yeast, a by-product of beer making at the pioneer firm, Gabriel Sedlmayr in Munich, was developed by the Danish entrepreneur J. C. Jacobsen into a commercial article in the 1850s, and a Danish export article in the next decade (details are given in Poul Strømstad's article). Among the customers were Norwegian town households. After a period as import agents for Danish manufac-turers, enterprising local merchants in Norway set up local yeast facilities in competition with the Danish imports, in what amounted to a simple case of import substitution.[6] The potential market was large. The main consideration for native imitators was price, their asset was nearness to the market. Hence the significance of tariff duties and marine transport costs for the timing of the transfer, since, after all, the technology was not very complex. In

5. Erling Petersen and C. J. Arnholm, *Frydenlund Bryggeri 100 år 1859–1959*, Oslo, 1959; Eilert Sundt, *Norges Handel og Industri*, Christiania, 1907 (index for Hansa, Schou and Frydenlund); S. Schmidt-Nielsen, 'Nærings-og nytelsesmidde-lindustrien', in *Norges Ingeniørforening 1874–1924*, Oslo, 1924; Olaus Schmidt, *E. C. Dahls Bryggeri 1856–1956*, Trondheim, 1956, p.17.
6. August Schou, *Gjær-og spritfabrikasjonen i Norge (De Norske Gjær-og Spritfabrik-ker A/S)*, Oslo, 1947, pp.35–6, 42, 47–8.

Bergen, the old Hansa town, it was quite common at the time for merchants' sons to walk down to the customs house to inspect the tariff lists and import volumes, to get ideas for setting up a possible import-substituting business. According to family lore, the first yeast factory in Bergen owed its origins to just such a tour of inspection. In the case of Henrik Bolstad Pedersen, a Bergen apprentice taught by Danish yeast masters, the start-up as independent manufacturer occurred when, in 1881, young Henrik, after a visit to the customs house, compared the customs tariff, transport cost and volume of yeast imports, with the local price, and made out that there was a sufficient margin for him to try out on his own.[7]

Demand, while generally deemed decisive, implies the intermediation of the *skilled artisan*. But this raises the question: when do demand side considerations yield to those of supply? Obviously, the common link was the commercial traveller, that is the agent who in a given situation matched demand with supply. Since normally the deal was voluntary, it makes sense to say that the commercial agent was the proximate causal agent. The practice of importing skilled artisans to start up domestic industries is age-old. We find numerous illustrations in the present papers dealing with technology transfer to Scandinavia, over a broad range, both in the mercantilist era and the free trade era of the nineteenth century. The margarine story appears representative.

The combination of European population growth and rising butter prices opened up the possibility of a surrogate for butter. A world patent for an emulsion method, using ox lard, was obtained by Mège-Mouriés, a French chemist, in 1869 and again in 1872. One of his assistants was Auguste Pellerin. Apparently by chance the young assistant met with a Norwegian businessman, Ejlert Sundt, in 1875 in Paris. Ejlert Sundt, an experienced wholesale merchant and match manufacturer from Christiania, saw the market opportunity for margarine and persuaded Auguste Pellerin to come to Christiania and start margarine production on his own, with start-up costs paid by Sundt. From this encounter stems the first margarine factory in Norway and in Scandinavia, A/S Christiania Smørfabrik Aug. Pellerin & Fils & Co., established in 1876. Norway exported margarine from 1880.[8] Initiative and capital in this case lay with the Norwegian businessman rather than foreign entrepreneurship.

7. Given to me by Johan H. Mohr Jr., Bergen, together with the family archive.
8. Nils Vogt, *A/S Christiania Smørfabrik Aug. Pellerin & Fils Co. 1876–1926*, Oslo, 1926, pp. 16, 24–6.

Transfer in the nineteenth century was mostly geared to profit as judged by private entrepreneurs. Unexpected conflicts were bound to occur. Up until the 1890s tobacco had been consumed mostly in the form of pipe and chewing tobacco, usually made by local craftsmen, but in that decade the American cigarette machines revolutionised production cost and cigarette prices. Cigarettes, by appealing to women, promised to widen the market and the demand for tobacco. The potential market was large, the question was who would get in first, the local manufacturers or BATCO, (British American Tobacco Company), the worldwide tobacco trust. BATCO around 1910 started a campaign to capture the Norwegian market, with sales methods deemed unethical by the Norwegian business community. Local tobacco manufacturers retaliated by organising a long-term boycott of retail tobacconists who sold BATCO cigarettes. The skirmish ended in 1925 when J. L. Tiedemann, one of the largest Norwegian tobacco manufacturers, joined with BATCO to establish a new tobacco company, the Norwegian–English Tobacco Factory Ltd, on a 50–50 per cent basis, whereby BATCO relinquished its aim of gaining national control and the Norwegian firms dropped their boycott.

The rise of the tobacco industry indicates the role of new technology for new openings, as well as the role of demand for the timing of transfer decisions. As for initiative, clearly, the native manufacturers rather than foreign entrepreneurship explain the what, when and where. It remains to add that the technology entered the scene as fully developed end-products. The reason is discussed below. To the point here is that, afterwards, Norwegian firms entered foreign markets with their smoke tobacco brands, thus giving another illustration of the Vernon trading cycle.

Perhaps the best example of the importance of the timing of technology transfer in the consumer goods industries is offered by the textile industry. Norway's modern textile factory industry, the country's first industrialised sector, goes back to the 1840s. By 1850 there were nine cotton spinning factories and ten weaving factories, against nil in 1840, with a workforce of 782 workers, two-thirds of whom were women. Imports of raw cotton had been negligible in 1840; by 1850 they reached 797 tons, and 2,285 tons in 1856. Reflecting the level of economic development, the founders were wholesale merchants or their sons. They were mostly natives, but in at least one case a foreigner (Peter Jebsen of Broager, Denmark) made a mark. Echoed elsewhere, Peter Jebsen's career is an offshoot of the traditional wandering guild craftsmen in search of market openings for their skills. They all had behind them some months' stay in Manchester, the world's cotton textile centre. Both machinery, skilled foremen and some skilled English female operatives came along. Machinery remained an import article, but maintenance and repairs were provided locally. The operation was one of pure import substitution, in which nearness to the local market, a moderate tariff and transport costs decided the calculation to try.

But why did the industry get its footing in the 1840s? After all, the industry was much older, several earlier attempts are known that turned out failures, and already the textile industry had a history in Switzerland, Alsace, Normandy, Belgium and the Netherlands. The best-known Norwegian early start is that of Mads Melchior Wiel (see Trine Parmer's article).[9] A Dane by birth, Wiel was a versatile entrepreneur and timber merchant from Halden, a town near the Swedish border. He started in 1813, with the most recent machinery from Britain bough in defiance of the British ban on machine exports. During the next twenty years Wiel tried to obtain continuous production. He even bought new machinery in the 1820s, but failed to secure a profit. In 1835 he finally gave up the efforts at cotton textile production.

So why did his successors succeed where Wiel, the pioneer, failed? We may rule out adverse localisation. Equally, tariffs cannot have been decisive, for although the tariff in the 1840s had a fostering bias, Wiel earlier had enjoyed rates that gave more protection. Nor was Mads Wiel a poor businessman. His other activities were run at a profit, despite the slump following the end of the Napoleonic Wars from 1815. We are left with timing. Timing was against Wiel. He ran up against a depressed market after 1815, which prevented success. By the same token, his followers set up their English factories in a period of rising incomes, free trade and expanding markets that absorbed just about everything they could bring out. In other words, timing was decisive.

Summing up the story of consumer goods, three generalisations appear warranted. First, as predicted by the Vernon theory, we easily find confirmation of the trading cycle and its phases. Second, as predicted by the Hayami–Ruttan theory, we find confirmation of the primary role of end-products in the initial transfer periods, both regarding production equipment, products and processes. Hence the predominance of material transfers. Without exception the evidence pinpoints the intermediation of skilled personnel in

---

9. Trine Parmer, *Mads Wiels Bomuldsfabrik 1813–1835* (history thesis, Dept. of History. University of Oslo), 1979. Printed in parts in *Volund*, 1981, pp.7–77.

various roles. Third, as to source of initiative, the decision makers who carried risks and provided initiative, were merchants, mostly local businessmen, their nose on the market, socially connected in a way that enabled them to raise venture capital. Technical expertise was a minor consideration. Direct foreign investment may be ignored as a transfer channel in these industries in the period considered. All this suggests absence of pressure and the existence of options.

(b) The Capital Goods Industries

The general remark one could make here is that the distance between risk, capital outlays and final sale is a long one. Second, for that reason initiative tends more often to come from the supply side rather than from the demand side. Investments after all, can always be postponed. A third is that the size of the market and the technology in question often prevent the setting up of heavy industries in small countries. Painful experience taught the entrepreneurs this lesson (see the articles by Rolf Adamson and Martin Fritz for exceptions). On the whole small countries tend to specialise and to exploit natural, comparative or competitive advantages.

Skills, organisation, training and standards for the new technology tended to accumulate in the new mechanical works and also in agriculture over a very broad range. Their appearance in Norway in the 1840s, as in Sweden earlier, signals the decisive step up the ladder to an industrial economy.

We are struck by the pervasive importance of Britain in the early phase, 1840–80, of capital goods production. It was natural; it also underscores Scandinavia's geographical proximity to the most advanced industrial country at the time. Geography meant that their resources came to be utilised in accordance with tastes, incomes and demand in the advanced rather than backward economies of Europe. The majority of the findings in the initial phase fall straightforwardly under the heading of transfer of men and machines. This phase is normally followed by the transfer of designs and capacity transfers.

Examples abound. As for Norway, we note that the first two paddle steamers, ordered and paid for by the government, came fully equipped from Britain in 1827. The first railway, 68 km long, was designed and built between 1851 and 1854 by Robert Stephenson Jr., the son of the locomotive inventor, on the basis of a contract between an English group and the Norwegian government, which acted on behalf of a powerful merchant lobby in the capital. The line was built with British rails and operated by British locomotives. The first locomotive drivers were also British. Later locomotives came from Swiss, German and British firms.

Not till 1875 was a single, small station locomotive constructed by a domestic firm, but forty years passed till regular domestic production of locomotives got under way, on orders from the state railway, the sole customer. By 1900 the Thune Mechanical Works, a local company in Christiania, had made ten locomotives out of 160. In 1910 48 per cent of the locomotives of the state railway system stemmed from domestic suppliers; equally, all freight and passenger cars were now built by local firms, but rails remained import items.[10]

The technical knowledge that was concentrated in another early mechanical firm, the Myren Værksted, established in 1848 in Christiania, stemmed originally from a Scottish travelling millwright, John Wilson. He taught the founding brothers, Jens Jacob Jensen (1817–90) and Andreas Jensen (1821–74), the art of drawing and mill construction, which provided a starting-point for the two young mechanics. They went on to design sawmill machinery, water turbines and steam engines for customers in Norway, Sweden, Finland and Russia. Their daughter firm later became a leading shipbuilding yard in Norway.[11]

The firm Christiania Spigerverk (Christiania Nailworks), established in 1853, and now part of Elkem, an industrial conglomerate of sizeable proportions, started as the name suggests, by making nails, using the patented English Coates nail-making machine, for which Spigerverket paid a licence fee. A skilled Englishman, John Hurst, was called in to demonstrate and operate the machine.[12] Tower lathes of British design were imported as finished products by the mechanical works, and usually we find English managers in the initial periods. For example, the Trondheim Mechanical Works, established in 1843 by a group of wealthy Trondheim wholesale merchants, hired an Englishman, John Trenery (1803–64) as daily manager. By that time Trenery was already working in Norway, at the Kåfjord Copper mine in Troms County in the north. He began his new job by a trip to England, where he hired skilled workers and bought machinery, tools and iron plates

10. Fritz Hodne, *Stortingssalen som markedsplass. Statens grunnlagsinvesteringer 1840–1914*, Oslo, 1984, p.175.
11. Chr. Gierløff, *Et bruk ved Akerselven. Myrens Værksteds hundreårsminne*, Oslo, 1948, pp.61, 63.
12. Edgar B. Schieldrop, *Christiania Spigerverk 1853–1961*, Oslo, 1961, pp.75, 83.

with a view to building steamships in Trondheim.[13]

His cousin, Thomas Trenery, took over the daily management of the Bergen Mechanical Works in 1866 in similar circumstances.[14] At Laksevåg Skipsverft, another Bergen shipyard, it was Daniel G. Martens Jr. (1856–1927), a Norwegian citizen, who stepped in as reorganiser in 1883. But the record has it that Martens junior had been trained as shipping engineer at the W. Armstrong & Mitchell Co., Newcastle. As technical director he hired James Bower, a Scot, who remained a member of the board from 1892 till 1924. Several contemporary Oslo yards tell a similar story, enough to form a pattern. Why buy the parts rather than the finished steamships? Apparently shipbuilding required assembly operations of a complexity that made building locally the most suitable solution. Other motives were present. Local business circles felt the need for local engineering facilities, for repair and maintenance purposes, quite apart from the expertise required for shipbuilding; hence shipbuilding foreshadows the second stage of pattern transfer, accomplished through the transfer of skilled personnel. The mechanical works taught three generations of workers the basic skills of an industrial economy. This is inferred from several papers in the present collection.

Moving to a different area, we note that it was a Briton, James Malam, who built the country's first gasworks in Christiania in 1847–8. A fellow countryman of his, James Small, constructed the gasworks in Trondheim in 1851–3. The gasworks in Bergen was built at about the same time by Oluf Pihl (1822–95), who had acquired his expertise in Britain at British gasworks. He later constructed the gasworks in six other Norwegian towns, while at the same time working as managing director of the Christiania plant. A busy fellow, Oluf Phil reminds us of the narrow base, the particularity, of the new skills.[15]

Also the textile plants were British in design and operations. At the Nydalen Spinderi Compagnie, in Christiania, one of the earliest cotton spinning mills, at one time the country's largest industrial establishment with around 1,000 workers in 1885 and 1890, we

---

13. Olaus Schmidt, *Aktieselskapet Trondhjems mekaniske Verksted 1843–1943*, Trondheim, 1945, pp.13–15, 30; *Amtmennenes Beretninger 1846–1850*, Trondhjem Amt, p.16 (county governor's reports).

14. Kaare Fasting and Henrik Myran, *Herfra går skibe. A/S Bergens Mekaniske Værksteder 1855–1955*, Bergen, 1955, p.50.

15. O. Pihl, *Norges gasverker og de til samme knyttede gasteknikere*, Christiania, 1913, pp.5, 8, 25, 29.

observe two generations of Jamesons as technical managers.[16] At the Dale Ullvarefabrik, established in 1879 in the west fjord country, the largest wollen mill in the country, we meet the same phenomenon. The British dynasty here was John Hartington and his son.[17]

The material transfer aspect is particularly well documented in the start-up of the mechanical pulp industry in the late 1860s. With large spruce forests at hand it was only natural that once a commercially viable technology was offered, Norwegian timber merchants should seize on the possibility of making paper out of spruce trees too small to be made into beams or boards. The new paper-making methods came, as we know, from Germany. One of the earliest cases of transfer is associated with Hendrik Fasmer. Of Dutch descent, Hendrik Fasmer (1835–1920) operated a many-sided family business outside Bergen at Alvøen. One activity was paper making. Hendrik Fasmer produced high quality paper in the age-old way, out of cloth and rags. He had obtained his education and training in Dresden and elsewhere in Germany, not Britain. Being short of rags, he got into correspondance with Herr Voelter in 1865, a German inventor of a patented wood-grinding machinery. This was followed up with a visit when Hendrik Fasmer next year journeyed down to Voelter's factory in Heidenheim. Apparently satisfied, Fasmer signed a purchasing contract for machinery for a pulp factory, based on Voelter's designs, and paper and cardboard production started the following year at Sævareid, a hamlet south of Bergen, where a waterfall provided direct transmission energy and local forests provided the timber raw material.[18] The material transfer story is echoed in the rise of the Danish clay and pottery industry (see the article by Ole Hyldtoft).

Flour milling technology illustrates the effects of market size on the technology transfer process. It also throws light on the location of initiative. Grain milling on the farm quern had been a seasonal household activity for centuries. In the 1860s, however, commercial mills, powered by water turbines, using imported stones and new sieving techniques, promised to turn the old seasonal activity into a large-scale capital-intensive industry, able to capture vast economies of scale. It was a common European wave. Several milling enterprises appeared in the 1860s in the environs of Bergen,

16. Ejlert Sundt, 'Norges Handel og Industri', *Farmand*, 1907, pp.3, 57.
17. Bernt Lorentzen, *A/S Dale Fabrikker 1879–1959*, Bergen, 1960, p.61.
18. Chr. Gierløff, *Sævareid. En vestlandsk treforedlingsbedrift og kultursaga*, Bergen, 1959, pp.47–8.

among them Tøsse Mølle, owned by Wollert L. Hille, Storemøllen Bruk, owned by W. F. Meyer, and Fotlands Mølle, owned by M. G. Risøen. Vaksdal Mølle, established in 1866 by Gerdt Meyer of Bergen at Vaksdal in the west fjord country near Bergen and in operation from 1871, offers a good illustration of the steps leading to a technology transfer.

Of German extraction, Gerdt Meyer (1817–97), a Bergen merchant, was engaged in the old northern fish export trade to the Baltic countries, to which Bergen owes its existence. The return articles had always been grain and flour. Not surprisingly Gerdt Meyer had dabbled in commercial flour milling earlier, but merely as a sideline, until 1866 when his knowledge of markets and technological progress in milling convinced him that the time was ripe for a large-scale undertaking, on the basis of imported grain. For his mill at Vaksdal Meyer bought machinery from the firm Gebrüder Seck of Dresden in Germany. German engineers set up the machinery, grinding stones, belts and water turbines.

His son Rasmus Meyer (1858–1919), after a fire in 1899, had the mill rebuilt on the basis of electricity and the newest milling technology, consisting of corrugated, high-speed steel rollers, that combined crushing, scraping and shearing, to produce a flour of previously unknown fineness and evenness. His own hydro-power station provided energy for the automatic rollers. The new grinding machinery of the 1880s was offered by Gebrüder Seck of Dresden and Ganz & Cie of Budapest in Hungary, the international market leaders in grinding machinery.[19] The entire mill being imported, the new Vaksdal was a case of 'turn-key' technology transfer, so familiar after World War II in Third World countries. Thus equipped with the latest automatic grinding machinery, including elevators, Vaksdal towered as the largest flour mill in Norway and Scandinavia, measured by turnover, with sales in excess of 10,000,000 kr. by 1910, and customers all over Scandinavia. The Norwegian agent for the Dresden firm was Hans H. Heyerdahl of Christiania, an engineering graduate from the Dresden Technische Hochschule, who had kept up contact with his class mates through the years.[20] Also Bjølsen Valsemølle A/S, another large mill, located in Christiania, in 1895 gave Gebrüder Seck the

---

19. Gunnar Nerheim, 'Fra bekkekvern til automatiske møller', in *Rent mel i posen. Bjølsen Valsemølle A/S 1884–1984*, ed. Helge Nordvik, Oslo and Bergen, 1984, pp.72, 74–7, 84.

20. Bernt Lorentzen, *Vaksdal Mølle 1866–1966*, Bergen, 1966, pp.154, 166, 169 and *passim*; Gulbr. Gulbrandsen, 'Kornmaling gjennom tidene', *Volund*, 1969, p.116.

contract for its new automatic grain mill, and again in 1912–13.[21] Similar changes in commercial milling took place in other Norwegian towns in the same period, with the result that by 1900 Norway boasted a modern milling industry of standard European sophistication.

(c) Market Size and Technology Transfer

The formula involved in the transfer of milling technology thus was: *local* initiative and capital, *foreign* skills and technology. The case points to a general lesson: market size usually prevents the rise of capital goods industries, or heavy industries, in small economies, which explains the relapse to the stage of material transfer in the Norwegian milling industry. Though in principle the Norwegian mechanical works could have constructed the new mill machinery, they would have been unable to achieve competitive prices. Unable to reap the necessary economies of scale on the home market, the domestic mechanical works usually avoided large-scale machinery and plants. Hence their tendency to specialise, hence the importance of *niches* in the world market.[22]

The problem remains though, for why did Ganz & Cie in Hungary, another small country, achieve an international position as market leader in milling machinery? Or why did L. M. Ericsson of Sweden, or Philips of Holland, both companies in small European nations, rise to become world market leaders in telephones and household electronics? An exhaustive analysis of these questions would take us too far afield, and would *inter alia* involve detailed examination of alternative rates of capital returns in formative periods. Shipping would document the role of relative returns on capital for business specialisation in Norway in the period.[23] Here consider the pulp industry. This industry rapidly grew into the major Norwegian industry in the lean years of the 1880s, in terms of employment, exports and value creation. For that reason it offered adequate incentives to domestic mechanical works to start import-substituting production of water turbines

21. Gunnar Nerheim, 'Fra bekkekvern til automatiske møller', pp.88–9.
22. S. B. Saul, 'The Economic Development of Small Nations', in *Economics in the Long View. Essays in Honour of W. W. Rostow*, ed. Charles P. Kindleberger and Guido di Tella, New York, 1982, vol.2, pp.125–9.
23. A. N. Kiær, *Statistiske Oplysninger vedkommende den norske skipsfarts Økonomi*, Christiania, 1871 (also published in the newspaper *Morgenbladet*, 1871), and Ole Gjølberg, *Økonomi, teknologi og historie. Analyser av skipsfart og økonomi 1866–1913*, Bergen, 1979, pp.214–18.

and pulp grinding machinery. Some firms got into exports. Even paper-making machinery was occasionally made on order. In other words, the transfer story turns on ordinary business calculations, involving likely returns on capital, in the near-to-middle time perspective.

Capital for industrial undertakings stemmed from private sources, mostly local sources, though Borregaard, the biggest paper and pulp undertaking, saw heavy investment of foreign capital from 1889, when the company underwent a major modernisation programme. So did mining and the new hydro-power industries around the turn of the century. Here British and German capital were heavily involved.[24] The main exception was the railways: they were financed mostly by the national government. By 1910 a total of 87.8 per cent of construction costs had been provided by the central government, as against 12.2 per cent from private individuals and local government.[25]

Agriculture illustrates how market considerations mingled with political efforts in the process of modernising 'the mother industry'. The diffusion of the horse-driven hay reaper deserves special attention. In the latter half of the nineteenth century agriculture in western Europe turned away from grain to concentrate on dairy products. The shift was quite drastic in Norway. Therefore the reaper holds a key place, since by 1890 hay represented 70 per cent of the fully cultivated farm area in Norway. Hay cutting by the old long-armed scythe was a particularly labour-intensive activity. Here, if anywhere, the savings derived from new technology were bound to have a substantial impact on farm productivity. Sources permit us to follow the technical changes and routes of diffusion in photographic detail. The American McCormick and Obed Hussey reapers were first shown in Europe at the Great Exhibition in London in 1851. The Norwegian Storting (Parliament) voted stipends for two Norwegian 'mechanics' to visit the world exhibition. Four years later the Royal Society for the Welfare of the Kingdom (Selskabet til Norges Vel) established in 1809, bought a British hay reaper. The idea was to demonstrate it at local farm fairs. These fairs became a regular feature in that decade, serving as channels of information on new farm tools and machinery, and as temporary markets for sellers and agents of farm machinery. Newspapers and magazines also reported on new implements and

---

24. Arthur Stonehill, *Foreign Ownership in Norwegian Enterprises*, Oslo, 1965, pp.32–44.
25. Hodne, *Stortingssalen*, p.283.

methods in agriculture. Three models of the hay reaper, all imported, were shown at the Christiania exhibition in 1868 as against fifty models in 1877, which also offered two domestically manufactured reapers for sale.

By this time several sales agents for imported reapers and other farm implements had appeared; indeed, they were the most generous sponsors of the farm shows. Their printed brochures, showing the latest farm machinery in the 1880s, offered examples of hard-nosed arithmetic, including discussions of the critical threshold farm acreages in regard to costs of purchasing and servicing the various reapers. The brochures apparently were translations from English and American originals, with the figures adjusted to the local topography. The version driven by one horse, was best suited to the hilly Norwegian terrain. The issue of lowest threshold size for the new hay reapers was regularly debated in the local newspapers and periodicals in the 1890s.[26]

Where the government had sown, the sales agents stepped in to reap the rewards, among them Ulrik Rosing & Co. and S. H. Lund, both in Christiania. Not coincidentally, Ulrik Rosing was a relative of Anton Rosing, co-founder of Akers Mek. Verksted, the country's first mechanical works in the capital, (established in 1841), and of Polyteknisk Forening (The Polytechnic Society), whose journal *Polyteknisk Tidsskrift* (established in 1854) he edited. The Polyteknisk Forening rose to become the nation's central debating forum. The journal, under its present name *Teknisk Ukeblad*, is an indispensable weekly for an estimated 20,000 readers. As in France, the technical novelties were passed on by a limited group of science-oriented families, who represented the repositories of technical expertise, going beyond one generation.

The statistics on the reaper (1,299 in 1875 against 49,277 in 1907 and 87,862 in 1917) suggest that its breakthrough came in the fourth quarter of the nineteenth century. We lack information on how many of these were made locally and how many were imported, but we do know that several mechanical works came to specialise in making farm machinery and spare parts. With reference to the Vernon trading cycle, it should be noted that one of these, the Kverneland Co., established in 1873 in the Stavanger area, has been a substantial exporter of farm machinery from the

26. Stein Tveite, 'Teknologisk diffusion og økonomisk tilpasning. Slåmaskinens gjennombrudd i norsk landbruk', paper at Utstein Symposium, October 1980; Fartein Valen Senstad, *Norske landbruksredskaper 1800–1850-årene*, Lillehammer, 1964, pp.357–67; Paul Borgedal, *Norges jordbruk i nyere tid*, Oslo, 1968, vol.3, p.235.

1950s. Its eight-blade ploughs have been a success in the American Midwest.

3 Pattern Transfer of Technology

Since pattern transfer, hinted at above, always involves the manufacture of capital goods, the distinction between consumer and capital goods may be ignored in the following. The transition from material to pattern transfer normally occurred when a new product, after having been imported and sold for a while by a sales agent, spurred an enterprising native to bring out a competing product. We also hear of an intermediate stage of licence production. Notably in agriculture the government played a part in the transfer, at times directly, but more often indirectly. In most cases a precondition for the orderly pattern transfer was patent legislation, designed to protect the property rights of foreign inventors and owners. Safeguarding patent legislation in Norway came in 1839, followed by new detailed legislation in 1885 and 1910, in addition to the international patent rights convention, signed in Vienna in 1883, to which Norway acceded.[27]

Alfred Nobel (1833–96), the Swedish inventor of dynamite, set up a subsidiary, Nitroglycerine Co. at Lysaker outside Christiania as early as 1865, after having secured a ten-year patent right the previous year, duly published in *Polyteknisk Tidsskrift*, which served as patent journal till the 1890s.[28] His British patent dates from 1867, and his American from 1868. Isaac Singer, who created one of the first large internationals, established a subsidiary in Scotland in 1867, and from 1881 he sold his sewing machines through Bruun & Co. of Christiania. Ten years later the mother company set up a sales agency in Norway under the name the Singer Manufacturing Co., Inc.[29] There was no production or assembly of sewing machines in Norway. This echoed the situation in textile machinery; both remained import articles. In 1899 the majority of the shoe machinery patents were pooled in the American Shoe Machinery Co.[30] In 1910 this international company registered a sales agency in Christiania under the name Aktieselskabet

---

27. Norwegian data in Ot.prp. no.7, 1885.
28. *Polyteknisk Tidsskrift*, 1870, p.77.
29. *Handelsregister for Norge 1891*, Christiania, 1891, p.597.
30. R. M. Robertson, *History of the American Economy*, 2nd edn, New York, 1964, p.335.

United Shoe Machinery Co., Ltd.[31] The same year the existing patent legislation of Norway was revised to include protection of trade marks and brand names as well.

Consider Thune, an early modern mechanical works in Christiania. The family business dated back to the village smith Anders Thune (1787–1874), who in 1817 established himself in Drammen. His son Andreas Lauritz moved the business to the capital in 1851 in order to tap the more promising market there. The firm kept in touch with the latest developments. Thus around 1890 the firm obtained the agency and production licence for the patented steam boiler by Babcock & Wilcox.[32]

To continue the round, Graham Bell, the American inventor of the telephone in 1876, had a daughter company in Christiania already in 1880, called the International Bell Telephone Co.[33] In the same year Kristiania Telefonselskab, a rival Norwegian company, was registered. This company used the patents by the Swedish inventor L. M. Ericsson. The two companies merged in 1885. The managing director of the new firm was Knud E. Bryn (1855–1941), one of Norway's pioneering engineers, educated at the Munich Technische Hochschule in Germany. Bryn already held the sales agency for an electric bell, patented by the German firm of Schuckert & Co., Munich. In 1885 Knud Bryn was appointed executive director of Elektrisk Bureau, yet another electro-company in Christiania, established in 1882. Through the interlocking directorships of Knud Bryn the telephone companies in Christiania together had access to all the products of the leading telephone firms of the day. Under Knud Bryn Elektrisk Bureau proved a remarkable engineering success. By 1914 it had moved into exports of telephones and telephone equipment, and installed telephone systems in a number of countries, including the Vatican, and the firm employed more than 700 workers.

In some instances patent rights were bought outright. This was the case in 1883, when the Frognerkilens factory, established in 1873 in Christiania, bought the patent to produce the Wenström

---

31. *Handelsregisteret for Norge*, 23 August 1910.
32. E. Munthe Kaas, *Boken om Thune*, Oslo, 1965, p.16. According to Y. Hauge, *Boken om Thune*, Oslo, 1965, pp.16–17, Thune wrote to Babcock & Wilcox in 1890 to ask for the agency of the boiler; a formal agreement was signed in 1895. I am indebted to Kristine Bruland for this point. See Bruland's article in Even Lange (ed.), *Teknologi i virksomhet*, Oslo, 1989 and chapter 10.
33. Ths. Norberg-Schulz, 'Elektriciteten i Norge', *Norges Ingeniør Forbund 1874–1924*, Christiania, 1924, p.437 (footnote).

dynamo from the Swedish engineer of that name for 2,000 Norwegian kroner.[34]

Consider the transfer routes involved in the milk separator, another key invention, and the intermediating role of Nils Claus Ihlen (1855–1925), a central figure in the Norwegian transfer. After his diploma from the Eidgenössische Technische Hochschule in Zurich he returned to Norway, where he inherited a decaying family sawmill business. This he turned into a prospering mechanical works, Strømmens Værksted, situated on the eastern railway, just east of Christiania. In 1883 he obtained for a time the exclusive production rights in Norway for the Swedish Laval hand milk separator. The inventor was Dr Carl Gustav Patrick de Laval (1845–1913).[35] In December 1886 Laval's company, Alfa Laval, obtained a fifteen-year patent in Norway for its smallest version, called the 'Baby' hand separator. It was small enough to be carried on horseback up to the mountain summer farm, and operated by the dairy maids. The official Norwegian statistics on separators go back only to 1939. According to that year's census Norway had 124,643 separators in 1939, against 136,853 ten years later, in 1949.[36]

Since milk production by 1900 towered as the single most important income source in Norwegian agriculture, the separator story is of considerable interest. So it is elsewhere, notably in Denmark and in Sweden. In 1894 the Laval separator provided 59 per cent of Sweden's engineering exports, and still 34 per cent twenty years later, when 95 per cent of all those manufactured by Laval were exported. They supplied 54 per cent of the whole Russian market, 80 per cent of the Dutch, 75 per cent of the north German, and 90 per cent of the dairy separators of Denmark, the cradle of the creamery industry.[37]

I therefore wrote to the present firm, A/S Strømmens Værksted, in May 1981, and asked for information on the firm's old separator business. Mr Per Hauan, the company's managing director at the time, kindly answered my queries by mailing several photocopies of brochures and patent descriptions from the 1880s and 1890s. One brochure from November 1890 informs us that dairy machinery had been a speciality of the firm since 1883. Unfortunately, none of the brochures offers information on how many Laval

34. Tore Halvorsen, *NEBB i vekst 1908–1983*, Oslo, 1983, p.7.
35. Kaare Fasting, *Strømmen Værksted 1873–1973*, Oslo, 1973, p.14.
36. CBS, *Statistical Survey 1958*, NOS, XI, 330, Oslo, 1959, p.32.
37. Saul, 'The Economic Development of Small Nations', p.124.

Table 7.1 Norwegian sales of milk separators, 1886–90

1886	1887	1888	1889	1890
10	50	160	200	250

Source: Catalogues of A/S Strømmens Værksted from the 1890s (in the company's archive)

separators Strømmens Værksted had actually produced, and how many sets the company had sold as Laval's sales agent. According to the 1890 brochure there were 800 Baby Laval hand separators in Norway in that year and 20,000 sold worldwide, including New Zealand, and La Plata. In addition there were 150 horse-driven ones and 530 steam power-driven separators. The Swedish brochures from the mother firm supplement these rather desultory bits of information. According to one set of clippings from 1902, the mother company by 1900 had assigned its sales rights in Norway to two sales agents in Christiania, namely E. C. Due & Co. and F. Anker & Co. On the Norwegian market they had sold as shown in Table 7.1.

By 1885 sales in Norway amounted to 1,500 sets, and all in all there are 20,000 Alfa-Laval hand separators in use in Norway. The case of Strømmens Værksted, involving licensing, agencies and full customer treatment, may be taken as the achetype of pattern transfer. But that is not the whole story, of which more below.[38] Needless to add, it all happened in a market setting with customers normally paying cash or in instalments.

Pottery and clay, besides demonstrating the British connection, throws light on the process of going from material to capacity transfer. In 1847 Johan B. Feyer (1821–80) established Egersund Fayancefabrik outside the Norwegian south coast town of Egersund. At that point young Feyer had been a trainee at a pottery in Newcastle for a year. Clay came from Britain, and the factory according to family lore 'was set up completely according to English standards'. The manager, Holmes, also came from Britain. Feyer played at import substitution, the tariff giving a small margin for finished porcelain ware. In 1870, however, parliament reduced the tariff, which undercut Feyer's economic viability. He was forced into bankruptcy proceedings in 1878, after thirty years' endeavour. Holmes, his manager, went to the nearby town of

38. Letter to author from Per Hauan, dir., A/S Strømmens Værksted, 22 May 1981.

Stavanger to oversee the construction of a new pottery plant, Gann Potteri & Teglverk.[39] He got the job table commission fee. The transmission route in the industry is best described in Ole Hyldtoft's paper.

To sum up this section on pattern transfer, the central point is that in the free trade era of the nineteenth century new patents were acquired and utilised whenever domestic entrepreneurs and occasionally foreign companies, conversant with the technical developments in diverse fields, judged the home market to be ripe for a new venture. The ventures do not fall neatly into either the supply-led or the demand-led category of development, for obviously, though swayed by both long-term and short-term considerations, the risk takers crop up as 'causes' in both cases.

What about governments? Though tariffs offered some incentive to import substitution, it would be difficult on the basis of the evidence offered to make the case that tariffs had much influence one way or the other for the timing and direction of pattern transfer in this period, railways included. As for subsidies, they amounted to very little indeed. Only in agriculture was the government involved in support programmes that went beyond the ruling canons of private enterprise. The lack of any strong protectionist tendencies apparently was common to the small European economies in the nineteenth century. Governments provided protection for private property and contracts, and jealously guarded this reputation. For example, several disputes arose concerning alleged patent infringements. One involved the Laval separator and its rival made by Burmeister & Wain of Copenhagen. The dispute was settled in the courts in a way which upheld confidence in the impartiality of the law system. Who in the 1990s would take that for granted? Moreover, the government assumed responsibility, which included most of the expenditure on infrastructure, education, health, transport and communications. So did the other enlightened governments of Europe at the time. Again, how universal is that in the 1990s?

There was no technological determinism. Overall, the native commercial entrepreneurs, when the time came, possessed the necessary insights, and could draw upon sufficient capital backing to have technological options. Similarly, though some foreign companies established themselves in Norway in the period, initiatives stemmed mostly from local people, whose advantage was

39. Thor B. Kielland and Olaf Lorentzen (eds), *A/S Egersunds Fayancefabriks Co. 1847–1947*, Stavanger, 1947, p.20.

proximity to the local market, its taste and culture. As yet the mentality of the merchant rather than the engineer dominated business decisions, one sign of which was the desire for quick returns and impatience with long-term commitments.

4 Capacity Transfer

A third stage is reached when local people have acquired the capacity to compete successively against earlier imports on the home market, apparently without direct reliance on foreign experts or foreign capital. The cases above have indicated various routes along which technological maturity was reached. Technical journals have already been noted. In addition we must note the significance of the many international exhibitions and fairs for the diffusion of new technology. As mentioned, the Norwegian government had already in 1851 provided stipends for two Norwegian mechanics to visit the Great Exhibition in London. The motive in the case of Jens Jacob Jensen from the Myren Mechanical Works, Christiania, was 'to aquaint himself with factories and mechanical devices in England and Germany'.[40] This quite clearly pointed to the idea of capacity transfer. One finds precisely the same motive and examples in the Danish clay and pottery industry (see Ole Hyldtoft's article). In several instances the international exhibitions inspired later attempts to demonstrate domestic industrial products, notably at the Paris World Exhibition in 1889. The attempts were sometimes backed by the central government, which paid part of the participation costs.

Training abroad This is a key word. In addition to the examples offered so far, consider Peder Dekke. Of a shipbuilding family in Bergen, Peder Dekke spent the two years 1851–3 as a trainee at the world-famous McKay shipyard in Boston to acquaint himself with the latest techniques of clipper ship construction. On completion of his apprenticeship young Peder Dekke returned home with a letter of recommendation, and proceeded to introduce his own version of the clipper ship to Norwegian shipping circles. The letter, including the signature of John McKay, may still be seen among the exhibits in the Bergen Maritime Museum. The Dekke yard in Bergen served in turn as a standard and training centre for local carpenters up and down the coast for a generation.

40. *Departementstidende*, 1851, p.757.

Travel This is another key word. Numerous examples are recorded in the present collection. The consequences changed with increased technical maturity. The first planing machinery in Norway was set up at J. N. Jacobsen's sawmill at Fredrikstad in 1863, after the owner had seen such machinery during a visit to business customers in Scotland the previous summer. The incident was typical: it is echoed in the textile pioneers, the paddle steamers, the lathes, the steam engines, the gasworks, the first hay reapers. The point of this story, though , is that J. N. Jacobsen dispensed with foreign experts, in the belief that his own local mechanics could successfully duplicate the foreign machinery.[41] He was right. The importance of travel is confirmed over and over again by Danish, Swedish and Norwegian experience.

Studies abroad Another route to industrial self-reliance was studies abroad at technical colleges and universities, notably the *Polytechnische Hochschule*, set up in Denmark, Sweden, Switzerland and Germany from the 1830s onwards, but not in Britain. The first intermediate technical education in Norway, at the Trondhjems Tekniske Læreanstalt, was organised in Trondheim in 1870, to be followed by a dozen similar schools in the next twenty years.[42] By comparison the Chalmers Polytechnical University in Göteborg, Sweden, dates from 1829, and the Polytechnical University of Copenhagen from the same year. Higher technical education was finally offered in Norway from 1910, when the present Norges Tekniske Høyskole (NTH) started in Trondheim in that year. The University of Christiania may be ignored in this context. Government papers record that twenty Norwegian engineering students studied at German polytechnical universities in 1854, and fifty-five four years later.[43] The Darmstadt Technische Hochschule alone produced twelve Norwegian graduates between 1850 and 1885, and twenty-one during the years 1896–1900.[44] The secretariat of the Norwegian Association of Engineers keeps a roll of its members. From one of the lists it appears that ninety-eight Norwegians

---

41. A. J. Jacobsen, *J. N. Jacobsen, lastehandleren, samfunnsborgeren og hans tid*, 5 vols, Oslo, 1950–54, vol.3, pp.38–41.

42. See two recent articles that place engineering education in an international perspective: Håkon With Andersen, 'Trondhjems tekniske Læreanstalt 1870–1915', and Gudmund Stang, 'Ble det for mange ingeniører?', both in *Trondheim Ingeniørhøgskole 1912–1987. Festskrift til jubileumsfeiringen 1987*, Trondheim, 1987.

43. *Stortingets Forhandlinger*, 1857, St.prp. no.3, pp.219–20.

44. By kind permission of the archive librarian at Darmstadt Technische Hochschule, Dr. A. Müller, and the Rektor, Prof. Dr Helmuth Böhme.

studied at the Eidgenössische Technische Hochschule in Zurich in the period 1855–1900, and forty-six in the next fifteen years, making 144 graduates in all up to 1915. By comparison the new Technical University at Trondheim prepared 123 engineers in the initial years 1910–15. This figure should be set against the total number of 302 engineering students from Norway graduating from foreign polytechnical universities in the period 1901–15, 128 of them in the comparable period 1911–15.[45] Brain drain was a recurrent spectre. The advent of electricity around the turn of the century led to another exodus of Norwegian technicians, eager to pick up the latest innovations in generator and turbine technology (see Gunnar Nerheim's article).

In other words, as far as technological competence is concerned, Norway obtained the mass of its human capital abroad well into the twentieth century. Note, however, that increasingly the students going abroad had received their basic knowledge in science at the technical colleges at home mentioned above.

Training on the job   A commonplace in the diffusion of technical expertise is training on the job. Peter the Great of Russia, who learned shipbuilding in Holland, attests to its universality. Yet the impact of such training on the job is not easily apparent in the subsequent founding of new businesses at home. Poul Strømstad's article offers some striking Danish examples. One Norwegian exception is the story of Harald Schie. He worked for four years at a horseshoe-making establishment in Philadelphia. On his return to Christiania, he succeeded in getting together enough venture capital to set up a horseshoe factory, Kristiania Hesteskosømfabrik. His horseshoes out-competed those of his rivals. When Harald Schie had saturated the home market, he established subsidiaries in half a dozen European capitals, and made a career in international business. His horseshoes tower, with Møller's cod liver oil and Mustad's fish hooks as an early successful case of internationalisation. Note that Harald Schie was not a worker; prior to his stay in Philadelphia he had a well-rounded education from the Trondheim Technical College of 1872 mentioned above.[46]

Technological give and take   The travel of ideas was not only one-

---

45. By kind permission of Björn Slungaard, Norske Sivilingeniörers Forening, Oslo, who let me study a copy of the Zurich list, and the published *matrikkel* of Norwegian engineers.

46. *Teknisk Ukeblad*, 1924, no.50, p.466.

way. The Krag–Jörgensen rifle, a Norwegian patent, was adopted by the US army in 1883, and by the Danish army in 1889. Other examples include the Honeywell–Bull punching card machine, and the jet engine, both going back to patents taken out by Norwegian engineers. A punching card machine was patented by Frederik Rosing Bull (1882–1925), the jet propulsion engine patents may be traced back to Ægedius Elling (1861–1949).[47] Swedish inventors provide a proud list to underscore this point of international scientific mutuality, notably L. M. Ericson, Alfred Nobel and Gustav de Laval. The build-up of technical competence in engineering was such that by the 1870s Norwegian makers of sawing machinery and pulp machinery sold their products to Sweden, Finland and Russia.

Of two Norwegian specialities, one was the whaling steamboats, used with the Norwegian Svend Foyn grenade harpoon in whale hunting in the northern Arctic waters from 1870 till 1904. In the latter year whaling in Norwegian territorial waters was banned. By 1914 the Nyland shipyard in Oslo had built fifty-seven whaling boats, the Aker shipyard 135, and Framnæs in Sandefjord twenty-seven, for both domestic and foreign customers, all fully equipped.[48] It is on record that the American whaling industry was outperformed by the Norwegians in the period up to 1880.

Another speciality was canning machinery. An international path-breaking patent for a canning machine, was taken out in 1902 by Heinrich Reinert, a technical draughtsman at the Rosenberg Mechanical Works in Stavanger, the European sardine canning town. By pulling a lever, the operative sealed the sardine tin to its lid hermetically, to provide lasting preservation of the product without any quality deterioration. While skilled canners could seal 700 tins a day manually, the Reinert machine easily achieved 7,000 tins daily. It secured the entry of a Norwegian speciality on the world market.[49]

Turning to the advent of electricity, we note at first a regression to material technology transfer. In the electro-technical industry, most of the cables, telephones, generators, transformers and instruments at first came in through sales agents for companies in Germany and the USA, and occasionally Sweden and Switzerland.

47. Lars Heide, 'Fra opfindelse til produktion. Frederik Bull's og Knut Andreas Knutsens's udvikling af hulkortmaskiner 1918–1930', paper at Fifth Norwegian Technological History Seminar, Trondheim, 1987, and Aa. Svinndal (ed.), *Styret for det industrielle rettsvern 50 år*, Oslo, 1961, pp.190, 192.
48. Hans Borgen, *A/S Framnæs Mekaniske Verksted 1898–1948*, Oslo, 1948, p.253.
49. *Stavanger Handelsforening 1836–1986*, Stavanger, 1986, p.39.

Adoption was rather swift, however, as is shown in the case of Elektrisk Bureau, mentioned above. The other two high technology industries around the turn of the century were electrochemicals and electrometals. A study of the literature provides sufficient evidence to form an opinion on the transmission routes. With the notable exception of the Norwegian Birkeland–Eyde patent for synthetic chemical fertilisers, nearly all the new patents came in from abroad, i.e. from the USA, France, Austria, Czechoslovakia, Britain and Sweden. The patents provided the foundation for·the modern industrial take-off in Norway on the basis of hydro-power, resulting in exports of crude aluminium, nickel, carbide, chemical fertiliser and iron pellets, all before 1914. Equally, as is apparent from Gunnar Nerheim's article, though Norwegian machine companies obtained some orders, the majority of the big water turbines and generators that powered the new high-energy plants, were supplied by Continental firms in the period up to the First World War, among them J. M. Voith of Heidenheim (Germany), Escher Wyss & Cie and Brown Boveri (Switzerland). Though the Norwegian suppliers could compete technically, they lacked the commercial, marketing and management skills, necessary for large-scale undertakings at home, let alone abroad (but see the article below by Håkon With Andersen'on the diffusion of steamships and motor ships in the Norwegian fleet from 1870 to 1940).

We round off the story by reviewing the introduction of the marine diesel engine. The diesel engine was patented by Rudolf Diesel, a German inventor, in 1893, and developed by numerous imitators, well before the patent rights expired in 1908.[50] In the late 1890s the Norwegian government, through special grants distributed by the local fishery associations, provided travel money for local experts to visit fishery exhibitions abroad. Such exhibitions were held in Sweden, Denmark, USA, Britain and elsewhere. Here the visitors familiarised themselves with the diesel engine. In 1894 Danish firms demonstrated five types of four-cylinder diesel engines, Rap, Gideon, Alpha, Dan and Danette. In 1902 the Ocean Fishery Loan Fund, established in 1889, granted the first loan in order to speed up the introduction of the diesel engine into Norwegian fisheries. The Dan engine was chosen for experimental use. Close on the heels of the Danish firms came Swedish offers for two-cylinder versions of the new engine, the Bolinder, Lysekilen and Advance, in the years 1897–1900.

After a phase of material imports, there followed a period of

50. Eivind Thorsvik, 'Mekanisering av fiskeflåten', *Volund*, 1972, pp.9–132.

pattern transfer and capacity build-up. Around 1900 foreign firms, so far as the Norwegian market was concerned, faced the option, either of being copied, in which case they would get into legal wrangling on patent rights, or of signing licence agreements, called 'working agreements', with Norwegian mechanical works to let them build the diesels. The first such licence agreements stem from 1900 and involved A. Gulowsen Ltd of Christiania and Isidor Nielsen's Mek. Verksted of Trondheim.

As if at a signal, there now mushroomed a Schumpeterian swarming of motor works in the coastal towns, now boasting domestic capacity. The plants offered Norwegian versions of the diesel, as well as repair and maintenance jobs. The market counted upwards of about 20,000 units. The turning-point came in 1905. That year saw sixty-eight old-fashioned steamers against eighty of the new motor vessels in the west coast winter herring fisheries. Most of the diesels were of local design. A detailed census was taken in 1926. Of 15,679 participating motor vessels, about 11,500 were powered by Norwegian motors from at least a dozen manufacturers. Of the 3,704 foreign engines, 2,985 were of Swedish design and origin. Several Norwegian firms made their way into exports, but never in any way rivalling the Swedish motor firms. Thus the three phases óf the Vernon trading cycle are in view.

5 Summary

To account for the forms, patterns, timing, capital, sources and initiatives involved in the technology transfers observed in Scandinavia in the formative nineteenth century, it is necessary to distinguish between the aggregate level and the level of individual industries, as the shifts from material transfer to pattern and capacity transfer show up only in individual industries, not in the aggregate. Five points sum up the story.

1. As a broad generalisation the evidence is sufficiently varied to conclude that the diffusion of new technology in the period did in fact pass through discernible stages: first one of material transfer coupled with the movements of skilled personnel, followed in turn by pattern transfer and capacity transfer. The transfers were matched by a concomitant build-up of domestic technical competence. The findings thus bear out both the Hayami–Ruttan theory of technology transfer and the Vernon theory of trading cycles. The two theories are complementary.

2. Market profitability rather than government planning decided technological diffusion in the nineteenth century. Exceptions include the railways, the telegraph, the roads and other infrastructural investments undertaken by the government. Accordingly, the decisive factors and actors in the transfer decisions appear to have been business calculations by native businessmen. Few of them initially possessed more than a superficial technological expertise. After a transitory period of material transfers, there followed a phase of pattern and capacity transfer, which involved import substitution. The yardstick in individual cases was the prospect of reaping economies of scale. In the transfer decision both demand and supply factors intermingled. Each case involved the decision by risk takers with an eye on both. The greater the capital outlay, the greater the risk, which is why the supply side and the engineer appear to loom most important over time.

3. To what extent do the stages mark degrees of higher technical sophistication? The evidence gleaned from imports of new technology is not clear-cut, some of it pointing up the ladder of technical mastery, other instances pointing downwards. Examples of the former would be the machine imports, or the telephone, which soon led to independent production and exports. An example of the latter would be the manufacture of horseshoes, which clearly consisted of well-known components. All of this points back to the absence of overall planning.

4. The canons of economic liberalism were observed from the 1840s in Norway, from the following decade by Sweden and Denmark. Accordingly, apart from agriculture, government intervention, in the form of privileges, bans, tariffs, quotas, subsidies and grants, was of negligible importance for the forms and directions of technological transfer in the period.

5. As for sources of technology, Britain dominated in an initial period, notably in textiles, glass, clay, pottery and mechanical and metalworking industries. In the following period of electrochemistry and electricity Germany and the USA dominate the transfer story, in terms of both products, patterns and capacity transfers.

8

The Diffusion of Technology and Industrial Change in Sweden during the Nineteenth Century

Lennart Jörberg

Technical change, often regarded as the same as technical progress, has been one central factor in explaining economic development during the last few centuries. It is enough to mention Adam Smith, Karl Marx, Joseph Schumpeter or Simon Kuznets to apprehend the continuity in these explanations.[1]

During the last twenty-five years much of the academic discussion on the change and scope of the importance of technical development for economic growth has focused on two different activity models: economists and economic historians have almost exclusively discussed the economic prerequisites, i.e. how to reduce the costs of production, whereas they have taken little interest in the innovation of products and quality improvements which in the long run have played an important part. Those who have discussed the history of technology have focused their interest on internal technical problems. As Nathan Rosenberg pointed out, the former starting-point is a practicable simplification which makes it possible, with comparatively simple methods of analysis, to discuss a large number of problems, at the same time as it causes difficulties in understanding technological change.[2]

According to Schumpeter's terminology, only the person who, through innovations, promotes progress, is a genuine entrepreneur.

An earlier version appeared in *Festskrift til Kristof Glamann,*. Odense Universitetsforlag, 1983.

1. See e.g. N. Rosenberg, *The Economics of Technological Change*, Harmondsworth 1971, and 'Technological Progress and Economic Growth', in L. Jörberg and N. Rosenberg (eds), *Technical Change, Employment and Investment*, Lund, 1982.
2. N. Rosenberg, *Inside the Black Box: Technology and Economics*, Cambridge, 1982, p.4.

What is strange about his theory is that he presupposes that only once in his life is a man likely to make an innovation, and that every fresh innovation is likely to give rise to a new enterprise. Consequently Schumpeter's definition is too narrow to be applied empirically. Therefore it is necessary to give the concept of entrepreneur a definition which is wider and more in agreement with general usage.

The question is also what is to be called innovation. If the concept is given the meaning 'adaptability and ability to profit by relative novelties', i.e. to use the advances of technology without the entrepreneur necessarily being an inventor or even the first to use the novelty, the problem can be simplified. For the enterprises the introduction of such novelties can entail a definite element of risk and also bring about adaptation difficulties and adjustment or education of the market to use the innovation, which in itself presupposes a certain degree of enterprising spirit. It implies a short-period prognosis where concrete facts and development tendencies are analysed, as well as a long-period prognosis in order to establish the development which may follow. As a rule innovations do not occur intermittently or in clusters as Schumpeter's model predicts. The enterprises which became the leading ones, e.g. in the 1870s, built their existence on inventions made quite a number of years earlier: the steam sawmill, the Bessemer process and wood pulp production, to mention a few examples. Schumpeter's concentration on a small number of great and decisive innovations, creating discontinuity in the development, is therefore hardly tenable. What characterises technological change is instead technological interdependence, the leading role held by the sector of capital goods for the diffusion of technology, and socio-economic factors. In sharp contrast to Schumpeter's emphasis on the discontinuous aspects of technical progress – a view which has left marked traces in the minds of most economists – Rosenberg stresses the importance of the small steps, the role of the less spectacular technical changes. Consequently a main aim, when innovations are studied, will be to study the decisions which lead to technical change at business level.

One should also distinguish between the ideas that arise because the market has need of a product, and the pressure which a new technology causes. One way of increasing the profit was for the enterprise to try to monopolise the technology. Patent legislation was, for the state, a way of guaranteeing this monopoly position for a limited number of years.

Nor should the word 'innovation' be restricted to signifying only the technical innovations, but this concept also includes changes in sales organisation and service as well as changes in the credit market or in financing. In this paper, however, the technical aspect will principally be considered.

Through new technology new products or assets have been created with a given quantity of energy or material. What is meant by new technology, however, is not at all unambiguous and clearly definable. The fact that the technology is new can mean that it was never used anywhere before, it can be new to a certain country, or a certain branch of trade, or to an individual entrepreneur. In the same way the diffusion of technology is a vague term. Diffusion can take place between continents and countries, between branches of a trade or within one branch.

In this paper I am going to make some comments on how international technology was applied within Swedish industry and what problems those enterprises faced which had the will and possibility to apply the technology that could be imported.

As the industrial breakthrough in Sweden occurred relatively late, compared with Great Britain, the USA and western Europe, it is natural that Sweden was technically dependent on the industrially more advanced countries. New technology can be difficult to transfer between countries, principally because the new technology reflects national markets and factor prices, and these are of course not the same everywhere.[3] What is required is therefore a certain adaptation of the technology to changed market prerequisites and other factors, such as the differences in factor endowments, relative costs and the social and institutional framework within which the enterprise is working.

Through technical change and mechanisation industrial productivity normally increases substantially. The new technology is also often connected with indivisibilities. The increased productivity leads to a reduction of the relative prices of industrial products at the same time as the increase in quantities leads to a growth in the industry's share of total production.

For continuous industrialisation to occur agriculture also must increase its productivity, so that a decreasing number of farmers, at least relatively, can support an increasing non-farming population. Sweden coped with this adaptation successfully without resorting to imports. Instead Sweden became an export country for agricul-

---

3. Cf. C. A. Olsson, *Relativa faktorpriser och teknisk förändring*, Meddelande från Ekonomisk-historiska institutionen, Lunds Universitet, no.18, Lund, 1981.

tural products from the 1830s, and oats and butter were the main export products up to 1900.

The new industries had of course no industrial working class from which they could recruit their labour. There was no skilled labour within the new industries, whereas the problem was less pronounced in the older, established branches such as the iron industry and, to a certain degree, the sawmill industry.[4] The new branches were principally the textile industry, which started on a small scale in the 1820s, and the engineering industry, which was little developed before 1850. About 1850 the textile industry had some 600 workers,[5] whereas the engineering industry employed approximately 1,500, with one enterprise, the Motala Engineering Plant, accounting for 450. The remaining branches, established about the middle of the century, were carpentry shops, pulp mills for mechanical pulp, and some food industries.

The introduction of a new technology within the older branches required ability to adapt the new methods to an already existing industrial structure, often resulting in concentration of production and higher productivity per employee. The new branches, which did not build on old traditions, had to train a completely new body of workers. After about 1870, when rapid industrial expansion set in, the problem became somewhat different, when, above all, the engineering industry but also the pulp industry and the iron industry developed technically very fast. During the 1890s a further dimension of the diffusion of technology was added when the many, often complicated, inventions made by Swedish technicians began to be exploited on a large scale, which often meant that new enterprises were created such as separators, internal combustion engines, ball-bearings and telecommunications.[6]

Obviously the influence from Britain was the strongest about the middle of the nineteenth century. Through intense trade relations contacts had long been established. However, innovations were introduced above all through British technicians being asked to come to Sweden, and through the many journeys that Swedes made to British enterprises. During the latter part of the nineteenth century the USA became an industrially leading country, and Swedish technicians practised at American enterprises, learning

4. L. Jörberg, *Growth of Fluctuations of Swedish Industry 1869–1912. Studies in the Process of Industrialisation*, Lund, 1961, and literature quoted there.

5. L. Schön, *Från hantverk till fabriksindustri. Svensk textiltillverkning 1820–1870*, Lund, 1979.

6. Jörberg, *Growth and Fluctuation*, chap. 8.

above all the methods of mass production. Germany, which also became an industrially developed country during the latter part of the nineteenth century, contributed to the development of Swedish enterprises through export of machinery, but played a role of secondary importance as an inspirer of technical innovations.[7]

Naturally technical innovations first penetrated the iron industry. In 1802–3 Th. Svedenstierna went to Britain and reported on the innovations that threatened the traditional Swedish industry.[8] In later decades, above all during the 1830s and 1840s, English smiths were often invited to come to Sweden to teach the Swedes the new finery methods. These changes, however, were not of a radical nature, but could quite easily be adapted and carried through by the Swedish iron industry, and through the existence of skilled workers the adjustment became continuous and relatively free from problems. When the Bessemer process was introduced into Sweden, Swedish technicians played an important role in making it industrially practicable.[9] As early as 1857 Bessemer's patent was bought, and in 1858 the technicians had managed to solve the problems. Still it is of interest to point out that the Swede who made the process industrially applicable was not a technician but a wholesale dealer.

The open-hearth process and the basic process were also introduced very rapidly into Sweden after Swedish technicians had travelled in Europe to get information about these innovations. In 1888–9 one technician went to twenty-five open-hearth works on the Continent, and in 1890 trials with such furnaces were begun in Sweden under the guidance of this technician among others.

Within the rolling-mill technology, too, achievements were made, principally by technicians who had studied in England, and the same is true of the introduction of pressed nail machines and the production of bolts.

During the 1860s, however, more and more impulses came from the USA, and in 1866 two engineers went there by order of the Swedish Ironmasters' Association to study processing within the iron industry. One of these engineers urged Swedish enterprises to 'follow the Americans' example, namely that of attaining cheap and good production through the use of machines to as large an extent

---

7. This and the following section are based on T. Gårdlund, *Industrialismens samhälle*, Stockholm, 1942, if other sources are not given.

8. T. E. Svedenstierna, *Resa genom en del af England och Skotland åren 1802 och 1803*, Stockholm, 1804, and *Några underrättelser om engelska jernhandteringen*, Stockholm, 1813.

9. E. Hedin, *Sandvikens jernverk*, Uppsala, 1937.

as possible'. He also compared industrial conditions in England and the USA.

> It is necessary to try to bring about division of labour, but this division of labour ought to be done through *different machines* and not through *division of manual work*. The former is the American method, the latter the one usually practised in England so far . . . The division of labour through manual work has lately brought about such great difficulties for English industrialists that an introduction of the same methods in other places ought hardly to be considered today. Apart from other disadvantages, the employers' dependence on the workers alone and the higher production cost make the method in general useless, all the more so as, in addition, in such a case we would get no help from our motive power, but instead every disadvantage, apart from the fact that craftsmanship is not available within the country, and cannot be acquired without an enormous loss of time and at great costs.[10]

The strong American influence within the rolling-mill technology is evident from the fact that of the most distinguished technicians, according to a Swedish study, thirteen had been born after 1850, and ten of them had earlier been employed as constructors or engineers in the USA, several of them for periods of between five and ten years.[11]

Also within the mining industry and blast-furnace technology the influence from Britain and to a growing extent from the USA was marked. Above all, pneumatic drilling machines, whose construction from the start had been of American origin, were improved. In blasting technique there was domestic expertise in Alfred Nobel.

The first engineering workshops in Sweden were run by Britons. At the Bergsund Engineering Plant, Samuel Owen, who had been trained at Boulton and Watt's, was employed in 1806. In his autobiography Owen gives a description of the difficulties he experienced when, later on, he started a workshop of his own in Stockholm. What above all hampered the development, according to Owen, was the lack of skilled workers. As soon as the workers employed were half-trained, they were offered advantageous employment by competitors.[12] The Motala Engineering Plant, established to furnish the construction of the Göta Canal across Sweden with

---

10. Quoted by Gårdlund, *Industrialismens samhälle*, p.238.
11. C. Sahlin, 'Valsverk inom den svenska metallurgiska industrin intill början av 1870-talet', *Jernkontorets bergshistoriska skriftserie III*, 1934.
12. Owen's autobiography has not been published. The manuscript is to be found in the city library of Göteborg.

machines and equipment, became the principal nursery for training within the mechanical engineering industry. For twenty years this plant was managed by Daniel Frazer, who had been a machine-fitter on the Continent for Bryan Donkins of London. On his retirement in 1843 he was replaced by a Swede who had been trained in Britain for three years, working for its best machine constructors. Even at the Motala Engineering Plant, too, the lack of skilled labour was considered as an obstacle. Andrew Malcolm, another immigrant Briton, became exceedingly frustrated and, after having worked for twenty years in Sweden, he wrote a pamphlet in which he predicted that the engineering industry had no future in Sweden, and for the same reason as the others had pointed out, namely the lack of skilled labour.[13]

At the same time a large number of Swedes went to Britain to get practical experience, and it is likely that all of the important enterprises had managers who had either been born in Britain or, later, were Swedes who had been trained in that country.

From the 1860s the impulses from the USA became stronger and stronger. In 1867 a delegation was sent out to the USA to study the manufacture of rifles, and they returned with machines and a number of American workers. To visit the industrial exhibition in Philadelphia of 1876 fifteen technicians received state grants. In 1878, during the discussion in the Technology Association in Stockholm, it was pointed out that one of the reasons why Swedish machines were often clumsy was that they were copies of British machines which had begun to be superseded by American ones.

A problem with introducing American technology into Sweden was that American technology presupposed supplies of raw material considerably larger than what was available in Europe, i.e. the technology was comparatively wasteful with raw material. About 1850 the American technique for processing wood was looked upon as superior to the European, but the Europeans regarded the American technique as of little use solely because the quantities of waste became too large, and therefore hardly justifiable from an economic point of view. Considering the relative scarcity of factors in the USA, the American technology, according to which raw material-wasting technology took precedence over an ambition to economise on raw material, was perhaps justifiable and even optimal, but it was difficult to apply it within the Swedish

---

13. A. Malcolm *Factiska bevis att fabriks- & industriväsendet inom Sverige ej är lika tacksamt som i andre länder* (factual proof that the factory and industry system in Sweden was not as profitable as in other countries), Linköping, 1869.

wood industry, even if Sweden, in comparison with England had plenty of raw material in its forest trees. [14] The emigrants returning from the USA seem to have played an important role for the diffusion of technology. In 1900 a statistical investigation about employment was made, and 23,000 machine operators were interviewed. Of these more than 1,400 had been employed abroad for more than one year, over 400 of them in the USA, while the majority had worked in the other Scandinavian countries. Within applied engineering British influences played an important part. This was true of both the construction of canals, principally the Göta Canal, which employed British foremen, and the building of railways. The first railway engines were also made in Britain. The production of engines began in Sweden in the 1860s, and after 1876 all engines for the Swedish National Railways were ordered from Swedish manufacturers, except for twenty which were imported from the USA in 1899. [15] Within the sawmill industry the import of technology and machines was smaller. It is true that frame-saws of British construction were imported, and the technology depended on British impulses, but on the one hand the technical application was relatively simple, on the other there were, from the middle of the nineteenth century, Swedish enterprises which could supply the sawmills with saws. In planing mills, producing for export, the British impulses were greater, and the same was true of the joinery industry. A number of craftsmen studied the London exhibition of 1851 at the government's expense. From the 1870s the American influence became more pronounced.

The paper pulp industry gathered all its technical impulses from abroad. The mechanical pulp technology, which had been developed in Germany by Keller and Voelter, was further developed in Sweden; the chemical technology came from Britain where, however, it was undeveloped in the early 1870s. Swedish technicians were to develop it further, and they led within that branch during the latter part of the century. [16]

The British influence was of course dominant within the textile industry. As long as the British prohibition of machine export was

14. N. Rosenberg, 'Innovative response to material shortages', *American Economic Review*, May 1973, pp.111–18.
15. H. Modig, *Järnvägarnas efterfrågan och den svenska industrin 1860–1914*, Stockholm, 1971.
16. E. Bosaeus, 'Utveckling av produktion och teknik i svensk massaindustri 1857–1939', in *Industrihistorisk skriftserie*, no.4, Uppsala, 1949.

in force, manufacturers tried to import British foremen and constructors, as well as machinery from Belgium, which had a considerable export, one of the reasons being that British industrialists had set up factories in that country to circumvent the export prohibition, which was not abolished until 1843.[17]

Within the food industry the German influence was considerable, especially within the brewery, flour mill, and beet sugar industries.[18]

The process of developing manufacturing industry in Sweden, an industry situated mainly in the countryside, involved substantial difficulties. Wages were probably lower in the countryside than in the towns, and that may be one reason why for example the textile factories at the beginning of the nineteenth century were located there. There was, however, hardly any labour to be had in the mostly small towns. Industrial production relied very little on urban labour, except in Stockholm and Norrköping. Furthermore, it was exceedingly difficult to get hold of the labour skills which were required.[19] Workers seldom left a craft to enter industrial work. Thus the workers recruited to industry came chiefly from the lower classes within the farming population, who lacked any notion of industrial methods. This is evident also from extant reports from the enterprises.

Exceptions to this general picture are to be found mainly within the iron industry, established long before. As mentioned above, the engineering plants constituted a school for the training of skilled workers. The Motala Engineering Plant got skilled workers through advantageous contracts, and a state loan in 1826 was made to enable employment of British workers and skilled Swedish workers, needed for rational production. The Swedish Ironmaster's Association also awarded grants to the Motala Engineering Plant, on condition that it accepted trainees, which continued until 1892. In its annual report of 1839 the company stated that there had been forty trainees working in its drawing office. Several of the most prominent technicians and industrial leaders in nineteenth-century Sweden had in their youth been trainees at the Motala Plant. It is also quite clear that not until the end of the century was it considered necessary to insist that engineers had a background in advanced training, and as late as the turn of the century, it was often the practice that foremen were in charge of technical management.[20]

17. L. Schön, *Från hantverk till fabriksindustri*, chap. 11.
18. E. Sylwan and G. Ohlsson, *Den svenska betsockerindustrin*, Malmö, 1932.
19. L. Schön, *Från hantverk till fabriksindustri*, chaps 8, 9.
20. G. Ahlström, *Engineers and Industrial Growth*, London, 1982.

As Swedish industrialisation came relatively late in comparison with western Europe and the USA, Swedish industrialists derived advantage from being able to choose the latest and best technology, and avoid experimental and other costs. The industrialists could exploit this advantage only if they knew how to choose expertly, i.e. if they had the ability to adapt the borrowed technology to the conditions of their own enterprise. Through a sceptial attitude to existing manufacturing processes, they gave an impetus to better and more rational methods of production. 'There is nothing more dangerous than using the word "impossible" in a wrong way', a manufacturer brought home to his technical production manager during the 1870s.[21]

The enterprises' sources of information were, as has already been stated, visits for purposes of study, as well as direct import of technicians. Besides there are in the correspondence archives of the enterprises references to journals, visits to factories, visits to exhibitions etc., which is what we would expect.[22]

The entrepreneurs were also aware that industrial espionage could be profitable. During the 1870s the production manager at the Kallinge Iron Works was sent to England, and he was instructed to

find out which ironworks in Staffordshire produced the best rolled bar iron for engineering works . . . It would be tremendously good if you could manage to get in and find out about the galvanisation of sheet metal etc. Try to find a worker or foreman who, on payment of a certain sum, is willing to come over to Sweden and set everything going. Get hold of the best journals on the iron industry and the best journals on the processing of metals . . . Try to make friends with the man at the works who is in charge of the construction of these rollers, invite him to late dinner at six o'clock at the Victoria Hotel, and give him one or two good nightcaps afterwards, hot, strong, sweet; that usually opens the Englishmen's heart . . . Apply for admission into the workers' so-called Union, pay the fee and apply for a job at some metal sheet rolling-mill, it may be more successful than expected.[23]

The Swedish Board of Commerce systematically sent out people to other countries, charged with the task of gathering information, and the Swedish Ironmasters' Association supported the publica-

21. L. Jörberg, *Svenska företagare under industrialismens genombrott 1870–1885*, Lund, 1988, p.86.
22. Ibid.
23. Ibid.

tion of technical literature, an activity which had started as early as the eighteenth century. During the latter part of that century, sixteen publications within the field of the iron industry were issued, during the period 1800–30 an additional forty, and between 1830 and 1866 a further eighty-five. The Swedish Board of Commerce has in its archives over 200 accounts of journeys from the period 1849–1900, handed in by holders of grants who had been sent abroad to collect new information of a technical nature.[24]

A more indirect way of supporting an improvement of the technology was achieved through the Manufacturing Discount, a state establishment which lent the enterprises money at low interest and with a long period of amortisation, especially aimed at supporting enterprises which started quite risky adventures. Among the borrowers the textile industry dominated, but the engineering enterprises became substantial borrowers from the 1840s. In several branches the loans from the Manufacturing Discount dominated. The activity ceased in the 1870s.[25]

It is quite clear that many of the new technical inventions could only slowly be made usable on a large scale. Consequently the state played a certain role as a risk taker and initiator by defraying the cost of journeys for purposes of study.

The technical transformation process can be said to consist of two components, namely one that is positive and one that is negative.[26] The positive one consists in the creation of innovations, and the negative one in the liquidation of old combinations of production factors. The difficulties experienced by an enterprise are caused by the pressure from the negative component of the transformation, i.e. the production consists of goods which are being superseded by new goods, or its methods of production are too costly in relation to the newer and better methods of other enterprises. Thus a certain method of production is located on the negative side of the development.

By differentiation between types of development, it is possible to distinguish between advancing, stagnating or regressive industries. Among the advancing enterprises it is also possible to distinguish between those which are subjected to a demand pull or a supply push, i.e. an increase in consumption without any discernible

24. *Jernkontorets historia*, Stockholm, 1955, vol.3, no.2.
25. T. Gårdlund, 'Manufakturdiskonten och fabriksindustrins finansiering', *Studier i ekonomi och historia tillägnade Eli F. Heckscher*, Uppsala, 1940.
26. E. Dahmén, *Entrepreneurial Activities and the Development of Swedish Industry 1919–1939*, Homewood, Ill., 1970.

measures from the enterprises, either because there are no substitutional products or because the production of the enterprise can replace other products. Subjected to the supply push, enterprises actively try to enlarge their production through, for example, reduced prices, new products or advertising. The difficulty in distinguishing between these elements consists in their often occurring simultaneously. Supply push can also take place even if no measures have been taken from the enterprise in the form of innovations and/or simultaneous price reductions. The course of events can even have started some decades earlier, and it is therefore necessary to try to discover concrete developments within the industrial branches and the enterprises so that it may be possible to study causality in the diffusion of innovations. Often one branch is too large a unit for these studies. The branches are heterogeneous, and to trace the course of development a more detailed level is required, perhaps specific products.

A method which has been used to measure technical change in relation to economic change is to analyse the patents which have been taken out, both foreign and Swedish ones.[27] There are usable series from about 1840. They show that Swedish applications increased slowly up to the early 1870s, by 2.6 per cent annually, and by 7.7 per cent annually up to about 1910. For foreign patents the percentages were 5.4 and 9.9 respectively per year for the periods 1840–50 and 1850–1900, i.e. foreign patents led the series by about twenty years. In relation to the changes of the volume of industrial production it appears that the applications for patents on the whole follow them. When the export rate increases, there is also an increase in the relative importance of foreign technology in Swedish industrial growth, and the opposite is true when the export rate decreases, i.e. an increase up to the middle of the 1890s and a decrease up to the beginning of the 1910s. The importance of foreign technology in the relation between domestic and foreign applications for patents becomes greater during the 1870s than earlier, at the same time as the short fluctuations of the foreign applications precede the industrial production which precedes the fluctuations of the Swedish ones. Evidently there are time patterns in the transfer of the foreign technology, but this does not explain the phases in Swedish industrial development.

A characteristic feature in Swedish industrial development was

27. O. Krantz, 'Teknologisk förändring och ekonomisk utveckling i Sverige under 1800- och 1900-talen. Iakttagelser från patentstatistiken', Meddelande från Ekonomisk-historiska institutionen, Lunds Universitet, no.26, Lund, 1982.

that the new industries only exceptionally grew out of the old guild system. This is seen clearly if the innovations and the building up of enterprises are studied. To be sure the new industries sometimes had their roots in the trades or in the so-called cottage or home industry; the textile industry, for example, came to be located in regions where such industrial activities already existed.

Another feature is the strong connections that existed between commerce and the establishment of new industrial enterprises. Engineering plants, sawmills, pulp mills, joineries, textile enterprises etc. were more often than not founded by people connected with the wholesale trade. This also explains the number of newly founded enterprises being much larger than the number of founders of enterprises during the first period of the industrialisation. Some twenty names reappear in at least about 100 newly established enterprises. This indicates of course that the basis of recruitment was very narrow, which in turn is a consequence of the class of capitalists being very small. The foreign element among the industrialists is also clearly discernible. Through this entrepreneurial recruitment foreign impulses to technical change were facilitated. After the 1870s, when the merchant houses lost their dominant position as both financiers and founders of enterprises, the foreign import of technology changed, which is also evident from the applications for patents by foreigners.[28]

The limited Swedish market also made it more difficult for manufacturers to import and apply the most modern technology, while the prerequisite for the introduction of an innovation was that the market was enlarged, i.e. that they could export. This was probably the case in several branches of the engineering industry. When for example the production of bolts started in Sweden and the enterprise analysed the advantages of importing a modern foreign bolt machine, future large exports were looked upon as essential for the purchase, 'for within the country such a large consumption of machine-made bolts is hardly likely to be sold'.[29] Within the cement industry, too, similar problems arose. The new techniques which were introduced in the 1870s would, according to the entrepreneurs, have the result that 'the storehouses become full and the plant must stop'.[30] During the depression in the late 1870s one industrialist was of the opinion that export to Russia, at any

28. L. Jörberg, 'Structural Change and Economic Growth: Sweden in the 19th century', in F. Crouzet et al. (eds), *Essays in European Economic History*, London, 1969.

29. Jörberg, *Svenska företagare*, p.87.

30. Ibid.

price, was necessary, as the fixed costs for the newly installed machines necessitated their continuous use. Besides it was impossible to dismiss the workers as it would be impossible to employ new unskilled workers when the work could start again.[31]

In this connection an analysis of development blocks, in Dahmén's terminology, can be informative. This term, among other things aims at directing attention to the fact that technical progress in a specific development process of production often cannot be used profitably, as long as certain other changes in other steps of the developmental process have not been implemented. This means that if the changes between the different parts of the process are not synchronised, structural tensions within these unfinished or incomplete development blocks will arise.

Both when there were tendencies to overproduction and malinvestments and when series of inertia factors were encountered, problems of equilibrium in the development arose, which remained until the new enterprises had driven the older enterprises out of the market, or until the market was enlarged so that a new branch structure with fewer structural tensions had been obtained. These tensions can also be regarded as a type of driving force in the development.

The problems were, to some extent, accentuated by the developmental phase of communications, in the technical and organisational development, and because quite a long time often passed between a technical invention and an innovation, and between an innovation and its general diffusion. If it was not profitable to introduce a new machine so long as wages were low or labour too unskilled, mechanisation in one branch could be intimately dependent on the development within other branches whose expansion contributed to a general rise of the wage level.[32]

This also implies that the diffusion of new technology is related to the market, i.e. that a new, more effective technology was used as soon as relevant costs and prices made the innovation profitable.[33] When there was a delay in the diffusion of technology, this can usually be accounted for by unfavourable cost and/or price relations rather than by an inability to master the new technology. Consequently the scope and the time for the introduction of a new technology are correlated with the labour's and the entrepreneur's level of training and market orientation before the introduction of the new technology.

31. Ibid., p. 71.
32. W. E. G. Salter, *Productivity and Technical Change*, Cambridge, 1960.
33. Cf. G. Mensch and R. Schnopp, 'Stalemate in Technology. The Interplay of

This may also explain the relatively slow industrial growth during the middle of the nineteenth century, at the same time as one finds a clear tendency to try to apply the new foreign technology within the branches which were growing, e.g. the textile, iron and engineering industries.

This capacity for adaptation to the market can also contribute to explaining why the tentative industrial start led on to a rapid cumulative industrialisation.

A cumulative element was that the recruitment basis for workers and entrepreneurs was enlarged. Even if merchant houses were very prominent in the industrialisation process as late as the 1880s and 1890s there were considerably more technicians among the entrepreneurs than earlier, above all within the engineering industry. The increasing contacts with other countries also contributed to creating a cumulative element in the development. Of importance was also the extension of credit and banking systems, above all when complete development blocks were built up, where the difficulties were often to be found at the financing level.

The fact that certain combinations of development and technical change were required for rapid expansion, created difficulties during the introductory period, as long as the entrepreneur's capability was undeveloped, the ability to survey the development blocks was insufficient, the labour unskilled and the market insufficient for advanced technical solutions.

But the new industries during the early phase of industrialisation before 1870 helped not only to enforce new technical solutions and innovations but also to activate a number of latent innovations in the older branches.

It is evident that industrial development was facilitated by the diffusion of technology, but it is also obvious that, more often than not, a new technology could not be applied until a certain development had taken place. Economic development implies an interplay between different categories: population, resources, technology and social institutions. To find explanations for this interplay is one of the main tasks for economic historians.

So far there is no generally accepted explanation of how technology is diffused between different parts of an economic system, whether this is the world economy as a whole or some part of it, national, regional, or between entreprises.[34]

Stagnation and Innovation', W. H. Schröder and R. Spree (eds), *Historische Konjunkturforschung*, Stuttgart, 1981.

34. A. Cairncross, 'Reflections on Technological Change', *Scottish Journal of Political Economy*, vol.19, no.2, June 1972.

9

Foreign Technology and the Danish Brick and Tile Industry, 1830–1870

Ole Hyldtoft

The years from 1830 to 1870 can be characterised as the founding period of the Danish brick and tile industry. The making of bricks and tiles was introduced to Denmark in the Middle Ages from the south, and recent studies show that during that period a varied and well-developed kiln technique had been mastered.[1] However, it was not until the end of the 1830s that the brick and tile industry started to spread widely. In the course of the 1840s and 1850s it developed into one of the largest Danish industries, sustained by a tremendous rise in demand for bricks, tiles and drainpipes. As early as 1855 there existed at least 178 brickworks in Denmark with more than five workers. In total, these brickworks employed 2,256 workers in a season. From 1838 to 1855 total production rose from about thirty million bricks to more than 130 million. In 1838 the largest brickworks produced approximately 600,000 bricks; in 1855 we find fifteen or sixteen brickworks with an annual production of more than one million bricks and drainpipes. Growth continued after a minor setback connected with the crisis of 1857. In 1872 there were at least 266 brickworks employing in total 4,162 workers, with an estimated production of approximately 227 million bricks, tiles and drainpipes. Almost half of the workers were employed in enterprises with more than twenty workers, and each of these firms had an annual production capacity of at least one million units. These estimates, and in particular the 1838 figures, are probably too low. However, there is no doubt that a dramatic expansion occurred in brick and tile production.[2]

1. B. A. Hansen, 'Middelalderlige teglovne', *Bygningsarkæologiske studier*, 1985, pp.7–16. A sketch of the development in the following decades is given by N. B. Josephsen, 'Med ler og tegl', in *Årbog for Den gamle By*, Århus, 1984, pp.117–73.
   2. Based upon the industrial censuses in 1838, 1855 and 1871–3 held in the

Ole Hyldtoft

In the standard work on the industrial history of Denmark, Axel Nielsen emphasises that the model for the Danish brick and tile industry during this period was the famous brick- and tile works on the Flensborg Fjord in the Duchy of Slesvig. Slesvig, together with the Duchies of Holstein and Lauenberg, was ruled by Denmark until the war in 1864, when they were annexed by Prussia. In particular, H. H. Dithmer's works at Rendbjerg by Egernsund won the reputation of being the best in Scandinavia. According to Nielsen, the inspiration for the increased use of machinery in the Danish brickworks form the 1850s onwards also appeared to come from the Duchies. In general, Nielsen's account gives the impression that Danish brick technology during this period was fairly backward.[3]

However, Axel Nielsen almost exclusively takes his examples from medium-sized estate brickworks in central and southern Zealand. If instead we look at the large, commercially run enterprises, the picture becomes more complex. For example, P. F. Lunde, the owner of a Copenhaguen iron foundry, built an extensive brickworks, Sorthat, on Bornholm. Besides being Denmark's first steam-powered brickworks, Sorthat was supplied with extensive railway tracks and advanced kiln and drying installations. Even the Prussian consul-general, R. Quehl, who visited the works in 1856, expressed enthusiasm, talking about 'eine grosse Ziegelei' (a magnificent brickworks), which bore witness to 'das Genie und eine Regsamkeit der Unternehmer, wie sie in Dänemark nicht eben häufig sind' (a genius and a business dynamism most unusual in Denmark).[4] A couple of years later the firm of J. Owen & Sons built Aldersro Steam Brickworks on the outskirts of Copenhagen. The ambitions here were equally high. According to the British construction engineer, the new works, 'regardless being of a smaller size than some of the works abroad . . . [was] the most complete brick and tile works, and in every respect based on the newest and most advantageous innovations'.[5]

In what follows, the main aspects of the technological development of the Danish brick and tile industry from 1830 to 1870 will

---

Rigsarkivet (hereafter RA (National Archives)). In 1872 the production per worker is estimated on the basis of information from fifty-two works with a total of 462 workers.

3. A. Nielsen, *Industriens Historie i Danmark*, Copenhagen, 1944, vol.3.1, pp.162–91.

4. R. Quehl, *Aus Dänemark*, Berlin, 1856, pp.263–5.

5. RA (National Archives) Indenrigsministeriet, 1. Kontor, Jnr. 2440, 1857. The firm got a patent for the drying installations and the clay-cleaning machine.

Brickworks producing more than 1 million units in 1855

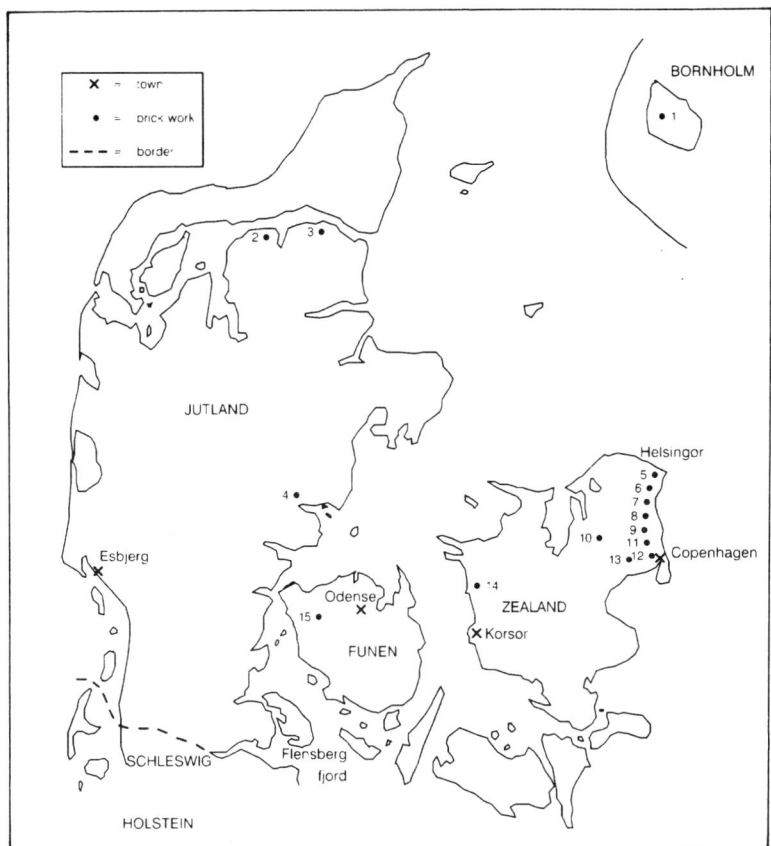

Key to brickworks:

1. Sorthat	6. Strandmosegård	11. Guldborg
2. Kyø	7. Niverød	12. Aldersro
3. Godthåp	8. Nivågård	13. Hakkemose
4. Bygholm	9. Lyngby	14. C. Collstrup
5. Klostermosegård	10. Bremeke	15. Sellebjerg

be examined. The emphasis will be on the larger works with production exceeding one million units. Most of these enterprises were situated in the vicinity of the capital, or had good transport facilities to the important Copenhagen market. Our interest will focus on possible foreign models and on typical methods of transferring technology at that time. The source material comprises fire insurance valuations, patent material, industrial censuses and contemporary technical literature. As a consequence of the slightly uneven source material the results will inevitably be of a provisional nature.

I Brick Manufacture by Treads, Hand Moulding and Periodic Kiln Technology

At the start of our period, the season of brick and tile manufacture normally lasted from after Easter until Michaelmas Day (29 September).[6] The first stage in the process consisted in digging the clay from the clay pits with special-purpose spades or forks. The best way to do this was to dig up the clay in the autumn, allowing it to 'winter', so that any lumps could be disintegrated by the frost, and moisture would be evenly distributed within the clay. From the clay pit, or 'the winter beddings', the clay was transported in wheelbarrows or by horse and cart to a 'tread' for kneading. The tread was a circular container with a radius of approximately 5 m, made of bricks and wood. In the middle of the tread a big post stood upright, and from this a bar protruded which was pulled by one or two horses outside the tread. To this bar was fastened a loaded treading cart, which gradually moved out and in along the bar. During the treading, water and possibly sand were added, if the raw clay was too greasy.

After treading, the clay was transported to the moulding table where the moulder took two lumps of clay and slapped them into the moulding form, a wooden frame with room for two bricks. He then smoothed the surface with a stick and placed them in the moulding yard. A hand moulder could in this way make about 4,000 bricks per day. After a couple of days, the bricks were turned on their sides and smoothed with a wooden knife, a task performed by women and children. After a few days the bricks were transported to the drying sheds. When sufficiently dry, depending on wind and weather, they were carefully stacked in the kiln.

---

6. Nielsen, *Industriens Historie*, pp.162–91; L. Vincent, *Vejledning i rørlegning*, 3rd edn, Copenhagen, 1872, pp.127–66. About traditional brickmaking technique see also T. T. Kragelund, *Gamle vestjyder fortæller om tegl og kalk-brænding*, Copenhagen, 1953; J. T. Christensen, 'Fra de små teglværker', *Historisk Samfund for Sorø amt*, 1935, pp.29–56; O. Nørregaard, 'Helligsø Teglværk 1875–1906', *Historiske Årbøger for Thisted Amt*, 1965, pp.9–26. About the technique of making bricks and tiles in general see B. Kerl, *Handbuch der gesamten Thonwaarenindustrie*, 3rd edn, Brunswick, 1907, and O. Bock, *Die Ziegelei*, 2nd edn, Berlin, 1898.

The firing was carried out in so-called chamber kilns or blast furnaces, which in their simplest form were just rectangular boxes made of raw or baked bricks. A good large kiln with room for 50,000 bricks would be something like 7.5 m long, 3.8 m wide and 3.8 m high, but both size and shape varied. A common characteristic for chamber kilns was that they were fired from below, and that the baking process was intermittent, processing one batch at a time. Peat, wood or coal were used as fuel. One baking lasted a good week, but a second week went by for the kiln slowly to cool down before the bricks were taken out and sorted. The whole process lasted at least three weeks, and the works therefore rarely managed more than ten firings in one season, and often had to be content with fewer.

This method of production rested on old traditions and required considerable practical skills. The training and judgement of the moulder and especially of the brick-burner were decisive for a good result. When baking, it was not only necessary to take into consideration the properties of the local clay, but also the weather, type of raw bricks and moisture content, the stacking, the kiln and the fuel. The baking procedure was judged by the emission of smoke and the sinking of the topmost layer of bricks.

During the period 1830–70 the Danish brick and tile industry experienced a complex process of change and development. Partly the process consisted in the expansion of the established method of production described above. This entailed the near-extinction of the older method of firing bricks in provisional clamps. Under the pressure of strongly increasing demand, the period of production was extended beyond the three months from sowing to harvesting, and the number of bakings increased. The brickworks increasingly stood out as an independent industrial sector, unable to manage only with surplus labour from the agricultural workforce. However, since manufacture during the winter months proved difficult to establish, a large part of the workforce were for a long period migrant workers. In the early phase of the expansion many of these came from Lippe-Detmold and surrounding areas in Germany. This process was then both an extension and a development of established methods, and it involved increased use of new technologies and machines. This will be examined in detail in the following sections.

The established method of working had several advantages. A skilled brick-burner could make a high quality product with an attractive appearance. The technology was relatively simple and familiar. The costs of construction could be kept in sight, and producing in batches made it possible to adapt production to changing demand.

However, there were also serious limitations and problems. Problems especially arose when demand increased sharply or new products were introduced, such as thin drainpipes. In addition, the method was strongly dependent on natural conditions, needed reasonably good quality clay, and was also dependent on the weather. Furthermore the production was discontinuous, both in the preparation of the clay and in the baking. This meant under-utilisation of fixed capital and easily led to accelerating transport costs. Finally the constant heating up and cooling down meant that the chamber kilns utilised only a fraction of the available energy. In other words, economies of scale were limited, and large-scale operation could instead easily generate difficult bottlenecks.

2 The Duchies and Germany

As Axel Nielsen pointed out, the Danish brick and tile industry was greatly influenced by the enterprises on the Flensborg Fjord.[7] From the 1820s, H. H. Dithmer developed the Rendbjerg brickworks into a model firm. To extend his knowledge, Dithmer travelled, with the king's patronage, on a study trip to Holland in 1824; Dutch brickworks provided a model for the Duchies' brick and tile industry. Partly based on this experience, Dithmer specialised in new and difficult products such as cornice bricks, decorative bricks, monuments, flagstones and, above all, clinkers. Rendbjerg's reputation encouraged many people to seek Dithmer's advice about modernisation or establishment of new brick- and tile works. For example, in 1832, the Holsteinborg estate corresponded with Dithmer in connection with a thorough modernisation of the Fuirendal brickworks in Zealand. The Egernsund brickworks were also used as educational establishments. The engineer Magnus Jespersen, after he had obtained his master's degree in 1854, had plans for a career in the brick and tile industry, and immediately travelled to Egernsund. Here he was trained in the working techniques of brick production.[8]

Unfortunately we know very little about the technical develop-

---

7. Nielsen, *Industriens Historie*, pp.162–91; A. Korse, 'Teglbrændingens historie på Broagerland', *Historisk Samfund for Als og Sundeved*, 1958, pp.75–92; I. Adriansen, et al., *Teglværker ved Flensborg Fjord*, Gråsten, 1984.

8. A. Garboe, *Bornholmer-geologen Magnus Jespersen*, Copenhagen, 1931, pp.53–4.

ment of the works on the Flensborg Fjord during this period. We do, however, have information about the development of a new kiln which, from an international point of view, was a significant technical advance. The main problem with the chamber kilns was their immense fuel consumption. An early, radical attempt to reduce fuel costs came in 1840, when the owner of a Flensborg brickworks, Hans Jordt, together with an iron foundry owner M. H. Holler, took out a patent on a 'channel kiln'.[9] Jordt's kiln, built on quite different principles from the chamber kiln, is the earliest tunnel kiln recorded. In the new kiln the firing was continuous. The raw bricks were stacked on small wagons, which were sent through a long tunnel. The actual burning took place in the middle of the tunnel at two fixed furnace points. The chimney was situated at the entry to the tunnel, and the heat emitting from the baked bricks gradually warmed up the raw bricks moving towards the firing places. Possibly, the incoming air had also been preheated by already fired bricks. This saved considerably on fuel.

The experiments continued the following years with support from the Danish state. The new kilns, however, caused problems, even after various improvements. Above all, the intense heat meant that the carriage movement of the wagons broke down repeatedly. In spite of promising results as regards reduced consumption of fuel, these problems seem nevertheless to have led Jordt to discontinue his expensive experiments in 1847.

H. H. Dithmer also applied for and was granted several Danish patents during this period. In 1842, together with the Belgian Alois Milch, he was granted a five-year monopoly on a piston-driven brick press, and in 1847 he acquired a five-year patent on a hand-powered tile press.[10] Simple presses of this kind were not, however, decisive in subsequent technological developments, but were used solely in the production of moulded bricks and tiles, and in after-pressing of bricks.

Several Danish brickworks ordered some of their machines and equipment from the Duchies. For instance, three brickworks stated in 1855 that they had bought their drainpipe machines from Schweffel & Howaldt's iron foundry in Kiel, which specialised in this branch of production.[11] When the Sparresholm brickworks in

9. C. Singer, et al. (eds), *A History of Technology*, Oxford, 1958, vol.5, pp.666–70; A. Olsen, 'Kanalovnen – tunnelovnen', *Lerindustrien*, 1924, pp.96–9; A. Olsen, 'Kanalovnen', *Lerindustrien*, 1933, pp.241–51.

10. Patents of 15 June 1842 and 14 July 1847.

11. Statistisk Bureau (hereafter SB (Statistical Bureau), industrial census of 1855;

Zealand planned to modernise in 1869, they obtained price quotations for machines from the engineering firm of Clausen in Broager near Egernsund.[12]

Another important channel in the transfer of brick and tile technology from the Duchies to Denmark were the newcomers from the Egernsund area, who founded or built up several of the period's most important Danish brickworks. As examples we can mention H. H. Dithmer's son, G. F. Dithmer, who in 1846 took over Klostermosegård in Zealand. In the following years he developed it into one of the leading brickworks in Denmark. Dithmer junior brought with him Fritz H. Friedrichsen, who in 1861 started up for himself, also in Zealand. Another arrival was H. P. Heide, who in 1832 became manager of Fuirendal brickworks and later leased and modernised the Bygholm brickworks in Jutland.[13]

Despite numerous examples it is difficult to evaluate more precisely the importance of the Duchies in the technological development of the Danish brickworks during this period. It is difficult to pinpoint specific machines or firing methods which spread to Denmark via the Duchies. To a very considerable extent it seems that the Duchies primarily acted as conduits, channelling an advanced method of traditional brick and tile production which had been developed in Holland.

The same applies to the early German influence, which was channelled to the Danish brickworks through the many Lippe-Detmolders who every summer worked in the Danish brick and tile industry. They were often employed as brick-burners, who, together with the men they had brought along with them, took over brick manufacture for one season by a kind of leasing arrangement. Several of these brick-burners, such as the long-standing manager of the country's largest brickworks at Hakkemose, west of Copenhagen, settled permanently in Denmark.[14]

It was not until the end of the 1850s that German technicians seriously began to make their mark internationally in the technological development of the brick and tile industry. In 1858 F. Hoffmann and A. Licht patented the circular, or ring, kiln, which in the following decades was to revolutionise the European

E. M. Holst, 'Meddelelser om dræning', Tidsskrift for Landøkonomi, 1853, pp.241–51.

12. Nielsen, Industriens Historie, p.187.

13. Ibid., p.179; G. Brask, En teglværkshistorie. Sølyst i Nivå, Karlebo, 1982, pp.7–18.

14. A. Larsen, 'Husmænd og teglværksfolk', Fra Københavns Amt, 1936–7, pp.195–250.

brick and tile industry. Almost at the same time came the breakthrough of continuous brickmaking machines, connected with
names such as C. F. Schlikeysen in Berlin and the Sachsenberg
Brothers in Rosslau. Even though patents were not secured for
Hoffmann's circular kiln or for the new brickmaking machine in
Denmark, knowledge about the machines rapidly reached the
country. In 1863 the *Journal for Håndværks- og Fabriksdrift* (Journal
for Craft and Industry) published a lengthy article from the *Deuts-
che Industriezeitung* (German Industrial News) about the 'organisation and running of brickworks', giving a detailed account of the
various innovations.[15]

3 Pug Mills and Drainpipe Machines

A closer look at the new machines and firing methods employed in
the brickworks shows, however, that the influence from the
Duchies and Germany was not exclusive. Even though most of the
production was carried out by hand, as earlier, some new machines
began to win acceptance in the Danish brickworks in the 1840s and
1850s. The most important of these were the pug mills and drainpipe machines.

The pug mill, which replaced the old horse-driven treading mill,
usually consisted of a cylinder-shaped container which had a vertical iron shaft in the middle with knives attached. This blended the
clay, and then pressed it out at the bottom.[16] Like its forerunner,
the machine was driven by a horse-gin, and the advantage of the
machine lay especially in the more continuous preparation of the
clay, whereas the treading mill blended one batch of clay at a time.
The earliest known pug mills in Denmark were installed in 1848 at
the Strandmosegård works north of Copenhagen and at the Hakkemose works west of the capital.

Strandmosegård brickworks was built in 1847 for Edvard Collin,
the Permanent Under-Secretary in Copenhagen, by J. C. Thygesen, Master of Engineering, who was himself manager of the
works until 1854.[17] Hakkemose was established in 1847–8 by J. P.
Langgaard from Copenhagen, who was originally apprenticed as a

15. Vol.4, pp.533–6, 550–4, 605–8 and 619–22.
16. Vincent, *Vejledning*, pp.132–5; C. I. Hansen, *Praktisk anvisning til dræning samt til drænrørs fabrikation*, Copenhagen, 1852, pp.22–3.
17. O. J. Rawert, *Kongeriget Danmarks industrielle forhold*, Copenhagen, 1850, p.294; C. Nyrop, *Bidrag til den danske industris historie*, Copenhagen, 1872, pp.280–81.

clockmaker and in the 1830s had founded the first orthopaedic institution in Denmark . The plans were extensive from the start, and the Hakkemose works developed into Denmark's largest brick- and tile works during the next twenty-five years. The central element comprised a giant double kiln with a capacity of two times 150,000 bricks. In addition, there were a couple of drying sheds and as many as four horse-driven pug mills, each consisting of 'a box of timber and planks with a cross in the bottom, to which is attached a cast iron shaft with five iron arms furnished with knives'.[18]

The pug mills were cheap to obtain. C. I. Hansen in 1852 estimated the price to be 150rd, and the four pug mills at Hakkemose were each valued at 120rd.[19] The use of pug mills spread rapidly in the larger brickworks. In 1852 C. I. Hansen recorded that pug mills were already employed at different places in Denmark, and in 1855 there were pug mills in at least a score of Danish brickworks.[20] This did not mean, however, that pug mills completely supplanted the old system of treads. Crusts and lumps had an unpleasant habit of passing straight through the pug mills, and in the following decades pug mills and treads were often used side by side in the brickworks; for example, the Klostermosegård works in Zealand used four treads alongside three pug mills in 1855.

The earliest known pug mills, made of wood and with wrought iron knives, appear in the Dutch tile industry in approximately 1800. In the 1840s and 1850s the source of inspiration for the Danish pug mills could have come either from Germany (Holland) or Britain.[21] It is, however, very rarely possible, on the basis of the available material, to make closer identifications. C. I. Hansen gives a description of the British type with iron vessel (developed by the London engineering firm of Clayton), and also the pug mills at Hakkemose seem to have been British-inspired with small knives at cross angles to the main knives. The fact that pug mills to a large extent spread together with the new drainpipe machines, as we shall see, points in the same direction.

Drainage came to Denmark from England, where this soil-improving technique started to spread from the 1830s. The prerequisite for a wider diffusion of drainage was relatively cheap, that is machine-made, pipes. The first drainpipe machine was constructed

18. Landsarkivet (hereafter LA (State Archives)), Zealand, Brandtaksationer, Kbhs. amt, Smørum herred, 24 July 1848; C. Langgaard, *En hundredeårsdag*, Copenhagen, 1911.
19. References as in notes 16 and 18.
20. References as in notes 16 and 18, and industrial census of 1855.
21. Kerl, *Handbuch*, pp.245–51; Vincent, *Vejledning*, p.135.

by the British machine maker Whitehead, and his machine set the pattern for most of the later types. It was a hand-driven machine with a horizontally placed iron box, inside which a piston operated to press out the clay through three or four circular openings. By the use of this machine a brickworks was, it was estimated, able to produce approximately 6,000 drainpipes in one day. Another widely use drainpipe machine came from Claytons; this too was a hand-driven machine with a piston, but the construction was different from the Whitehead type in that it had a vertical and cylinder-shaped box.[22]

From about 1850 drainage spread quickly in Denmark, energetically backed up by the Royal Agricultural Society. A good ten years later, in 1861, there were about 33,000 hectares of drained fields, and in 1872 the area had increased to 200,000 hectares. As early as 1848, the Strandmosegård works was the first Danish works to start producing drainpipes. During the first three years production was essentially experimental and did not exceed 10,000 pipes; in 1851, however, the works increased production to about 75,000. In his efforts to keep abreast of technical developments, the manager, J. C. Thygesen, visited the 1851 World Exhibition in London, where he purchased, *inter alia*, one of the Whitehead drainpipe machines on behalf of the Danish Industrial Society. This machine was installed at A. C. Bock's pottery in Copenhagen.[23]

The machine manufacturer M. P. Allerup in Odense, Funen, was quick to copy the new machine. With support from the Royal Agricultural Society the estate proprietor Bruhn, from Jutland, bought a drainpipe machine from Allerup in 1853, as did an estate manager named Schmidt, also from Jutland. In 1855 Allerup delivered a double-acting drainpipe machine to the Cathrineholm brickworks in Zealand. Almost at the same time P. F. Lunde, the iron foundry owner in Copenhagen, began to manufacture drainpipe machines.[24] Other brick works got their drainpipe machines, as already mentioned, from the Duchies. Because the investment was moderate (C. I. Hansen puts the price at 220rd) and the market

22. Hansen, *Praktisk anvisning*, pp.24–5, Vincent; *Vejledning*, pp.137–47; Kerl, *Handbuch*, pp.829–36; P. J. Winstrup, 'Whiteheads maskine til formning af lerrør og hule mursten', *Qvartalsberetninger fra Industriforeningen*, 1852, pp.37–41.
23. J. Christensen, *Landbostatistik*, Copenhagen, 1985, p.72; Nyrop, *Bidrag* and 'Jordudtørring ved underjordiske lerrør', *Årsberetninger om det kgl. Landhusholdningsselskabs Forhandlinger*, 1851, pp.32–41; *Qvartalsberetninger fra Industriforeningen*, 1851, p.256.
24. Industrial census of 1855 and 'Dræning', *Årsberetning fra det kgl. Landhusholdningsselskabs Forhandlinger*, 1853, pp.33–4.

promising, several brickworks rapidly entered drainpipe production, and in 1855 at least fifteen works in Zealand alone made drainpipes. A good fifteen years later, in 1872, drainpipes were manufactured at a minimum of ninety-eight works, and at that time drainpipes accounted for between 10 and 20 per cent of the Danish brick and tile industry's total production.[25] The manufacture of drainpipes had an important side-effect on the Danish brickworks since the thin-walled pipes required a more thorough blending of the clay; for this reason drainpipe production promoted the acquisition of pug mills and blunging machines. Blunging was used, presumably according to the English method, in cases were the clay was of inferior quality due to considerable amounts of gravel or sand. In 1855 two horse-driven blunging machines were in operation, one at the Gisselfeld brickworks in southern Zealand and one at the Brochdorffs works in south-east Jutland. The first steam-powered blunging machine was installed in 1857 at the Aldersro brick works in Copenhagen, where the clay was of very low quality. This machine was of British origin, as was most of Aldersro's machinery. The new acquisition was described thus: 'a blunging machine with an upright iron shaft with four wooden arms and two fork-shaped iron harrows plus two cast iron rollers, which move along a paved track which is enclosed by a wooden vessel'. Apart from the two rollers, which might suggest that the fire valuation officers had confused a grinder with iron wheels with a blunging machine, the construction described matches a type of blunging machine which was later widely diffused. In 1862 the Hakkemose brickworks also acquired a blunging apparatus supplied with a steam engine.

4 Steam Power and Brick Machines

Denmark's first steam-powered brickworks was the Sorthat works on Bornholm, built in the years 1853–5 by P. F. Lunde.[26] An

---

25. Industrial censuses of 1855 and 1871–3.
26. LA (State Archives), Brandtaksationer Bornholm, Vester Herred, 10 December 1853 and 5 March 1857; M. Jespersen, 'En skitse af Sorthat Kulværk på Bornholm', *Indbydelsesskrift til de offentlige eksamina i Rønne højere Realskole*, 1866, pp.3–38; B. Tornehave, 'Teglproduktion på Bornholm', in R. Egevang and J. Thoms (eds), *Rønne – købstad i 650 år*, Rønne, 1977, pp.107–24. At the other coal works on Bornholm, the Hasle Kulværk, the steam engine was from 1853 powering not only the mine pumps but also a vertical pug mill at a small, adjacent brickworks. See LA (State Archives), Brandtaksation Bornholm, Nørre herred, 5 January 1854.

8-horsepower steam-engine pulled the clay preparation machinery, which comprised a rolling mill and a crushing machine, and also two drainpipe machines. All the machines were made at Lunde's own machine works in Copenhagen. The preparation of the clay was no longer carried out by a pug mill. The crushing machine was presumably a grinder with iron wheels; this machine would grind clay by driving two cast iron wheels around inside a fixed cast iron bowl, or by rotating the bowl. The process of shaping the bricks, however, seems to have been carried out in the traditional method of hand moulding. The steam-engine and the drainpipe machine clearly point towards Britain, but the rolling mill and the clay-crushing machines may equally have been constructed after British models.

A couple of years after Sorthat, mechanisation gradually started at other Danish brickworks. In 1857 the Zealand works of Nivågård and Aldersro both acquired steam-engines.[27] At Nivågård the engine was used for kneading the clay, at Aldersro the steam-engine pulled, in addition to the blunging machine mentioned above, one clay-cleaning machine and four brick machines. The clay-cleaning machine and the most powerful of the brick machines came from the London machine makers G. Fletcher, whereas the three other brick machines came from Claytons. All four brick machines were presumably piston presses of a type similar to drainpipe machines. It was possible to adjust these machines to make 'hollow bricks', and one of them was described as a combined brick and drainpipe machines. In 1859 the Zealand works of Klostermosegård also acquired a steam-engine, and in 1862, as we have seen, the Hakkemose works followed suit. From the middle of the 1860s the pace quickened, and by 1872 as many as twenty-two Danish brickworks used steam power, with a total of 213 horse power.[28]

The machines so far discussed could perform only single operations. By combining the single processes and making them continuous, further savings could be achieved, and many attempts were made particularly to combine the preparatory process of

27. Industrial census of 1872; LA (State Archives) Indenrigsministeriet, 1. Kontor, Jnr. 2440, 1857, and Brevbøger K 1310, 2739, 3440, 3074; LA (State Archives) Brandtaksationer Copenhagen, Ubd. Klædebo, 10 November 1857, 7 September 1858, 6 May 1859, 31 July 1862, 6 June 1865, 28 May 1867, 19 November 1867, 9 May 1868.
28. RA (National Archives), Toldinspektørernes Årsberetninger, 1859; LA (State Archives), Brandtaksationer Kbhs. Amt, Smørum herred, 6 November 1862; industrial census of 1871–3.

Figure 9.1 Clay cleaning machine. (Made by G. Fletcher, London, for Aldersro, 1858.) *Below* Grinder with iron wheels.

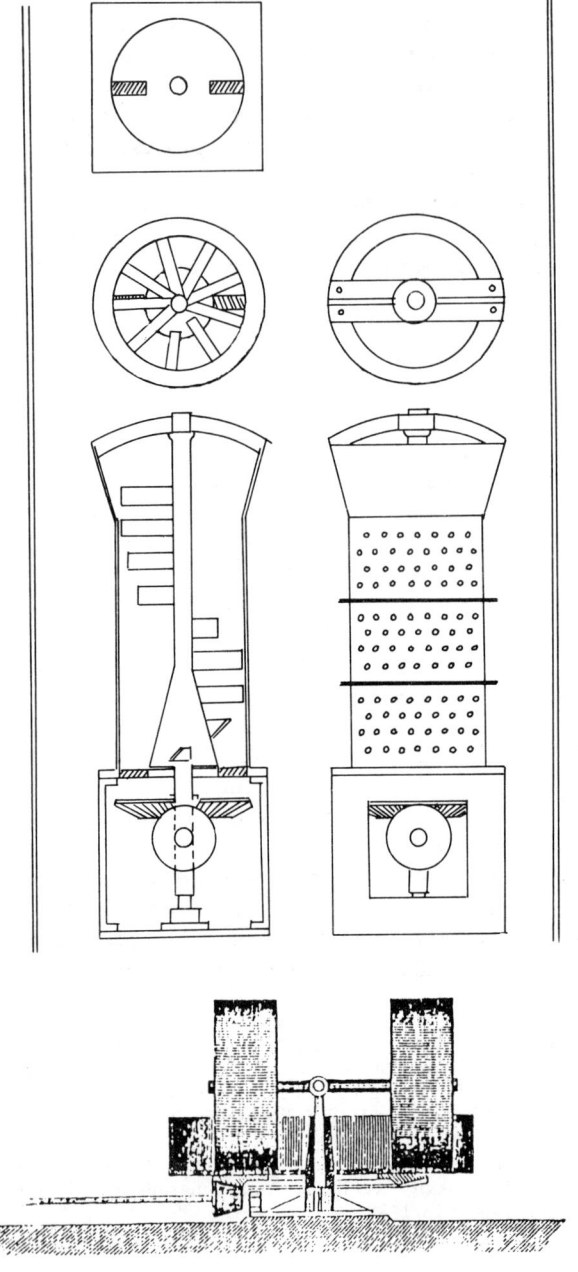

kneading with the moulding operation in one single machine. An early step was taken in 1850 by the Hakkemose brickworks, which was in the process of changing towards 'machine-powered' production. At Hakkemose the pug mill was driven by a horse gin which was placed outside the machine house. Underneath the pug mill was placed a hand-operated brick moulding machine made of iron and with wheels and rollers. The clay passed through the moulding machine, was turned into raw bricks and then transported outside the building by means of a screw mechanism and a horizontal belt. According to his son, the owner J. P. Langaard had himself constructed this brickmaking machine. From the description this new machine seems to have been similar in construction to the moulding machine with rollers for which the Briton I. Hunt in 1843 obtained a five-year monopoly.[29] The combined kneading and moulding machine at Hakkemose was not, however, a success. In 1855 the works reported somewhat laconically to the industrial census that in former days they used to employ machines.

Towards the end of the 1850s the German firms Sachsenberg Brothers and C. F. Schlickeysen succeeded in constructing a combined kneading and moulding machine which was of practical application. The machine gradually gained acceptance during the 1860s and put its mark on the development of the next few decades. This innovation initiated widespread conversions of both pug mills and rolling mills into continuous brickmaking machines, in which the clay was squeezed out like a sausage through a mouthpiece and thereafter cut into suitable sizes by means of a steel wire arrangement.

The development of rolling mills into brick making machines, combining clay preparation with brick moulding, was pioneered by Sachsenberg Brothers at Rosslau, near Berlin. At the beginning of the 1860s their machine, powered by a 3–4-horsepower steam engine, produced between 8,000 and 10,000 bricks in the course of a working day of ten hours. The firm did not take out a Danish patent. Instead, in 1866, Anker Heegaard's firm at Fredriksværk in Zealand was granted an eight-year monopoly for a very similar machine. This firm also delivered mobile steam engines and various brick manufacturing equipment to a number of brickworks in west Zealand.[30]

29. LA (State Archives) Brandtaksationer, Kbs. amt, Smørum herred, 24 June 1850, 19 November 1856, 30 March 1858, 6 November 1868, 3 May 1869, 4 October 1872 and 13 December 1872; RA (National Archives) Generaltoldkammer og kommercekollegiet, Industrifaget, Jnr. 154, 1843; Chr. Langarrd, En *Hundredeårsdag*, Copenhagen, 1911.
30. As note 15, RA (National Archives) Klasseinndelte patentsager, 1864–94,

However, as a consequence of the poor blending ability of the rolling mills, it was the pug mills, further improved by C. F. Schlickeysen in 1855 (see Figure 9.2), which spread in great numbers to the brickworks under various designations such as brickmaking machines, moulding machines and auger machines. Schlickeysen's first machines were mounted upright and designed for horse power. In the beginning of the 1860s, a 2-horsepower machine could produce 4,000 bricks in one day.[31] We lack more detailed information about the pace of diffusion of brickmaking machines in Denmark. This would be difficult to reconstruct because of the unclear terms. We know that in 1866 Klostermosegård used a moulding machine, and in the same year a brickmaking machine was valued for fire insurance purposes at Gustav Kähler's works in Zealand.[32] The Sandberggård brickworks in south-east Jutland reported to the Industrial Exhibition in Copenhagen in 1872 that they used two to three horses to drive the 'Schlickeysen machine'. Employing fourteen workers, this was one of the smaller enterprises; this information may therefore indicate significant spread, even though it was quite exceptional that such small brickworks were represented at industrial exhibitions in the capital.

Even at the beginning of the 1870s it appears that the process of brick production in the main remained separated into two distinct processes; preparation of the clay in pug mills, followed by hand moulding. This seems to have predominated also at the larger works. At the Hakkemose brickworks in 1872 there were five mechanically driven pug mills operating in the various drying sheds, and in the same year the Fredriksholm brickworks in Copenhagen employed as many as eleven mechanically driven pug mills.[33]

5 Artificial Drying and All-year Operation

The limitations of seasonal production constituted a bottleneck for increasing production in response to the growth in demand. All-

kl. 80b, 24 July 1866; F. Borup, *Teglværksindustri*, Copenhagen, 1886, pp.21–7; Kerl, *Handbuch*, pp.545–63.
31. Ibid. and Singer, *History*, p.669.
32. RA (National Archives) Toldinspektørernes Årsberetninger, 1866; LA (State Archives) Korsør Rådstue, Brandforsikringsprotokoll no.279, April 1866.
33. Erhvervsarkivet (National Archives of Trade and Economy) in Århus, In-

Figure 9.2 Two types of continuous brick making machines

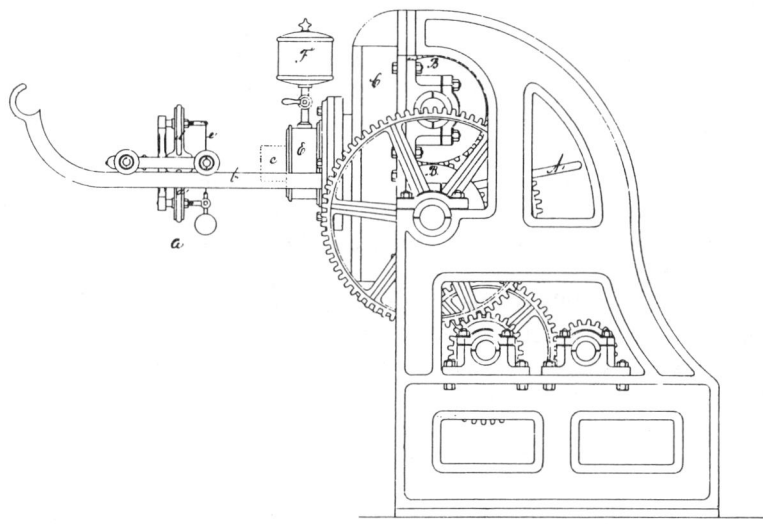

Rolling mill from A. Heegaard, Frederiksværk, 1866

C. F. Schlikeysen's vertical horse-driven machine, developed from the pug mill

---

dustriforeningens archives; Bedømmelseslister (judges' reports) for the industrial exhibition in 1872; LA (State Archives) Brandtaksationer for Sokkelund herred, 5 November 1872.

year production, which was the norm in most other industries, would give considerable advantages for plants with increasingly expensive capital equipment. Furthermore, the costs of transporting to and from the drying sheds would also be reduced, and each brickworks could build up a permanent well-trained workforce. We find that a couple of enterprises tried to pursue this line of approach.

The main obstacle to all-year production was in the drying of the bricks. It was difficult to dry the raw bricks in the damp and cold autumn, and with ordinary open air sheds a single frosty night could ruin the production.

From the middle of the 1850s the pioneering works at Sorthat and Aldersro both attempted to solve this problem, and were, to some extent adapted to all-year production. At Sorthat the common, open air drying sheds were supplemented with a large brick-built winter drying house with a tiled roof. Inside were four brick-built vaults, of which two had steam drying pipes of cast iron, whereas the other two had brick-built pipes with fire doors of iron. The brick-shelves travelled on iron rail tracks which ran through the building. From the description, it is not clear how the drying house functioned. Apparently, two sources of heat were used; one in the form of direct heating from inside the house, the other by the use of surplus heat from the steam engine. Four blowing machines in the engine house blew the steam to the drying house where it was dispersed by another two blowing machines. On the whole, the construction seems to be very similar to central heating installations of that time, with which P. F. Lunde was already acquainted.

In contrast, at Aldersro (see Figure 9.3) the system was based on surplus heat from the brick kilns. Four large brick-built drying houses were erected close by the kilns. Through each of the drying houses ran brick heating pipes which were connected with the kilns through pipes embedded in the ground. The required draught was supplied by one large chimney. When the kiln reached the stipulated temperature, the warm air was channelled into the drying houses and then returned back through the kiln and up the chimney. This British construction was an expensive investment; the four drying houses alone were estimated at 17,600rd.

From an international point of view, these two drying installations were remarkably early. The engineer O. Bock, who was born in Denmark but mostly worked in Germany, wrote in 1901 that the oldest winter drying installations known to him were built in Germany in 1858, and several handbooks refer to an 1867

Figure 9.3 Artificial drying installation at Aldersro, 1857 (Vertical and horizontal views.)

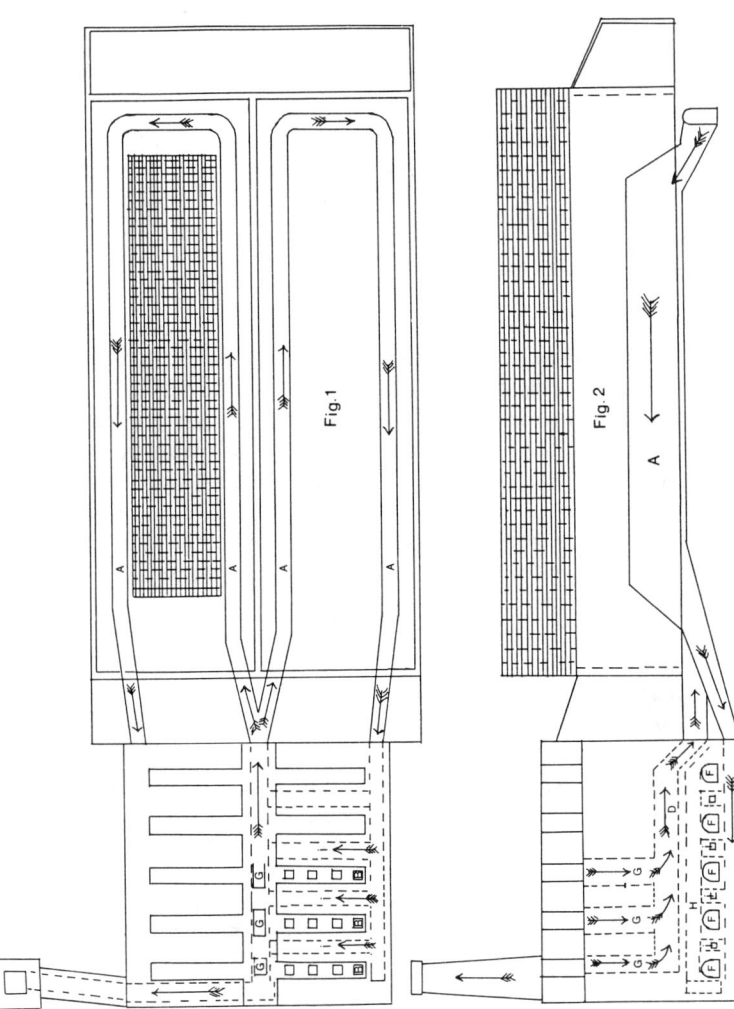

Fig. 1   Fig. 2

Key:
A: heat pipes
B: smoke holes
F: fire hatches
G: connecting pipes

construction built by the German Mensing as the first. Mensing's installation was, in common with the system at Aldersro, also based on surplus heat from the periodic chamber kilns.[34] However, the two early Danish attempts were not successful. Sorthat was closed down in 1858. Although savings were made by Sorthat being self-sufficient in coal, nevertheless it seems unlikely that the method of drying, being essentially a kind of central heating system, would ever be cost-effective. At Aldersro the estimated value of the drying houses was halved in 1865; the heating installations presumably no longer functioned. Two years later the buildings were replaced by four open drying sheds. Even so the drying principle at Aldersro was a considerable advance. It must, however, have been very difficult to get a more or less steady supply of heat to the drying houses, as the heat was generated by periodic kilns running intermittently.

6   Kiln Technologies.

Fuel supply proved to be another serious bottleneck hampering expansion of the established method of production. The chamber kiln used considerable amounts of fuel, largely as a consequence of running intermittently. Cold air was sucked in from outside, and most of the energy supplied subsequently vanished with the smoke or during the cooling of the kilns. In 1866 accounts from thirteen brickworks in the Helsingør district in Zealand show that normal expenditure for fuel amounted to between 30 and 40 per cent of the value of the finished bricks at the larger works, and even exceeding 50 per cent at many smaller peasant brick works in the countryside. As a rule, the baking of 1,000 bricks consumed 1,500 pieces of peat or about 400 kg of coal.[35] The large fuel consumption not only made the bricks more expensive, but was also a socio-economic burden in a country with only small energy resources. With a total production estimated at 227,000,000 bricks in 1872, the process of baking in chamber kilns required about 91,400 tons of anthracite coal, the equivalent of approximately 28 per cent of total Danish imports of anthracite coal in 1871–2.[36] The fact that some of the

34. O. Bock, 'Om den kunstige tørring og dens udvikling', *Nordisk Tidsskrift for Lervare- og Sten-industri*, 1901–2, pp.23–32; Bock, *Ziegelei*, pp.44–67; Kerl, *Handbuch*, pp.597–631, especially pp.612–13.

35. RA (National Archives) Toldinspektørernes Årsberetninger, 1866; Borup, *Teglværksindustri*, p.5; *Nordisk Tidsskrift for Lervare- og Sten-industri*, 1900–1, no.11.

36. See p.201 and *Statistisk Tabelværk*, 3.20, pp.16–17, 48–49.

fuel consisted of domestic peat and firewood did not make the consumption less. This was a cause of concern at the time. In 1843 the Technical University made a statement to the effect that fuel consumption in brick production was excessive, and that in some parts of the country the level was alarmingly high.[37] As mentioned above, Hans Jordt and M. H. Holler sought to solve this problem by developing a kind of tunnel kiln using less fuel in the 1840s; however, the attempt to develop this into a profitable technology did not succeed (see above, p.207).

Instead, in the 1840s and 1850s the leading brickworks concentrated on developing the old chamber kilns. Kilns with increased capacity, exceeding 100,000 bricks per baking, were gradually introduced. Furthermore, improved fuel economy and operational dependability were achieved partly by building vaulted kilns and double kilns, and partly by constructing brick-built firing channels and permanent fire grates. The use of firing channels and grates was linked with the gradual change-over to coal firing.

It is difficult to point to definite foreign models for these less radical improvements. The building of vaulted kilns probably followed German models, while the use of fire grates was probably inspired by Britain. Inspector Friedhling, who managed Gisselfeld's three brickworks in the southern Zealand, reported in the 1850s that he had ordered fire grates from Britain.[38]

The breakthrough for a new, fuel-saving kiln came with the circular, or ring, kiln which the Germans F. Hoffmann and A. Licht patented in 1858.[39] The circular kiln consisted of a circular or oval-shaped brick-built channel, divided into twelve to twenty-four chambers. Each chamber had an entrance gate and a smoke channel which led into a joint smoke channel and from there onwards into a tall chimney. It was fired through firing holes at the top of the kiln. The baking was continuous since the firing was moved on, about every twenty-four hours, to the adjoining chamber. Considerable fuel savings were achieved partly by preheating the incoming air and partly by preheating the 'green bricks' (see Figure 9.4). Whereas the baking of 1,000 bricks previously used about 400 kg of anthracite coal, the circular kiln managed the same quantity with approximately 160 kg. In addition, the continuous operation made it possible to increase production, and production required a

37. Olsen, 'Kanalovnen – Tunnelovnen', pp.96–9.
38. Nielsen, *Industriens Historie*, p.188.
39. Bock, *Ziegelei*, pp.87–129; Kerl, *Handbuch*, pp.686–707; A. Olsen, 'Ringovnen og dens anvendelse', *Lerindustrien*, 1907, pp.20–3.

Figure 9.4 Circular kiln with fifteen chambers

Operation of the kiln: (a) entrance gate, (b) smoke channel, (c) chimney, (d) moveable iron plate. Raw bricks are placed in chamber 15, while the baked and cool bricks in chambers 14 and 13 are unloaded. Firing takes place in chambers 6 and 7, while bricks in chambers 8 to 12 cool down. Cold air directed through chambers 8 to 12 warms up and is first directed to the firing chambers; then onwards to the chambers 5 to 1 it pre-heats the raw bricks for finally being passed up the chimney. After approximately 24 hours the iron plate (d) is moved on to separate between the chambers 15 and 14 and the process repeated.

smaller area. A further advantage was that a number of bricks could be dried on the top of the permanently warm circular kiln.

The iron-founder P. F. Lunde maintained in 1888 that in 1855, or three years before Hoffmann, he had built the world's first circular kiln at Sorthat.[40] According to Lunde, the invention was subsequently transferred to Germany by German workers returning home at the expiry of their employment at Sorthat. On the basis of a fire insurance valuation in 1857 and Lunde's own descriptions from 1888, it is possible to describe the general outlines of the kiln. It consisted of six small 'French kilns' built together in two rows. These kilns, which could each take about 10,000 bricks, were interlinked by a system of vents. From each kiln a second vent led to a joint smoke funnel and onwards to a 19 m tall chimney, which was situated to the north-east of the building. When kiln no.1 was

40. P. F. Lunde, *Hovedtræk af et levnedsløb*, Copenhagen, 1888, pp.26–9.

fired, the vent between nos. 1 and 6 was shut, while those between the others were open. Thus, five kilns were heated free of charge. Clearly, this installation had several features in common with Hoffmann's circular kiln, for example the adjoining and interlinked chambers, and the joint outlet through a smoke funnel and chimney. On the other hand there were also obvious differences. With only six kilns, continuous operation was hardly possible, and consequently recirculation of hot air more limited; the heat from baked bricks could not be used to preheat the incoming air, for example. In addition, Lunde's kilns were interlinked by means of vents, and they were not, as were Hoffmann's, designed in the shape of one large channel. Finally the firing would not have taken place through the top, which was a characteristic feature of the circular kiln. The Sorthat kiln was thus not a fully developed circular kiln and could not match the later Hoffmann kilns in fuel economy. It must be considered as a forerunner of the circular kiln, just like a number of similar contemporary constructions in Britain, France and Germany. Even so, it was a considerable technical achievement.[41]

From the beginning of the 1860s Hoffmann's invention became generally known among Danish specialists. In 1863 the circular kiln was described in detail in two Danish trade journals, based on German and Swiss sources, and in the following years it frequently figured in the technical literature.[42] The first circular kiln in Denmark was erected during the winter of 1866–7 at Gustav Kähler's brickworks at Korsør in Zealand. It was an impressive construction; the kiln measured 26 m in diameter and comprised twelve chambers, each containing about 20,000 bricks. In the middle of the kiln was a 34 m tall chimney. One surprising feature was that the drawings for the kiln came from a British architect, while a local mason built it.[43] As early as the spring of 1867 a second kiln was built as part of modernising the Aldersro brick works. This kiln may also have been of British origin.

Circular kilns were subsequently acquired by Hakkemose in 1868, by the Klostermosegård, Tårnborg and Svenstrup works,

41. P. Loeff, 'Ueber die Entstehung ringförmiger Ziegelöfen', *Zeitschrift für Praktische Baukunst*, Berlin, 1871, pp.119–34.
42. See note 15; 'Ringovn for teglstenfabrikation eller kalkbrænderi', *Journal for Håndverks- og Fabriksdrift*, 1863, pp.99–103, 116–17, 180–1; A. T., 'Ringformige teglovne', *Tidsskrift for Physik og Chemi*, 1863, pp.11–14.
43. LA (State Archives) Brandtaksationer Korsør Rådstue archives, 19 September 1867; 'Teglværksdriften ved Korsør', *Berlingske Tidende*, 18 January 1867; G. Kähler, *Ungdomserindringer*, Copenhagen, 1907, pp.42–3.

Ole Hyldtoft

near Korsør, in 1869, followed by Nivågård in 1870; thus, in 1870 there were at least seven circular kilns in operation in Zealand. During the first years of the 1870s the pace quickened, stimulated by a nearly explosive rise in coal prices. What was characteristic of this group of circular kilns was that it represented a breakthrough for German influence. The installations at Tårnborg, Svenstrup and Nivågård were referred to by contemporaries as 'Hoffmann kilns'.[44] Hoffmann did not take out a Danish patent for his kiln. The earliest Danish patent for a circular kiln was acquired by the Swede C. V. Kull from Höganäs in 1867. A distinctive feature of his kilns was that they were arranged on two floors. Presumably Kull's kiln was never built and operated in practice in Denmark. In Sweden the first circular kiln was constructed as late as 1873 by the Skåne Cement Company (Skånska Cement Aktiebolaget) in Lomma. In this important area, the leading brick works in Denmark were also more advanced than in the Duchies; in 1874, when the first circular kiln was installed in the Egernsund area, circular kilns were employed at something like twenty Danish brickworks.[45]

7 Summary

The technological development of the Danish brick and tile industry in the years 1830–70 was a combination of developing and extending the method of production already established at the start of our period, in response to strongly increased demand, and the introduction and development of new techniques and machinery. The brickworks became increasingly distinct from agriculture, and established as an independent industry. From the end of the 1840s some new machines were introduced at the larger works. This was primarily a matter of relatively simple machinery, such as pug mills and drainpipe machines, where each machine managed one single process.

From the middle of the 1860s modernisation took new and more radical forms at the leading works. Brickmaking machines, com-

---

44. LA (State Archives) Brandtaksationer Helsingør Rådstue, 25 June 1869; Lynge-Krone herred, Hørsholm distrikt, 13 April 1870; industrial census of 1872; P. Strømstad, *Ringovnen ved Nive å*, Karlebo, 1985.
45. RA (National Archives) Klasseinddelte patentsager, 1864–94, kl. 80c, 19 September 1867; W. de Shårengrad, 'Bränning i ringugn', *Nordisk Tidsskrift for Lervare- og Stenindustri*, 1904–5, pp.84–5. I am indebted to S. O. Christensen for information relating to Egernsund.

bining kneading and moulding, were introduced, as were the fuel-saving circular kilns. Both technologies pointed towards more continuous production. At the same time more works changed over to steam power.

Particularly during the first decades of the period the Danish brick and tile industry was influenced by the works near the Flensborg Fjord in the Duchies. However, this influence appears to have been a matter of the Egernsund works passing on a particularly highly developed form of traditional production modelled on the Dutch brick and tile industry. This was to a large extent also true of the German influence, which was carried to Denmark by the Lippe-Detmolders who worked in Danish brickworks during the summer seasons.

While the southern influence is generally recognised in the literature, the simultaneous, strong British influence has largely been ignored. That A. Nielsen overlooks the British influence is partly due to the nature of his source material, which in the main includes small and medium-sized estate brick works in Zealand, and partly because his treatment in effect concentrates on the period between 1820 and 1850. The British model is clearest in connection with drainpipe machinery and production, but some pug mills, piston presses and blunging machines can also be traced back to British techniques. Furthermore, one of the largest works in the 1850s, Aldersro in Copenhagen, was built under the supervision of a British engineer and supplied with predominantly British machines. British and French brick and tile technologies still occupied prominent international positions in the 1850s. In 1857, when Magnus Jespersen went abroad to further his education, having left his position as manager of the Sorthat works, he travelled not to Germany, but to Britain and France. He visited, among other places, Paul Borris's brick and tile works in Paris, the porcelain manufacturers at Sèvres, and Clayton's famous machine factory in London.[46] Towards the end of the 1850s German technicians seriously began to set their mark on the brick and tile industry by the introduction of the continuous brick machines and the circular kiln. Even so, and despite the traditionally very close links between Danish and German professional circles, it was British drawings which were used in the construction of the first, and possibly also the second, circular kiln in Denmark in 1866–7.

The Danish brick and tile industry was not at the frontier of technological development. Nevertheless it seems that the Danish

---

46. Garboe, *Magnus Jespersen*, pp.64–70.

works were operating at a reasonably high technological level. The leading works, that is the larger commercial works with good communications with Copenhagen, were quick to exploit the most important of the new technologies coming from abroad. Subsequently, and within a fairly short period of time, Danish engineers succeeded in copying the machines which were still of fairly simple construction. Among the more independent contributions were Hans Jordt's tunnel kiln from 1840 and P. F. Lunde's brickworks at Sorthat in the mid-1850s. These early attempts at applying advanced techniques were not, however, as is often the case, economically successful.

The transfer of foreign technologies to Denmark took place through a number of different channels. Information about current technological developments was sought through technical journals, patents and foreign machine suppliers. Even more important than this was, it seems, direct personal contact. Personal contact was established by Danish engineers visiting main international industrial exhibitions and travelling and studying abroad, and by recruitment, for longer or shorter periods, e.g. of foreign brickmakers and engineers to work in Denmark.

Considering that Denmark possessed a good system of international contacts, it may seem surprising that the diffusion of continuous brickmaking machines and the circular kiln did not begin until the mid-1860s, that is, after a ten-year lag. A number of factors contribute to an explanation of this perhaps surprising situation.

Firstly, the brickmaking machines and the circular kiln suffered from 'teething problems' which it took a long time to solve. The Sachsenberg brickmaking machine, or rolling mill, combining clay preparation with brick moulding, was relatively inefficient in blending the clay. The brickmaking machine developed from pug mills had a tendency not to disintegrate lumps and stones, which therefore passed straight through the machine. Furthermore, on both machines the cutting devices caused problems. Quality problems were also connected with the circular kiln which frequently produced impure and stained bricks; the reason for this was that the preparatory process of 'smoking' the raw bricks was not separated from the firing in the original version of the kiln.

Secondly, the crisis in 1857 far from encouraged expensive experiments on top of the investments already made during the preceding years. From the mid-1860s onwards demand again showed an upward-moving tendency. Nevertheless brick prices remained at a low level until 1872. In the years 1862–6 the problems intensified as coal prices rose sharply.[47]

Thirdly, difficulties arose in the supply of skilled German labour, owing to strong anti-German feelings in Denmark after the war in 1864. At least in theory, the new and developing technology was, so to speak, tailor-made to solve these difficulties – the rising coal prices, increasing demands and problems with the supply of German labour – which culiminated in 1865.

47. J. Pedersen and O. S. Petersen, *An Analysis of Price Behaviour during the period 1855–1913*, Copenhagen, 1938, pp.246–9.

10

The Norwegian Mechanical Engineering Industry and the Transfer of Technology, 1800–1900

Kristine Bruland

1 Introduction

From the early nineteenth century Norwegian engineers were able to establish a dynamic and solidly based mechanical engineering industry, which grew to become a core industry of Norwegian industrialisation. But how did they do it? What kinds of activities lay behind the construction of this new industry? In this chapter we argue that technology transfer – that is, the acquisition and adaptation of foreign technologies and engineering skills – was a central element in the development of this Norwegian industry. Norwegian engineers and entrepreneurs were alert to the extraordinary development of foreign engineering technologies in the early nineteenth century, and to the opportunities these technologies provided. With great energy they learned the key techniques and skills being pioneered abroad and successfully implanted them in Norway. What was the main source for this import of technology? As we shall see below, Britain was overwhelmingly important as a technological exemplar for Norway. Moreover it remained so to the end of the nineteenth century, despite a German challenge which seems to have emerged only slowly. This technology transfer from Britain involved a wide range of activities: foreign travel, purchases of foreign equipment, foreign education, contacts with foreign firms, agency relationships and so on. The technology transfer process was thus a complex one. But it was, and remained, a core element in the success of Norwegian mechanical engineering as a whole.

2 Technology Transfer and Industrial Development

The modern industrial era began in Britain during the later part of the eighteenth century, with the emergence of a mechanised manufacturing industry based on continuous technological change and continuous productivity growth. Since then the world economy has been characterised by the existence of technological 'leader' and 'follower' economies, and by the transfer or diffusion of technologies from the former to the latter.[1] Technological leadership has, of course, changed hands several times during the past 200 years, often with momentous economic and political results: by the early twentieth century, British dominance had been supplanted by that of Germany and the USA, which in turn have been challenged by the economic rise of Japan and the Far East. For follower economies, especially small economies such as Norway, the identity of the leader at any particular time is perhaps less important than the follower's ability to imitate, adapt and utilise the technological innovations and advances on which leadership is based. Small countries can never hope to match the levels of investment, scientific activity and applied research and development of the major economies. Whether or not they become industrialised and remain at the forefront of advanced industrial performance, therefore, depends in large part on whether they can develop the skills to use technologies developed abroad. This is not just a matter of buying machines. It involves the development of 'technological capability', which Fransman has defined as follows:

> By a technological capability [we] mean the ability, embodied in people, to select the appropriate technology; to implement it; to operate the production facilities so implemented; to adapt and improve them, and possibly to create new processes and products. A technology may be purchased, but a technological capacity is operated only through the build-up of human capital.[2]

---

1. A. Maddison, *Phases of Capitalist Development*, Oxford, 1982, chap. 1.
2. M. Fransman, Introduction, in M. Fransman (ed.), *Machinery and Economic Development*, London 1986, p.xiv. Myra Wilkins presents a similar point of view, but refers instead to 'absorption' rather than 'capability': 'It is worth considering the difference between mere transfer and the absorption of technology within the host country. A company can export capital goods. In one country the machines installed might be allowed to break down and eventually fall into disrepair; in another country, the same machines might be used efficiently in modern industry, copied, adapted or produced locally' (M. Wilkins, 'The Role of Private Business in the International Diffusion of Technology', *Journal of Economic History*, (vol.39, 1974, p.171).

In this chapter we treat 'technology transfer' or 'technological diffusion' as the ability to use such technological capability to deploy and develop technologies developed abroad. In fact relatively few countries have proved able to develop advanced technological capabilities or capacities in the sense used by Fransman, which in turn suggests that the development of such capabilities is a complicated and difficult process. The apparently simple process of technological diffusion, like that of technological development itself, appears to involve a particularly complex interweaving of cultural, educational, legal, economic and political factors, which is perhaps why few if any really adequate histories of technological change and development in the West have yet been written.[3] This chapter concentrates on one particular aspect of this process, namely the acquisition of skills in mechanical engineering, and the role played by the mechanical engineering industry in developing what we might call 'diffusion capability' within a small economy.

Mechanical engineering is a particularly important industry for industrial, technological and economic development as a whole. One key reason for this is that *product* innovations by mechanical engineering firms, for example in new machines and equipment, become *process* innovations when they are put to work by user firms; this can have a wide impact on technological change and productivity growth within the economy. Competition among mechanical engineering firms in terms of products thus embeds process innovation within the wider economy. The immense historical significance of this was emphasised strongly some years ago by Nathan Rosenberg:

> In both the US and the UK in the nineteenth century, technological change became institutionalised in a very special way – that is in the emergence of a group of specialised firms which were uniquely oriented toward the solution of certain kinds of technical problems. The rapid rate of technological change was completely inseparable from these capital goods firms. In fact I would regard the emergence of such firms as the fundamental institutional innovation of the nineteenth century from the point of view of the industrialization process.[4]

The development of specialised mechanical engineering firms

3. Thomas P. Hughes, 'History of Technology as Modern History', *ISIS: Journal of the History of Science Society* (forthcoming).
4. N. Rosenberg, 'Economic Development and the Transfer of Technology: Some Historical Perspectives', in idem, *Perspectives on Technology*, Cambridge, 1977, p.152.

was clearly an important component of accelerated industrialisation in Britain during the early nineteenth century. But it had an international impact as well, for British capital goods enterprises increasingly sought markets abroad. Particularly after 1843, when prohibitions on some types of machinery exports were removed, this market-seeking by British engineering firms opened up possibilities for 'follower' economies to develop a range of industries using British techniques. In Norway, for example, a mechanised textile industry rapidly developed from the mid-1840s on the basis of 'packages' of technology – comprising technical information, equipment, skilled labour and managerial expertise – assembled by British textile engineering firms.[5] This was but one component of a general diffusion of technology from Britain at that time. Another important part of the spread of industrialisation was the spread of mechanical engineering industries themselves into those countries which succeeded in industrialising in the middle and late nineteenth century. Since mechanical engineering industries 'undertake technological change and adaptation as a matter of routine',[6] as Rosenberg puts it, this spread was of central significance in European industrialisation. But how did this happen, and how did the mechanical engineering industry facilitate the further spread of industrial technologies into growing economies during the nineteenth century? These are the fundamental questions which are asked in this paper.

In discussing the mechanical engineering industry from the point of view of technology transfer, we need to distinguish between two quite distinct phenomena. The first concerns the historical origins of the industry, while the second concerns its wider economic impact. These technology transfer issues are:

1. What role did technology transfer play in actually establishing the mechanical engineering industry in Norway? For example, how did embryo European (in this case Norwegian) engineers acquire both the general technical knowledge and the specific technical skills associated with the new technology developed in Britain, how did they acquire the competence to run specific techniques, what particular types of equipment were imported from abroad, and what types of labour, skill and managerial input

---

5. K. Bruland, *British Technology and European Industrialisation: The Norwegian Textile Industry in the Mid-nineteenth Century*, Cambridge, 1989.
6. N. Rosenberg, 'The Direction of Technological Change: Inducement Mechanisms and Focussing Devices', in idem, *Perspectives on Technology*, Cambridge, 1977, p.99.

were imported from abroad? In examining this question below, we shall be principally concerned with Norwegian absorption of technologies from Britain; later sections will deal with Germany.
2. What were the implications of the new mechanical engineering industry for the wider technological development of Norwegian industry? Specifically, how did the mechanical engineering industry shape the diffusion of foreign technological practice into other industries internally, by acting as an 'entry point' for foreign technologies?

The principal objective of later sections of this chapter will be to answer these questions empirically. Before doing so, however, we move to a brief description of the development of mechanical engineering in Britain, since it was Britain, as world technological leader, to which early Norwegian engineers looked first, and it was the British industry which was the source of virtually all of the early industrial technology deployed in Norway.

3   The Development and International Activities of Mechanical Engineering in Britain

The development of a substantial mechanical engineering industry in Britain in the early nineteenth century was perhaps the most significant long-term effect of the industrial revolution of the late eighteenth century. Early British industrialisation rested on a very narrow industrial base, with three 'staple industries' of coal, textiles and metal manufacture being the first to develop. Outside these core industries, output and productivity growth was definite but slow, and both investment and the use of new techniques lagged significantly. But within the core industries, growth was spectacular. In textiles, where expanded input supplies and falling input prices combined with new spinning techniques to produce high output growth, the industry grew from 1770 at a long-run average compound rate of approximately 3.5 per cent. This rate is sufficient to double output every twenty years. But in the first half of the nineteenth century textile growth accelerated: between 1810 and 1842 the growth rate averaged almost 5.5 per cent per year, and between those years annual output increased by over 500 per cent.[7] Over the same period, the capital equipment of the industry

7. These growth rates are calculated from P. Deane and W. A. Coles's estimates, *British Economic Growth 1866–1959*, 2nd edn, Cambridge, 1978, p.191.

increased at an even faster rate: both the number of cotton spindles
and the total capital stock increased at approximately eight per cent
per year.[8]

These growth rates were unprecedented, and were associated
with a massively increased demand for machinery, power equipment and tools. This demand played a crucial role in the development of the fourth 'staple industry', namely engineering. 'In the
long run', as Landes has pointed out, 'the diffusion of mechanized
manufacture called forth major improvements in tool design.'[9]
During these years a combination of advances in iron manufacture,
new techniques in precision engineering pioneered by a small but
influential group of engineers[10] and a flow of labour from such
occupations as instrument and clockmaking, led to the establishment of a large number of engineering firms. Since the textile
industry was by far the largest market, a large part of the industry
was concentrated in Lancashire: by 1841 Lancashire had 115 engineering enterprises, employing over 17,000 workers.[11] By 1851,
according to the population census of that year, over 63,000 workers
were employed in making machinery or boilers. Employers' returns in the same census indicated over 800 'engine or machine'
enterprises; most were very small, but 155 firms employed over
twenty men each, and 34 firms employed more than one hundred
men each. The major firms in the industry were very large indeed:
'By the early 1840s, Fairbairn's were employing a total of between
1,000 and 2,000 men in their Manchester and Millwall works; in the
Atlas works of Sharp, Roberts and Co there were nearly 1,000 by
the early 1850s, while Nasmyth, Wilson and Company were by
then employing 1,500 in their Bridgewater foundry.'[12]

Naturally these enterprises looked not only to domestic but also
to foreign markets. A major constraint on machinery exports,
however, was the existence of laws prohibiting the export of a wide
range of tools and equipment, especially for textile and iron and
steel manufacture. Such prohibitions had existed, in various forms,

8. M. Blaug, 'The Productivity of Capital in the Lancashire Cotton Industry
during the Nineteenth Century', *Economic History Review*, 2nd Series, vol.13, 1961,
p.32; and *Select Committee on the Exportation of Machinery*, vol.7, 1841, Appendix 2,
p.230.

9. D. Landes *The Unbound Prometheus. Technological Change 1750 to the Present*,
Cambridge, 1969, p.105.

10. K. R. Gilbert, 'Machine Tools', in Ch. Singer et al., *A History of Technology*,
1958, vol.4, p.418.

11. K. Bruland, *British Technology*, p.33.

12. A. E. Musson, 'The Engineering Industry', in R. Church *The Dynamics of
Victorian Business*, London, 1980, p.95.

Table 10.1   British exports of machinery to 1856 (£000)

	1814–16	1844–6	1854–6
Steam-engines	–	319	753
Other machinery	28	614	1537
Total	28	933	2690

Source: R. Davis, *The Industrial Revolution and British Overseas Trade*, Leicester, 1979, p.27.

for many centuries, but were significantly strengthened in the 1780s. The regulations did not, in practice, stop the flow of machines and equipment to Continental Europe and North America: smuggling, espionage, emigration and so on led to a steady flow out of Britain.[13] None the less, the regulation constrained the sales of British engineering firms, and this became an increasing problem after the first quarter of the nineteenth century, as the engineering industry ran up against the constraints of the domestic market. A major political and economic debate ensued, as engineers and machine makers sought to repeal the prohibitions: as Mathias has put it, 'Once specialised as an industry in its own right, its leaders claimed the right to export markets of their own. Engineer after engineer argued thus before the Parliamentary Committees on the Export of Machinery in 1824 and 1843.'[14]

The prohibitions were repealed in 1843, and from that time British machinery exporters actively sought foreign markets. The growth of the trade can be seen in Table 10.1. These exports formed only a tiny part of British commodity trade; indeed they made up less than 10 per cent of exports in the 'metal and metalware' category. None the less they were of very great importance in non-British industrialisation, for these exports were a primary source of equipment for continental enterprises. In this sense, Britain genuinely was 'the workshop of the world'. But the trade in machinery ought to be seen, in my view, in two ways. On the one hand, there was the direct supply of equipment, and associated specific 'know-how' which enabled particular plants or production processes to be established. On the other, however, there was a less tangible effect, as foreigners acquired general technological

13. D. Jeremy 'Damming the Flood: British Government Efforts to Check the Outflow of Technicians and Machinery 1780–1843', *Business History Review*, vol.51, 1977, pp.1–34.
14. P. Mathias, *The First Industrial Nation*, 2nd edn, London, 1983, p.110.

capabilities as a result of trade and other relationships with British engineering firms. It was these capabilities which underlay the spread of mechanical engineering itself, and this is the focus of subsequent sections of this chapter. We turn now to the international trade background to the development of mechanical engineering in Norway; later sections will examine the technological flow between British and Norwegian engineering directly.

4 Norwegian Industrialisation, Technology Imports and the Growth of Mechanical Engineering

Even before industrialisation began during the 1830s and 1840s, Norway was in some ways an unusual economy by virtue of its openness to foreign trade: fishing, timber and shipping were major export industries.[15] Two of these, shipping and timber (including especially sawmills), were also industries which, as the nineteenth century progressed, increasingly required the application of industrial techniques and thus engineering products. Even before Norway began the industrialisation phase of the mid-nineteenth century, therefore, there was significant scope for the activities of the mechanical engineering industry.

But fishing, timber and shipping also produced substantial foreign earnings, and Norway was therefore able to import a very wide range of manufactured goods, especially from Britain. Certainly this trade fluctuated according to economic circumstances, but by the mid-1830s Norway was importing large quantities of such products as cotton manufactures and yarn; woollen goods; canvas; coke and coal; pottery; soap; shoe leather; oil; paints; lead; ammunition and explosives; anchors and chains; iron and iron plates, castings and manufactures; steel and steel goods and wire, and so on.[16] Growth of demand in many of these product groups in turn opened up the possibility of import-substituting domestic manufactures, and this probably lay behind the growth of a number of important new industries in Norway. Growth here was interwoven with that of the traditional export industries.[17]

Naturally the industrial sector was, in absolute terms, small, but

15. See E. Hovland, H. V. Nordvik and S. Tveite, 'Proto-Industrialization in Norway, 1750–1850: Fact or Fiction?', *Scandinavian Economic History Review*, vol. 30, no.1, pp.45–56, for the structure of the pre-industrial economy and foreign trade.
16. A. Schweigaard, *Norges Statistik*, Christiania, 1840, pp.164–5.
17. F. Sejersted, *En Teori om den Økonomiske Utvikling i Norge i det 19 Århundre*, Oslo, 1973, p.37.

Table 10.2   Development of Norwegian industries: number of firms

	1865	1870	1875	1879
Metal industry	25	22	56	65
Chemical industry	47	60	141	136
Textile industry	80	107	122	145

Source: *Statistik over Norges Fabrikanlæg* (Kristiania, 1881), p.vii

by 1850 a significant part of the industrial workforce was employed in such industries as textiles, chemicals, pottery, iron foundries and engineering.[18] During the 1860s and 1870s some older industries began to decline in terms of the number of enterprises within them: quarrying, paper, timber etc. all contracted.[19] But in the newer industries growth in the total number of firms continued. As Norwegian industry grew from the 1840s, so in consequence did its demand for machines and equipment. At first much of this was imported. From the early 1840s, imports of machinery, in particular from Britain, accelerated, as Tables 10.4 and 10.5 indicate. Table 10.3 should probably be interpreted as a process of steady growth for the falls in exports in the late 1840s were the result, firstly, of financial crisis within Norway, and secondly, of recession and the profound political upheaval which swept Europe in 1848 and 1849. As Europe recovered from this crisis, from 1850 trade grew strongly, and this too is reflected in Norwegian imports of machinery. From 1850 imports of machinery increased sharply,

Table 10.3   Machinery and millwork exports from Britain to Norway, 1843–50 (£)

1843	1,392
1844	2,483
1845	9,449
1846	15,518
1847	5,270
1848	5,727
1849	4,187
1850	12,175

Source: British *Parliamentary Papers*, 1854–5, vol.52, p.226

18. NOS, *Historisk Statistikk, 1978*, Oslo, 1978, table 42. See Table 10.2
19. See *Statistikk over Norges Fabrikanlæg*, Christiania, 1881, p.vii.

Table 10.4 Imports of machinery by Norway, 1841–64 (thousand kr.) (from 1851, 3-year moving averages)

1841	28
1844	72
1847	93
1850	322
1851	142
1852	171
1853	363
1854	503
1855	549
1856	499
1857	408
1858	218
1859	409
1860	497
1861	467
1862	318
1863	280
1864	328

Source: Statistisk Sentralbyrå, Historisk Statistikk 1978, table 159, pp.276ff.

in value terms, until the rate of growth slowed noticeably towards the end of the decade.

These two series cannot be compared directly, but with an exchange rate of approximately 16 kr=£1 they are consistent with a British share of Norwegian imports of machinery varying between 60 and 90 per cent during the 1840s. The important trend here, however, is stabilisation and then decline of machinery imports from around 1860; since the Norwegian industrial sector continued to grow at this time, Table 10.4 implies that engineering needs were increasingly being met from domestic sources, that is that domestic engineering firms were successfully competing with foreign producers.

This growth of Norwegian mechanical engineering can be seen in Table 10.5, which traces employment from 1850 through to the end of the century. In response to general industrial growth and hence increasing demand for machinery, the Norwegian mechanical engineering industry expanded rapidly from the mid-1840s, and continued to do so until the turn of the century. But this occurred through a sharp upturn during the 1860s, which was subsequently maintained, with fluctuations in employment reflect-

Table 10.5 Growth of employment in Norwegian mechanical
engineering, 1850–1900

	Total engineering employees	Eng. as % of industrial employment
1850	1,368	11.1
1860	1,608	7.8
1865	4,999	17.6
1870	7,161	20.6
1875	10,927	22.6
1879	7,929	18.3
1885	9,570	20.1
1890	13,663	21.4
1895	12,626	20.4
1900	16,790	21.1

Source: NOS, *Historisk Statistikk* 1978 (Oslo 1978), table 42, p.79

Table 10.6 Structure of the Norwegian mechanical engineering
industry (number of enterprises)

	1865	1870	1875	1879
Mechanical workshops	35	23	28	44
Coach manufactories	4	5	8	7
Railway carriage Manufactories			1	1
Shipyards	72	197	200	112
Rifle manufactories			1	1
Musical instruments	3	6	8	5

Source: *Statistik over Norges Fabrikanlæg* (Kristiania, 1881), p.viii

ing cyclical factors. Employment in the industry rose sharply from
1860 in absolute terms but also as a percentage of the industrial
workforce, reflecting the increasingly important place of engineering in the industrial structure of the country.

In terms of the number of enterprises, this engineering industry
remained strongly oriented, during the mid-nineteenth century, to
the shipbuilding and repair industry which was so important to
Norway's foreign earnings. The structure of the engineering industry during the 1860s and 1870s is given in Table 10.6. The

239

material presented in Table 10.6 suggests that the growth of Norwegian industry was closely connected with foreign trade, in particular machinery imports. But domestic mechanical engineering came to play an increasingly important role. This suggests questions about whether the decreasing reliance on machinery imports and the growth of domestic engineering, also have a foreign dimension. How did this growing industry establish its technological basis? In particular, what role did foreign influences play as Norwegian entrepreneurs, managers and engineers acquired the skills of the modern industrial era? We turn now to an empirical examination of these questions.

5 Transfer of Technology into the Norwegian Mechanical Engineering Industry

In this and following sections we examine the role of the transfer of foreign technology, primarily British, in the establishment of Norway's mechanical engineering industry. 'Technology' should not be confused simply with equipment, nor 'technological diffusion' with the purchase of machines (although of course it usually does involve machine acquisition). Rather, technology and diffusion capability consist of a complex combination of:

1. Information, which is in turn a complex phenomenon including, for example, (a) general information on the scope, range and structure of available technologies and on the main lines of technological advance at any particular time, (b) specific information on available techniques, (c) knowledge relevant to the construction, setting-up and operating of equipment.
2. Skills, both labour and managerial, in the construction, operation, supervision, maintenance and management of equipment. Training in all of these areas is thus an important component of technological diffusion.
3. Equipment acquisition, operation, adaptation and development.

In examining the details of technological diffusion within the Norwegian mechanical engineering industry we shall be looking closely at the practical ways in which information, skills and equipment were acquired. In particular, the following sections will investigate:

1. The role of foreign travel and foreign training in the develop-

ment of Norwegian engineers and engineering firms
2. The nature of contacts between Norwegian and foreign engineering enterprises
3. The acquisition, adaptation and development of foreign equipment
4. The role of foreign workers in the development of the Norwegian engineering industry

During the nineteenth century the development of the mechanical engineering industry was extensive in Norway, with vigorous creation of firms. The following examination will range fairly widely over this often dispersed industry, but concentrate in the main on the technological histories of eleven firms. They are:

Mesna Works (founded 1814)
Thune Mechanical Workshop (founded 1815)
O. Mustad (founded 1832)
Aker's Mechanical Workshop (founded 1841)
Trondhjem's Mechanical Workshop (founded 1843)
O. Jakobson's Machine Workshop (founded 1845)
Myren's Mechanical Workshop (founded 1848)
Christiania Nail Manufactory (founded 1853)
Kværner Works (founded 1853)
Nyland's Works (founded 1854)
Kampen Mechanical Workshop (founded 1865)

These firms had a wide range of primary and secondary activities, but in general we can say that they cover the whole spectrum of activities of mechanical engineering in Norway.

6 Technology Transfer (1): Foreign Travel and the Development of Norwegian Engineering

Britain's industrial development, from the inception of its industrialisation in the late eighteenth century, was of very great interest to foreigners. The interested parties included those with a scientific or technical concern over what was happening in Britain, those who desired to become entrepreneurs on the British model, and – perhaps most importantly – European governments who were worried about the economic, political and military implications of Britain's emerging industrial dominance. Thus by the middle of the nineteenth century visits to Britain's industrial areas were a standard part of European entrepreneurial practice.[20] Like many European governments, Norwegian legislators were anxious for close contact with British industrial development, and accordingly the Norwegian parliament discussed, in 1836 and again in 1854, the desirability of promoting visits to Britain by Norwegian businessmen and engineers. It was decided to subsidise such visits with official stipends, which became extensively used to support visits to Britain. Since these stipends were intended to support visits which would not otherwise have been made, there were of course many visits other than those which were officially supported. But we shall begin with a description of travel stipend applications as a way of getting some idea of the numbers and interests of Norwegian engineers who were visiting Britain.

Appendix 10.1 describes principal stipend-funded visits by Norwegian engineers against the background of overall travel stipend applications for the forty-five years from 1850. It can be seen that in any particular year engineers formed a significant proportion of those whose applications were granted. A total of 163 engineering stipends were approved during these years. Of the recipients, 101 (or 61.9 per cent) went to Britain; most of the remainder went to Germany, with a small proportion going to Denmark or Sweden.

A typical example of early stipend-backed travel might be that of the mechanic and machinist P. Nørbech of the Nylands Works. He had worked for a year in various mechanical engineering establishments in Britain, and spent five months in New York. In 1853 he received a travel stipend of 150 specie daler to visit several big workshops in Britain and France. He applied for a further 50 specie daler in 1855, and was clearly very familiar with British technical practice. The important engineer O. Jakobson received a similar amount for visits to English and Scottish workshops and the London Exhibition, on which he was expected to write a report for the Department of Domestic Affairs (Indredepartementet).

Many engineers travelled, however, without relying on stipends. Some of these visits are listed in Appendix 10.2. The visits covered a wide variety of activities. A. Jensen, of Myren Mechanical Workshop, for example, who specialised in turbine construction, acquired his expertise in Germany, where he had 'studied turbines with first-class German experts'.[21] His brother, J. Jensen, was

20. See e.g. E. Robinson, 'The Transference of British Technology to Russia, 1760–1820', in B. M. Ratcliffe (ed.), *Great Britain and Her World 1750–1914*, Manchester, 1975, p.3. For a description of visits undertaken by Norwegian textile entrepreneurs see Bruland, *British Technology*, chap. 5.
21. K. Anker Olsen, *Kværner Brug 1853–1953*, Oslo, 1953, p.71.

rejected for a travel stipend in 1842, but apparently then used the resources of the University of Oslo Library for his informational needs; he subsequently travelled to England and Germany in 1851, 1857 (to study steam-driven saws in anticipation of the repeal of the 'sawmill privilege' which occurred in 1860) and 1860.[22] Subsequently Paul Holmsen, the manager of Myren's Fredrikstad filial, travelled in 1870 in Britain and Germany.[23] For Nyland's Works, the works manager Bang travelled to England in 1888 and 1890 'to study the progress made in England's best-known workshops'.[24] Later, in 1893, A. L. Thune in his correspondence with Babcock & Wilcox about business promotion in connection with the Chicago exhibition referred to at least fifty Norwegian technicians planning to visit the exhibition, and remarked that 'more will go than just those with stipends'.[25] Where firms themselves funded travel, conditions were sometimes attached: when the directors of the Trondhjem works gave their English works manager John Trenery 200 specie daler to visit the London Exhibition in 1862, he had to agree not to leave their employment for two years after the trip.[26] The firm subsequently engaged A. N. Olsen as works manager 'by telegraphing to England', and subsidised three visits to Britain for other employees.[27]

As in the textile industry,[28] foreign travel sometimes appears to have provided the initial impetus for enterprise formation itself: Steenstrup, of Aker's Mechanical Workshop, 'must have got the idea to establish the mechanical engineering firm from his stays in England and Sweden', where he had studied both mechanical engineering and shipbuilding.[29] Similarly Halvor Thune worked in Scotland in 1838, in a Glasgow workshop, and began boiler production in Norway on his return.[30] Steenstrup first visited England in 1834–5, then twice in the 1840s (during which time he visited the Maudsley works, perhaps the most important engineering shop in England at that time); later Aker documents record visits to

22. C. Gierløff, *Et Bruk ved Akerselven: Myrens Verksteds Hundre Års Minne*, Oslo, 1948, pp.69, 133.
23. Ibid., pp.158–9.
24. Nyland archives, Maritime Museum, Oslo, Styreprotokoll 1B; 11/11–88 and 8/8–90.
25. Thune archives, Norwegian Technical Museum, Oslo, Brevkladdebok; 24/4–93.
26. O. Henmo, *Trondhjems Mekaniske Verksted 1843–1918*, Trondheim, 1919, p.16.
27. Ibid., pp.24, 64.
28. See Bruland, *British Technology*, chap. 4.
29. L. Egge and H. Sandsbråten, *Gamle Akers Verksted*, Oslo, 1982, p.2.
30. Y. Hauge, *Boken om Thune*, Oslo, 1965, pp.16–17.

England by Steenstrup to visit the Exhibition at the Crystal Palace in 1851,[31] then again in 1856 to order machinery, with reports of the visit and accounts of expenses. He visited London again in 1861, and visits by directors and foremen at Aker occurred in 1860 and 1882.[32] In fact, Aker exhibited at the London Exhibition in 1862, and won a prize.[33]

We note the beginnings of a reorientation among the Norwegian mechanics and engineers travelling abroad; the number of different destinations increased and were situated further afield as more countries, in particular Germany and the USA, developed into advanced industrial economies. In the 1870s H. G. Stub, for example, a director of Kværner, visited the United States, and the Kværner foundry-manager, W. Bergh, visited the USA around 1895: 'The trip resulted in several technical improvements in the production process.'[34] However, based on available material, Britain remained the dominating destination. It should also be noted that Norwegian firms often received visits from foreign engineering firms, which no doubt facilitated technology flow: Nylands, for example had visits from the English companies Foxwell & Co. in 1883, William Reid & Co. in 1884, Kenyon & Co. in the same year, and Turton & Sons in 1885.[35]

7 Technology Transfer (2): Foreign Education and Training of Norwegian Engineers

Foreign training was a frequent element in the development of early Norwegian mechanical engineering firms. This could involve either practical workshop training or formal academic education or both. Some patterns of formal education abroad are summarised in Table 10.7.

Foreign training was widespread through the firms studied here. For example, Mustad sent 'a man to learn in England' about 1847.[36] A director of Kværner trained at the Gewerbe Akademie in Berlin, and another – the technical director H. M. Smith – after

31. Aker archives, Norwegian Technical Museum, Oslo, Forhandlingsprotokoll, 1843–52, pp.49–52.
32. Aker archives, Norwegian Technical Museum, Oslo, Styreprotokoll, 1854–71, pp.69, 74; Styreprotokoll, 1871–1900, 27/4–82.
33. H. P. Lødrup, *A/S Akers Mekaniske Verksted 1841–1951*, Oslo, 1951, p.72.
34. Olsen, *Kværner Brug*, pp.163, 227.
35. Nyland archives, Maritime Museum, Oslo, correspondence in IIE 2316, 14/4–83; 2319, 13/5–84; 2320, 15/8–84; 2322, 28/7–85.
36. O. Wicken, *Mustad gjennom 150 År, 1832–1982*, Oslo, 1982, p.32.

Table 10.7   Education and work experience from abroad

Firm	Person	Visit
Thune Mechanical Works	H. Thune	Employed at workshop in Glasgow, 1838
	S. Thune	Technical education in Germany, c. 1900
Aker Mechanical Works	Steenstrup	Studied engineering and shipbuilding in England and Sweden; 1834, 1835, 1840, 1841
	Bronn	Technical education from England
	G. Swensen	Educated in England 1872–74
	J. G. L. Lie	Technical education and visits to England, Scotland, USA
	H. G. Stub	Educated in Berlin
	S. A. Weidemann	Visits to England and Scotland
Trondhjem's Mechanical Works	H. B. Holmsen	Educated in Germany and England
	A. Nørbecholsen	Educated in Sweden, and three years in England
	H. J. Olsen	Educated in Denmark and Germany
	W. H. C. Swenssen	Educated in the USA
	Trenery	Born and grew up in Britain
	S. A. Weidmann	Stays in England and Scotland
	H. G. Jürgens	Visits to France
Jakobsons Machine Works	H. Jakobson	Educated in USA
Myrens Mechanical Works	A. Jensen	Stays in Germany
Kværner Works	G. Onsum	Technical education from France and Germany
	H. G. Stub	Educated in Berlin, and visits to USA
	H. M. Smith	Worked in England for nine years, 1862–71
Nylands Works	Morterud	Studied in England, $2\frac{1}{2}$ years
	L. Rode	Stays in Sweden in the 1860s

Sources: secondary literature referred to in this chapter; firms' records

visiting Britain funded by a stipend in 1862, stayed to work as a draughtsman and foreman at Camell & Co. of Sheffield and Newton Iron Works of Hull. He returned to Norway in 1871.[37] Another important Kværner director, G. Onsum, was educated in France and Germany, visited German workshops in the 1870s and sent back (with permission) detailed technical drawings of a travelling crane to the firm.[38]

Within the Aker firm, technical training from England was regarded as a 'family tradition', and nineteenth- and early twentieth-century directors (such as Bronn, G. Swenson and J. G. Lie) received engineering education in Britain and the USA.[39] Sverre Thune, of the Thune enterprise, received his technical education in Germany in the early twentieth century.[40]

The foreign influence was not simply a matter of initial training, but also of further education in mid-career. Some firms were very active in this regard. The Trondhjem's works, for example, apart from actively subsidising travel for its employees, specifically funded education. In 1878 works manager Helseth, engineer Nyhus and foreman Nielsen were all given 400 kroner each to study abroad. In 1872 Trondhjem's employee H. J. Olsen was given money for travel 'if he accepts to stay on . . . and if he tries to make the trip as technically useful as possible and uses this knowledge to our advantage'; in 1881 he was given more money for further education in Germany, Britain and Sweden, and in the same year assistant Falck was to spend two years abroad for further education.[41] Falck was granted two years abroad on full pay on condition that he remained with the firm for five years after his return. In 1878 the foreman, J. F. Nielsen, spent at least three months abroad on a state stipend and with full wages. He travelled again to Britain the following year.[42] This process continued: in the late 1880s, W. H. C. Svenssen visited the USA; in the mid-1890s H. B. Holmsen spent two years abroad, in Berlin, then studying shipbuilding techniques in Britain and Germany; in 1900 H. G. Jürgens (who had been educated in France) went abroad to study new machinery and work methods, and bought a substantial quantity of machinery for the firm while away.

37. Olsen, *Kværner Brug*, pp.62, 227.
38. Ibid., p.120.
39. Lødrup, *Akers Mekaniske Verksted*, p.35.
40. Hauge, *Boken om Thune*, pp.16–17.
41. Henmo, *Trondhjems Mekaniske Verksted*, pp.66–70, and Thune Archives, Norwegian Technical Museum, Oslo, Forhandlingsprotokoll, 6/7–72.
42. Henmo, *Trondhjems Mekaniske Verksted*, p.64.

Although, as noted above, a significant number of Norwegian engineers were trained abroad, others were trained by foreign workers within Norway. Thus Andreas Jensen, the father of J. and A. Jensen of Myren, was taught in Norway by the Scottish mill builder John Wilson.[43] Similarly, one of Andreas's sons, J. Jensen, was trained by the textile engineer Gellertsen, who had himself trained abroad (at the Nordberg engineering firm in Copenhagen).[44] Others, such as Andreas Thune and O. Jakobson, studied English and German in order to be able to read the technical literature of the industry.[45] Jakobsons's son was trained as an engineer in the United States.[46]

Sometimes foreign training had a significant impact on the subsequent technological development of a firm. For example, Jens Jacob Jensen went on to be trained at a technical school in Zurich, 'the first of many young men from Myren', and this appeared to be an important factor in the switch by Myren from making turbines of the British type to new Continental turbine types.[47] Much later, foreign training extended away from purely technical questions into the area of engineering management: thus a Kværner manager, H. P. B. Lund, visited the Berwick Co. of Brooklyn, New York, in order to study the 'scientific management' techniques developed by F. W. Taylor.[48]

For most of the nineteenth century Britain was the dominant foreign source for technical education and experience. It remained important, but as the process of industrialisation spread, and in particular as German industry grew strongly, the choice of destination for further education and training open to Norwegian mechanics expanded. Furthermore, German industry was backed by a system of education – particularly within technical subjects – which at the time was unrivalled.[49] The material above, which indicates a gradual decline of Britain's dominance, does not reflect the dramatic changes which took place within the formal education of engineers. During the latter half of the nineteenth century aspiring Norwegian engineers increasingly sought technical education in Germany. Mainly based on source material from the educational

43. Gierløff, *Et Bruk ved Akerselven*, p.61.
44. Ibid., p.63.
45. Hauge, *Boken om Thune*, p.24.
46. *Jubileumsbok*, Oslo, 1946, p.32.
47. Gierløff, *Et Bruk*, p.164.
48. Olsen, *Kværner Brug*, pp.271–2.
49. Landes, *The Unbound Prometheus*, p.340.

institutions, this has been demonstrated by, among others, Håkon
With Andersen and Fritz Hodne.[50]

8 Technology Transfer (3): Contacts With Foreign Engineering Enterprises

The central institutions initiating and receiving technology transfer
are business firms.[51] A key indicator of the existence of international transfer is therefore a high level of contact among firms
internationally. Much depends, of course, on the quality of these
contacts, but we should remember that even superficial relationships between firms may involve important information transmission. In tracing the international contacts of Norwegian
engineering firms we are, inevitably, limited by the available
archive material, especially correspondence and invoice files. Unfortunately this material does not survive for all the firms studied
here, and we are therefore limited to examining Thune, Jakobson,
Aker, Mustad, the Christiania Nail Factory, Kværner, Kampen,
Nyland and Trondhjem's mechanical workshop. The traceable
foreign contacts of these firms are listed in appendix 10.3, which
gives details of the names, locations and main functions of foreign
contacts, transactions which took place, as well as dates of contact
and sources.

The evidence suggests a very high level of contact: at the very
least, several hundred foreign engineering firms had some form of
contact with one or other of the Norwegian firms. During the
period 1830–1900 the Norwegian engineering enterprises were in
contact with 342 foreign firms, although it should be emphasised
that not all of these were mechanical engineering firms. Of these
firms 194 were British, thirty-four German, two from the USA
and twelve from various other countries. The level of contact
varied considerably between the firms; Thune, for example, had
some kind of contact with fifty-six foreign firms, virtually all of
which appear to be within the engineering industry. Over half of
these were in Britain; next most numerous were contacts with
German firms, with some contacts in the USA. A similar pattern
can be found with other Norwegian firms. Jakobson had contacts

---

50. H. With Andersen, 'Germany and the Education of Norwegian Engineers',
in *Berichte über das 2. deutsch-norwegische Historikertreffen in Bonn*, May 1987, NAVF,
1987, p.100; F. Hodne *Norsk Økonomisk Historie 1815–1970*, Oslo, 1981, p.247.
51. See Wilkins, 'Role of Private Business'.

Table 10.8 Foreign contacts of Norwegian firms

	1841–50	1851–60	1861–70	1871–80	1881–90	1891–1900
Thune mechanical works				35	18	13
O. Mustad	1		5	1	1	9
Aker mechanical works	1	10			19	
Jakobsons machine works			15	42	5	7
Kværner Works				1		
Christiania Nail Works					5	8
Nylands works		15	51	10	68	1
Kampen mechanical works				1		17
Total	2	25	71	90	116	55

Source: appendix 3

with at least sixty foreign firms, Nylands with 119, Aker with thirty-one, Mustad with sixteen, the Nail manufactory fourteen, Kampen eighteen and Kværner one. This suggests a very high average number of contacts with strong variations around the median, but it must be emphasised that the numbers given above are lower limits. For some of the firms (especially Kværner) the available source material is very poor, so that the figures probably underestimate the real number of contacts – in Kværner's case by a large margin. As one might expect, general engineering firms, such as Aker, Nylands, Jakobson and Thune had a wider range of contact than firms producing a more limited product range (such as Mustad and the Christiania Nail Manufactory).

In Table 10.8 we have disaggregated the foreign contacts according to contact with Norwegian firms and time periods. We can trace the number of foreign firms active in the Norwegian market within each ten-year period. Because some firms were active during several ten-year periods, the table gives a higher total number, namely 359, than the total number of foreign firms operating on the Norwegian market during the whole period. The table shows clearly marked growth in establishing foreign contacts after the

1840s. It culminates in the 1880s, then to decrease sharply in the 1890s.

How are we to interpret this decrease? It may be connected with a problem of source material. In particular the Nyland sources seem to be better for the 1880s than the 1890s. Nyland had contacts with a substantial number of foreign firms in the 1880s, and our ability to establish the real number of contacts for the 1890s might be seriously impaired by a possibly incomplete set of data for this period. On the other hand the sources may be equally incomplete for all the ten-year periods. We do not know this, but if it is so, the table probably reflects a real decrease in number of contacts. Since the 1890s was not a period of depression in Norway (actually it was the period when Norway definitely overcame the crises of the 1870s and 1880s and of industrial expansion and reconstruction), we cannot explain the decline in the number of foreign contacts by a general economic recession. What we see, therefore, may be a decline in Norwegian mechanical workshops' technological dependence on foreign supply, and a growth in technological independence of Norwegian enterprises. We note that the Norwegian firms' number of contacts as well as the profile of contacts over time vary considerably, but even so the 1890s show a sharp decrease in contacts for the firms – Thune, Jakobson and Nyland – which had definitely the largest share of contacts in the preceeding periods.

In the 1890s the geographical distribution of the foreign contacts also began to change. This is shown in Table 10.9. British contacts still made up the largest share of the total decreasing contacts, but the German share grew rapidly, from 11 per cent in the 1880s to 36 per cent in the 1890s.

Much depends, however, on the *content* of these relationships: what exactly did they mean for the process of technology transfer? Broadly speaking, we can distinguish from the available records four types of transaction between foreign and Norwegian firms:

1. Purchase of 'basic' inputs such as iron bars, iron and steel plate, glass, coal, oil etc. Such items were very frequently obtained abroad; where they were purchased domestically, it was often through importers such as Henry Hutchinson of Drammen.
2. Purchase of relatively simple but 'engineered' inputs, such as screws, washers, pins, tubes, springs, knives etc.
3. Purchase of machinery, machine tools and general engineering capital goods. Such equipment might include, for example, lathes, rolling machines, milling machines, boring machines,

Table 10.9 Geographical distribution of foreign firms active in the Norwegian market

	1831–40	1841–50	1851–60	1861–70	1871–80	1881–90	1891–1900
Britain	1	2	15	60	42	86	31
Germany			3	1	12	12	20
USA					2	1	
Others					4	4	4
Total	1	2	18	61	60	103	55

Source: appendix 3

turbines, pumps, jacks and lifting equipment, planing machines and steam-engines. From available records it is difficult to be precise about the number of foreign firms supplying such equipment, but for Aker, for example, out of thirty-one foreign firms supplying an identifiable product, seventeen were supplying equipment in this category. Jakobson and Myren appear to be roughly comparable.

4. Supply of technical or economic information. A number of these contacts involved, for example, information on prices and availability of equipment, or the supply of drawings, or catalogues of equipment, or information regarding patents and licensing.

Clearly it is the latter two categories which are important for the development of technological capability in Norway. Information flows have an obvious role in acquisition of technological competence. But we should remember that the supply of equipment (which will be discussed in more detail in the next section) is firstly usually associated with technological information flows, and secondly is normally associated with competence building through 'learning by doing'. The considerable extent of transactions in these areas, therefore, is prima-facie evidence of significant technology transfer.

9 Technology Transfer (4): Acquisition and Adaptation of Foreign Technologies.

Although, as I have argued above, technology transfer can never be reduced simply to the purchase of machines and equipment, nevertheless it normally does involve equipment acquisition. Machine acquisition frequently involved the acquisition or development of new skills, and, in the case of the mechanical engineering industry, was central to the development of new capabilities and new products. Our ability to trace such acquisition is of course limited by the availability of source material, but extant invoice and other documentary material enables us to give at least an outline of the overall process of machine acquisition by the firms studied here. Those purchases which can be definitely documented by invoice or shipment records, principally for Aker, Nyland and Trondhjem's Mechanical Workshop, are described in Table 10.10.

This Table almost certainly understates, because of limited sources, the extent of acquisition: there are very many other references – in correspondence or secondary literature – to other machines. But it can be seen that these firms maintained a consistent programme of machinery acquisition from abroad from the early 1840s. About eighty major items were purchased, mostly machine tools of various types. The principal source was Britain, and the trade involved some of the major engineering firms of the time, such as Whitworths; it is only towards the end of the period that purchases from Germany begin to appear.

In fact most of the firms studied here engaged in more or less extensive machine purchase, especially in their early years, and this was presumably an important element in the industrial learning process in Norwegian engineering. Thus Myren's new workshop in 1855 contained a large English boremachine, 'distinct by its complete arrangement and self-acting motion';[52] Mustad purchased all its early nail machines from William Thompson & Co. of Birmingham, and subsequently purchased lathes, a stick machine, a slotting machine and a planing machine.[53] At approximately the same time the Nail Factory purchased six nail-making machines, as well as patent rights, apparently for the Coates machine, for Sweden and Norway.[54] From the early 1840s, Aker acquired lathes, saws and machine tools (of various types) from England, as well as a steam hammer from the British engineer James Nasmyth.[55] Customs duty on the largest machine acquired, a lathe costing 802 specie daler, was refunded by the state.[56] Another round of pur-

52. Gierløff, *Et Bruk*, pp.113–4.
53. Wicken, *Mustad*, pp.13–14, 28.
54. Christiania Nail Works, archives, Norwegian Technical Museum, Oslo, Forhandlingsprotokoll, 1854, p.1; Olsen, *Kværner Brug*, pp.20, 79.
55. Akers archives, Norwegian Technical Museum, Oslo, Forhandlingsprotokoll, 1843– , pp.4, 17, 51, 52; Lødrup, *Akers Mekaniske Verksted*, p.39.
56. Lødrup, *Akers Mekaniske Verksted*, p.39; Nasmyth archives, England.

Table 10.10 Documented machine and equipment acquisitions

Year	Firm	Purchase	Exporting country	Producer
1842	Aker	1 lathe		
1843	Aker	1 lathe		
	Trondhjem's	1 steam engine		
		1 blowing machine		
		1 lathe		
1847	Mustad	1 nail machine		Thompson & Co.
1850	Aker	1 bolt-boring machine		
		1 expanding iron chuck		Matthew Young
1851	Aker	1 self-acting lathe		
		1 shaping and planing machine		Whithworth & Co.
1854	Aker	1 steam hammer		Nasmyth
		1 planing machine		
		1 lathe		
		1 chuck		Parr, Curtis & Madeley
	Chra. Nail Works	6 nail machines		
1855	Myren	1 bore machine		
1856	Aker	[unspecified]		
	Nyland	1 planing machine		
		1 circular saw		Worsam & Co.
1858	Nyland	1 shaping machine		Whithworth & Co.
1859	Nyland	1 plate-bending machine		J. Buchton & Co.
1860	Aker	1 lathe		
1861	Nyland	1 steam hammer		R. Morrison & Co.
1862	Nyland	1 steam riveting machine		W. and I. Garforth
		1 steam governor		W. Sergant
1865	Nyland	1 [indecipherable]		
		1 slotting machine		
		equipment for lathe		

continued on page 254

Table 10.10 *continued*

Year	Firm	Purchase	Exporting country	Producer
1866	Nyland	1 punching machine 1 slide lathe bed 1 slotting machine 1 drilling machine 1 wheel-cutting machine 1 edge-planing machine 1 plate-bending machine		Collier & Co. J. Hulse Whithworth & Co. Smith, Peacock & Tanneth
1870	Trondhjem's	1 slotting machine		Hutton & MacDonald
1871	Mustad	1 lathe 1 stick machine		
1872	Mustad	1 lathe 1 planing machine 1 slotting machine		
	Trondhjem's	1 planing machine		
1873	Trondhjem's	1 riveting machine		
1882	Nyland	manometer control app.	Germany	A. Barber & Co.
		1 blower & duplex engine 1 steam winch 1 cylinder		Vulcan Iron works Clarke, Chapman & Guerney Hawkes, Crawshaw & Sons
1883	Nyland	[unspecified]		
1884	Nyland	horizontal punching, beam-bending and angle-cutting machine 1 steam winch 1 plate bending machine [unspecified] 1 mandrel (spindle)		Campbells & Hunter Clarke, Chapman & Co. Scriven & Co Foxwell & Son

Year	Firm	Purchase	Exporting country	Producer
1886	Mustad	[unspecified] 1 wire-drawing machine		Appleby Bros.
1892	Nyland	1 steam hammer		J. Rennie & Son
1893	Kværner	1 slotting machine	Chemnitz	
1894	Nyland	1 punching and cutting machine		
1898	Chra. Nail Works	13 nail machines	Germany	
	Kampen	1 transmission steel-wire pliers		
		[unspecified] Worthington pump	Germany	Naxos Union Worthington by Fischer & Son
		1 lathe	Germany	E. Sonnenthal, by L. Ewald

Note: unless stated otherwise machines and equipment are from Britain.
Sources: Aker: accounts, 14 December 1842; Protocol (Forhandlingsprotokoll); Nasmyth archives, Britain; Trondhjem's Mechanical Works: archives, box 1; Directors' Protocol (direksjonsprotokoll), pp.12, 20, 23; O. Mustad: O. Wicken, *Mustad gjennom 150 År 1832–1982*, Oslo, 1982, pp.13, 28; Christiania Nail Works: K. Anker Olsen, *Kværner Brug 1853–1953*, Oslo, 1953, p.20; Protocol (forhandlingsprotokoll), p.1; T. Parmer, *'Spigeren' som temmet jern og stål: Produksjon og arbeidsforhold ved Christiania Spigerverk 1860–1960*, Oslo, 1982, p.143; Myren's Works: C. Gierløff, *Et Brug ved Akerselven: Myrens Verksteds Hundre Års Minne*, Oslo, 1948, pp.113–14; Nyland Works: accounts (hovedbøker); correspondence; Directors' Protocols (styreprotokoller); Kværner Works: Olsen, *Kværner Brug*, p.162; Kampen Mechanical Works: correspondence.

chases – of similar but more modern equipment – was made in England in the early 1850s as Aker moved to Holmen, a harbour site, and formed a new company.[57] Purchases would sometimes be associated with a new product line: when Aker was considering whether to begin making coke ovens in 1851, Steenstrup went to England to acquire information and models, and to purchase appropriate

57. Egge and Sandsbråten, *Gamle Akers Verksted*, p.4.

equipment.[58] From the late 1850s, Nyland appears to have purchased a very large part of its machine stocks in Britain: at least a dozen major machine tools of various types.

Similar purchases continued to be made into the late nineteenth and early twentieth centuries. In 1898, 1901 and 1912, for example, the Nail Factory purchased equipment in Sweden and Britain (the Swedish purchase including men to operate the equipment).[59] Both Myren and Aker continued to buy rolling and lifting equipment from the English firm Smith Bros. in the last years of the nineteenth century.

Where it is possible to trace the machine and equipment stocks of firms, we find a large proportion of equipment to be of foreign origin. For example, the Jakobson copy book records, in 1903, the movement of a number of machines and lists their values: by far the largest part are foreign, in particular boring and grinding machines by Barnes and by the American firm Pratt & Whitney. The general picture which emerges here is that machine tool acquisition in particular was an important element in the development of technical capability by Norwegian engineering firms. The fact that these transactions decreased in number in the 1890s suggests that such capability actually had been developed.

10 Technology Transfer (5): Use of Foreign Technical Information and Imitation of Foreign Techniques.

The direct imitation or adaptation of foreign techniques, on the basis of foreign information, was an essential element of early Norwegian engineering. Norway's first Scotch turbine, for example, was constructed by the Myren firm in 1849, on the basis of British technical drawings.[60] Likewise, Myren's first planing machine, built in the early 1860s, appears to have been an imitation and perhaps development of a British machine imported to Frederikstad in 1860. Later, in the 1890s, Myren changed to an American model.[61] In 1870, Myren made a paper machine for Bentse Brug which was in fact of copy of a machine delivered twelve years earlier by the

58. Lødrup, *Akers Mekaniske Verksted*, p.52.
59. T. Parmer, '*Spigeren*' *som temmet jern og stål: Produksjon og arbeidsforhold ved Christiania Spigerverk, 1860–1960*, Oslo, 1982, pp.131, 143, 145.
60. Olsen, *Kværner Brug*, p.70.
61. Gierløff, *Et Bruk*, pp.138–9, 146.

Edinburgh engineers James Bertram & Sons.[62] Within Myren 'English books of instruction were the most used', and the firm was often in receipt of technical information, such as the technical drawings of spinning machinery sent to Myren by Bertram and Sons in 1879.[63] In the mid-1840s, the engineer O. Jakobson visited the William Thompson engineering firm in Birmingham, collecting a nail-making machine on behalf of Mustad. He subsequently built a copy of it for Mustad, which was installed early in 1849; his agreement with Mustad stipulated that he would not build a similar machine for others. This pattern was repeated thirty years later: in 1876 one of Mustad's employees, one Topp, studied the American horseshoe machine in Christiania after the world exhibition, and subsequently – in 1881 – made a copy.[64]

Nylands were frequently in receipt of detailed technical information from British engineering firms. For example, in 1882, Clarke, Chapman & Gurney sent tracings of plans of a furnace which they had supplied in England. In the same year, Palmers Shipbuilding Co. of Jarrow sent similar tracings of a furnace, and ten days later sent photographs.[65] They also received patent information from UK patentees, one of whom suggested that Nylands take out a Norwegian patent for one of his inventions.[66] Nylands, like other Norwegian firms, were often also in receipt of circulars, catalogues, prospectuses and so on from potential suppliers: these were in themselves important sources of technical information.

The process of imitation became more sophisticated over the years. In Norway, as in other nineteenth-century industrialisers, railway construction was a key development. The Norwegian Railways had acquired most of the equipment abroad, in particular through the British firm of Beyer, Peacock: between 1866 and 1883, Beyer, Peacock sold fifty-seven locomotives to the Norwegian market.[67] But the railway also promoted domestic production. Thus, in 1901, Thune's first railway locomotive was successfully constructed. But like his later early locomotives, this was constructed using British drawings which had been acquired for him by the railway.[68] Another and important form of imitation was the

62. Hauge, *Boken om Thune*, p.129.
63. Gierløff, *Et Bruk*, p.164. These drawings are kept at the Norwegian Technical Museum in Oslo.
64. Wicken, *Mustad*, pp.14, 39.
65. Nyland archives, Maritime Museum, Oslo, correspondence in IIE 2315, 6/11–82.
66. Ibid., IIE 2321, 12/2–85.
67. Beyer, Peacock archives, Glasgow University Library.
68. Hauge, *Boken om Thune*, p.103.

ability to use foreign technologies under licence. Aker, for example, in 1912, licensed the marine diesel motor design of the Danish firm Burmeister & Wain, receiving detailed drawings and construction plans.[69]

11 Technology Transfer (6): Foreign Engineers and Workers in Norwegian Engineering Development.

Expertise or skill is a central component of any technology, and equipment generally cannot be operated without it. One key way in which technical skills can be acquired is through the import of foreign workers, who may operate important equipment, transmit their skills by teaching, or manage and supervise local workers. The extent to which we can trace such workers in Norwegian engineering is, as with other aspects of technology transfer, dependent on surviving records. These are unfortunately limited in this area. But where records do survive, we find an important presence of foreign workers, usually British. Sometimes these were involved in the establishment of firms in the first place: thus John Trenery, of Trondhjem, was one of the founders of the firm. Unlike the other founders, he had no financial liability for the firm, which suggests that his role was primarily technological. Indeed one of his earliest tasks was to travel to Newcastle to purchase a steam engine and a wide range of other equipment which formed the basic fixed capital stock of the firm. Some played important roles in introducing major new technologies: William Stephenson (himself a son of George Stephenson, builder of early steam locomotives and a figure of great importance in world industrial history) constructed Norway's first locomotive. Others played roles less noticeable but perhaps no less important in the long run. Those workers who can be traced are described in Table 10.11.

Unfortunately, the sources from which Table 10.11 is drawn do not permit us accurately to describe either the lengths of stay or the technological functions of British workers. It also probably understates numbers of workers, in particular omitting short-stay workers associated with the setting up of new equipment. We can see however that all of the major Norwegian engineering firms at some time employed foreign workers, usually British. Although it is difficult to ascertain lengths of stay, we know that a significant number of workers stayed for periods of several years: of the

---

69. Lødrup, *Akers Mekaniske Verksted*, p.106.

Table 10.11 Foreign workers in the Norwegian engineering industry

Name	Period	Function
O. Mustad		
Hurst	1887–	steel drawing
Henry Haynes	1887–97	filer
Holloway	1880s	steel worker
John Croft	1880s	fish hook worker
Tom English	1880s	steel worker
Wm. Masters	1880s	polisher
'six English'	1891–93	start up needle production, machine purchase, management
'English women'	1899	fishing fly makers
Aker Mechanical Works		
Asmundsen (Danish)	1843	
Charles Morris	1846–48	iron worker
Mellwright Spickles	1855–56	
Pickles	1857	
J. J. B. (Bing?)	1884	
Trondhjem's Mechanical Works		
J. Trenery	1843–64	founder
Wm. Trenery	c.1849–60	
Wm. Stevenson	c.1850–68	foreman
B. Cook	1890s	
Stephenson	1860s	
(Danish)	1872	formers
Myren Works		
Rollowy (?)	1861	foreman
Anton Harris	1868–70	drawer
Wilh Wettergren (Swedish)	1869	mechanical worker
Christiania Nail Works		
G. Hudson	c.1895	
J. Hurst	1854–1865	
Nyland Works		
F. Ratcliffe	1859	turner
T. Jowsey	1861	foreman

continued on page 260

Table 10.11 *continued*

Name	Period	Function
T. Ratcliffe	1861	mechanic
Samuel James	1861–72	foreman
James Rippon	1862–67	foreman
James	1882	

Note: the workers are of British origin, unless otherwise stated.
Sources: firms' archives, taxation records, secondary literature referred to in the footnotes

approximately sixty workers involved, about twenty-five appear to have been in Norway for periods of longer than two years. In other Norwegian industries, in particular textiles, I have been able to show that foreign workers performed a number of key technological functions. In equipment acquisition, setting up and operation of new techniques, information flows, training, production management and so on, they played a role out of all proportion to the small percentage of the workforce which they represented.[70] This was possible, however, only through access to detailed correspondence records of a type which does not appear to have survived in the engineering industry. None the less we can presume that similar functions were carried out in mechanical engineering, especially since foreign workers normally commanded wages higher than those being paid to Norwegian workers at that time.

Norwegian firms actively sought British labour: Nyland, for example, asked for help in this regard from the Norwegian firm of Bodin & Co., who were operating in Glasgow. But they also received applications for employment from workers in Britain. Often, interestingly, these were Norwegians who had trained or worked in Britain. Nyland received a number of detailed letters of application from such workers from the early 1880s, sometimes from previous employees: 'Having been in England for nearly two years now, and having gained experience from a variety of shops and drawing offices I take the liberty of writing, in accordance with Nylands Works' kind suggestion before I left for this country, to ask if there is any chance of a position at the works in the near future.'[71]

Klouman had worked for six months at the North Eastern

70. Bruland, *British Technology*, chap. 8.
71. Nyland archives, Maritime Museum, Oslo, correspondence in IIE 2318.

Marine Engine Works, Wallsend on Tyne, in order 'to better acquaint myself with the English practical approach'. Subsequently Klouman worked as a draughtsman with the firm Cheesbrough & Roysden Engineers and Patent Agents, and spent nine months as an employee of Albion Foundry, Liverpool. Klouman in fact wrote three separate letters of application and forwarded references; the reference from Albion Foundry described Klouman as a 'good draughtsman, accurate in his work, punctual, steady and very obliging'.[72] Nylands received at least six separate similar applications during the early 1880s; a number had worked both in Britain and Germany, and for other Norwegian firms, as engineers, draughtsmen and so on. This kind of labour flow may well have been an important element, therefore, in the diffusion of technical skills. One such applicant, P. C. Pettersen, had in fact had his own firm in Christiania in the 1850s, but had sold up and gone to the USA, where he worked for an engineering firm in New York.[73]

Nylands actively sought British help in the management of labour. One of their letters, now lost, to the very important British firm of Whitworth & Co. apparently asked for such help, for it drew the reply that 'we shall be pleased to send out someone for the purpose you name – but in the meantime we shall be glad to know what class of work you are principally engaged with, and whether there is, or is likely to be any opposition on the part of your workmen to any reforms that may be suggested from time to time.'[74]

12 Technology Transfer (7): Norwegian Engineers and the Spread of Foreign Techniques in Norway.

One of the most important functions of Norwegian engineering firms was to operate as conduits for the flow of foreign techniques to other Norwegian firms. Myren, for example, in the mid-1850s offered to order machines from abroad for factories in Norway,[75] and in the late 1860s wrote to the textile entrepreneur Halvor Schou about his requirements for water pipes, steam engines and boilers,

72. Ibid., IIE 2318; 2319.
73. Ibid.
74. Ibid., IIE 2321 17/11–84.
75. M. E. Nord in *Skillingsmagasinet*, 1856, cited in Lars Thue, 'Framveksten av et Industriborgerskap i Kristiania 1840–1875', thesis, University of Oslo, 1977, p.32.

about which 'we have written to several of our acquaintances in Manchester'.[76] In the early 1880s they were again advertising that they could use their connections abroad for the purchase of foreign machinery for Norwegian customers.[77] The engineer O. Jakobson obtained information from England for Mustad in 1851; they also received technical information on English hardening and plating techniques from the Bergen merchant Wallendahl in the late 1870s.[78]

The role of Norwegian engineering firms in the diffusion of foreign technologies is intimately connected with their activities in licensing foreign technologies and using foreign patents; some examples of this have been given in section 9 above. Even more important was the transition through which Norwegian engineers became agents of foreign engineering enterprises, constructing, selling and therefore spreading foreign techniques. The best-documented example of this is the firm of Thune, which acted as agent in Norway for the power engineering firm Babcock & Wilcox. Thune purchased his first Babcock & Wilcox boiler in 1890, and within four years had supplied seventeen such boilers to Norwegian customers. A formal agency relationship was not in place until 1895, but this led to the construction of sixty-three steam-engines using Babcock boilers within the following two years. Thune played a continuing and important role in marketing, skill transfer and the general diffusion of information relevant to power technologies in Norway.[79] Such activity was widespread and of considerable importance. Perhaps the most important technology involved here was steam power. For example, the English firm Clayton & Shuttleworth sold eighty-eight steam engines to Norway between 1857 and 1896; eighty-one of these were sold through Jakobson, and a further three were sold through Myren.[80] In the very late nineteenth century Thune sold a substantial number of high-pressure boilers for power purposes, from the firm of Babcock & Wilcox.

76. Hjula archives, Norwegian Technical Museum, Oslo; Correspondence in 7/3–68.
77. Gierløff, *Et Bruk*, p.170.
78. Wicken, *Mustad*, p.38.
79. E. Lange (ed.), *Teknologi i Virksomhet, Verkstedindustri i Norge etter 1840*, Oslo, 1989, pp.62–72.
80. Clayton & Shuttleworth archives, Reading University Library.

Table 10.12 Stipends for foreign travel; Germany and Britain as % of all destinations

	Germany only	Britain only	Both Germany and Britain
1851–1860	14.2	40.4	9.5
1861–1870	3.5	67.8	5.3
1871–1880	10.0	45.0	10.0
1881–1890	21.0	31.5	31.5
1891–1900	33.3	16.6	16.6

Source: calculated from appendix 1.

13   The Rise of Competing Leaders

From the mid-nineteenth century, the spread of industrialisation began to produce challenges to Britain's economic and technological leadership; the most significant long-term development was the rise of Germany as a machinery-producing and exporting nation. It is important to note, however, that this was a gradual development, and only slowly became reflected in the international relations of Norway's engineering industry. But a changing focus on the part of Norwegian engineers can be traced through most of the technology transfer mechanisms which have been studied above. If we consider foreign travel funded by state stipends, for example, we can see a sharp increase in German destinations as a percentage of all destinations from 1880. Although the sample numbers are small, and there are wide fluctuations, Table 10.12 shows a clear trend increase in German destinations, a trend decline in UK-only destinations, and a steady increase in those visiting both Germany and Britain.

Similar changes can be noted in terms of machine acquisition, and in the general commercial contacts between Norwegian and foreign engineering firms. But the slowness of the transition ought to be emphasised: note, for example, that Table 10.10 above, tracing machine acquisition, shows the beginnings of German impact only towards the end of the nineteenth century. Table 10.13 shows ninety-eight larger machine and equipment purchases. The British dominance is obvious.

These figures do not, however, reveal that German influence did not begin until towards the end of the century, as we can see in Table 10.14. However, machine acquisition, as I have suggested, is

263

Table 10.13 Transactions (1831–1900)

	Britain	Germany	Other	Total
Machine parts	31	6	2	39
Machines	30	7	2	39
Machine tools	18	2	0	20

Source: appendix 3, firms' records.

Table 10.14 Foreign contacts by Norwegian firms

	1841–50	1851–60	1861–70	1871–80	1881–90	1891–1900
Thune Mechnical Workshop						
Britain			19	15	7	
Germany				12	1	6
USA				2	2	
Other				2		
O. Mustad						
Britain		5	1	1	9	
Germany						
USA						
Other						
Aker Mechanical Workshop						
Britain	7			19		
Germany		3				
USA						
Other					1	
Jakobson's Machine Works						
Britain		15	31	4	4	
Germany				8		3
USA				2	1	
Other				1		
Christiania Nail Works						
Britain				5	6	
Germany						1
USA						
Other						1

Table 10.14 *continued*

	1841–50	1851–60	1861–70	1871–80	1881–90	1891–1900
Nyland Works						
Britain	15	50	9	54	1	
Germany			1	1	10	
USA					1	
Other					3	
Kampen Mechanical Workshop						
Britain			1		5	
Germany						10
USA						
Other						2

Source: appendix 3

less important than other forms of business relationship. More significant are general business contacts. Table 10.14 is a development of Tables 10.10 and 10.11 above; it shows the countries of origin of the foreign contacts of seven Norwegian mechanical engineering firms during the period 1830–1900.

Within a general framework of sharply decreasing total number of contacts, Table 10.14 shows a clear general reorientation towards German contacts towards the end of the century. Even here, however, Britain remained overwhelmingly important to the end of the nineteenth century. Kampen Mechanical Workshop was strongly oriented towards German contacts, while Thune, Aker and Nyland appear to be shifting focus with regard to their foreign contacts.

None the less by the turn of the century German techniques were making an increasing impact. The German firm of Steinmüller were providing substantial competition to Thune's Babcock agency, and Jakobson would 'make all kinds of new machines of German construction'.[81] The decline of British engineering dominance clearly lies outside the time period of this study, though its beginning can be clearly seen.

81. Thune archives, Norwegian Technical Museum, Oslo, copy book, 1901–3, p.453.

14 Conclusion

When, in 1892, the directors of Nylands Mechanical Workshop decided that they required a new hydraulic riveting machine, they considered purchasing it from England. But they found that their own workshop 'could make it for 8,000 kr., whereas the price from England was 12,000'.[82] This is an index of the fact that during the nineteenth century Norway had developed a mechanical engineering industry which ultimately became of world standard. In the opening section of this chapter I argued that this effort, for any small country, must involve the emulation of larger technological leader economies: the import of the technological capabilities developed in the dynamic industrial leaders. In the nineteenth century this meant Britain and then Germany. Norway made the effort, but the technology transfer process was a complex one: we have seen that it included foreign travel, training and education; machinery acquisition; the use of skilled labour from abroad; agency relationships; and a wide range of contacts with foreign engineering enterprises. It is useful to contrast this technology transfer process in the engineering industry with that which occurred in another important emerging industry in nineteenth-century Norway, namely textiles. In the latter, Norwegian cotton and wool entrepreneurs acquired machinery, expertise, information and labour from abroad in 'packages' which were put together by British textile engineering firms. The entrepreneurs themselves required commercial and marketing skills: they could, and did, remain relatively lacking in technical expertise.[83] In the engineering industry, by contrast, skill development and competence building were central: this is because engineering is not so concerned with the production of standardised products, but is much more a matter of technical problem solving in which competence is of critical importance. For that reason the role of technology transfer in the development of Norwegian engineering is much more a matter of training and education, of access to information about foreign technical developments, and possession of the ability to use that information. Norwegian engineers clearly had certain advantages in this process: the economy was a particularly open one, and Norway was – in cultural and political terms – outward-looking. Norwegian engineering entrepreneurs made the most of these

---

82. Nyland archives, Maritime Museum, Oslo, Styreprotokoll, (directors' protocol) IB 2113.
83. See Bruland, *British Technology*.

advantages, deploying and developing foreign techniques to construct an industry which was central in Norway's transition from peripheral isolation to one of the richest economies of the advanced world.

Appendix 10.1

Norwegian state travel stipends

Year	No. applicants	No. granted	No. to mechanics	Mechanics to UK	Destinations
1851	25	7	3	3	1: England 1: England and Scotland 1: England and Germany
1852	67	29	6	2	2: England 1: Sweden and Denmark 1: Germany and France 1: unspecified
1853		25	4	2	1: England 1: England and France 1: Sweden
1854					
1855		23	3	2	1: England 1: England or Germany 1: unspecified
1856		16	3	1	1: Scotland 1: Germany 1: unspecified
1857	40	20	4	3	1: England

				Origin
1858	66	22	5	3	1: Scotland
					1: England, Belgium or Germany
					1: Germany
1859	49	22	6	5	1: England
					1: Sweden and Germany
					1: Germany
1860	45	27	8	2	5: England
					1: Sweden
					1: Scotland
1861	29	17	7	5	1: England and Germany
					1: Germany
					1: Sweden
					4: unspecified
1862 (London exhibition)	46	24	24)		4: England
	20	9	7		1: Scotland
					1: Germany and Switzerland
					1: unspecified
1863					6: England
					1: England and Germany
1864	77	30	7	7	5: England
					1: England and Scotland
					1: England and Germany

continued on page 270

Appendix 10.1 continued

Norwegian state travel stipends

Year	No. applicants	No. granted	No. to mechanics	Mechanics to UK	Destinations
1865	81	18	3	3	2: England 1: England and Scotland
1866	65	25	6	6	4: England 1: England or Scotland 1: England and Scotland
1867			6	1	1: England 5: unspecified
1868	93	27	5	4	4: England 1: Switzerland 1: unspecified
1869	94	28	6 (1 died)	5	3: England 1: England and North America 1: England and Germany
1870	90	32	7	4	2: England 1: England and Sweden 1: Germany 1: Belgium and France

Year					
1871	86	27	4	3	2: England 1: England and Germany 1: Belgium and France
1872		28	2	2	1: England 1: England and Sweden
1873		31	9	6	5: England 1: England and Germany 1: Sweden 2: unspecified
1874		16	4	0	2: Germany 1: Germany and France 1: Sweden
1875		25	5	2	1: England 1: England and Scotland 2: Sweden 1: Germany
1876		28	9	4	3: England 1: England and Germany 2: America 2: Sweden 1: unspecified
1877					
1878					
1879–80	110	19	4	2	1: England 1: England or America

continued on page 272

Appendix 10.1 continued

Norwegian state travel stipends

Year	No. applicants	No. granted	No. to mechanics	Mechanics to UK	Destinations
1880–81	103	24	3	3	2: England 1: England or Germany
1881–82	96	25	2	1	1: England 1: Switzerland
1882–83	104	25	4	2	1: England 1: England and Germany 1: Germany 1: Sweden
1883–84	146	25	2	2	1: Scotland 1: England and Germany
1884–85		26	3	2	1: England 1: England and Sweden 1: Germany
1885–86	117	25			
1886–87	108	25	2	0	1: Germany 1: Sweden and Denmark
1887–88	137	28	2	2	1: England or Germany 1: England, Germany and France

1889	26	3	2	1: England
				1: England and Germany
				1: Germany
1890	24	1	1	1: England and Germany
1891	27	2	1	1: Scotland, Germany and Austria
				1: Sweden
1892				
1892–93	27	1	1	1: Scotland
1893				
(Chicago exhibition	32)			
1894	30	2	0	2: Germany
1895	38	1	0	1: Denmark

Sources: Indredepartementets skriv on håndverkstipendier (1.2301, 2.2302, 3.2303 in the Riksarkivet, Oslo); *Departements-Tidende*

Appendix 10.2

Non-stipend foreign visits

Year	No. of visitors	Destination	Firm	Aim
1847	1	England	Mustad	learn to operate and collect machine
1848				
1849				
1850				
1851	1	England	Aker	buy models
1852				
1853	1	England	Jakobson	
1854	1	England	Nyland	
1854	1	England	Chra. Nail Works	buy machine, employ man.
1855				
1856	1	England	Aker	buy machines
1857	1	England	Aker	
1858	1	Scotland	Trondhjem	
1859				
1860	1	England	Myren	
1860	1	England	Aker	
1861	1	England	Aker	visit exhibition and workshops
1862	1	England	Trondhjem	visit exhibition
1863				

Year	No.	Country	Company	Activity
1864				
1865	1	England	Nyland	
	1	England	Nyland	order raw materials
1866	1	Sweden	Thune	
1867				
1868				
1869				
1870	1	Germany	Kværner	
	1	Germany, England	Myren	
1871				
1872	1	England	Chra. Nail Works	visit exhibition, buy models, employ moulders
	1	Denmark	Trondhjem	studies, buy raw materials and machines
1873	1	England	Trondhjem	
1874				
1875	1	Denmark, USA	Kværner	
1876				
1877				
1878	3	France	Trondhjem	
	1	Germany	Trondhjem	visit foundries, buy models
1879				
1880				
1881	1	Germany, England, Sweden	Trondhjem	visit exhibition, education, machine information
	1	unspecified	Trondhjem	further education
1882	2	England	Nyland	visit exhibition
	1	England	Aker	visit exhibition

continued on page 276

Appendix 10.2 *continued*

Non-stipend foreign visits

Year	No. of visitors	Destination	Firm	Aim
1883	1	England	Nyland	visit workshops
	1	England	Nyland	buy machines
1884				
1885	1	unspecified	Trondhjem	
1886	1	England	Trondhjem	
1887	1	England	Nyland	visit shipyards and workshops
1888	1	France	Trondhjem	
	1	England	Nyland	witness steamship building at works
1889				
1890	1	England	Nyland	study improvements
1891				
1892	1	Germany	Thune	
1893	2	Scotland	Thune	
	1	USA, Denmark	Kværner	education

Sources: Akers mechanical workshop: H. P. Lødrup, *A/S Akers Mekaniske Verksted 1841–1951*, Oslo, 1951; Directors' Protocol no.1; O. Mustad: O. Wicken, *Mustad gjennom 150 År 1832–1982*, Oslo, 1982; Jakobson: ibid.; Christiania Nail Factory: Forhandlingsprotokoller; taxation records; Trondhjem's mechanical works: O. Henmo, *Trondhjems Mekaniske Verksted 1843–1918*, Trondheim, 1919; accounts, Thune Mechanical Workshop: letter books; Kværner: K. Anker Olsen, *Kværner Brug 1853–1953*, Oslo, 1953; Nyland: Correspondence, Directors' protocols.

Appendix 10.3

Foreign firms in contact with the Norwegian mechanical engineering firms

Firm	Location	Type of product sold	Dates
Thune (1815)			
Altendorf & Wright	Birmingham		1855–79
D. Auld & Sons	Whitevale Foundry, Rochester Street, Glasgow	machine parts	1893–4
G. Aultman & Co.	Canton, Ohio, USA		1878–85
Babcock & Wilcox	London	machines	1891–4
H. Bamford & Son	Britain		1887
A. Barber & Co.	Hamburg		1877
J. Bedford & Son	Sheffield		1877–9
Bellis & Co.	Birmingham	(agent)	1891
Berliner Eissengiesserei und Verkzugmaschinen Fabrik	Berlin		1878–91
The Blackman Ventilation Co. Ltd	London	(agent)	1891
Clayton & Shuttleworth	Lincoln		1877–85
Wm. Cook & Son	Glasgow		1877–83
Croggan & Co.	London		1881–2
Wm. Crosskill & Son	Beverley		1884
Crownshaw, Chapman & Co.	Sheffield	machines	1878–80
Dicksley, Sims & Co.	Leigh		1877

continued on page 278

Appendix 10.3 *continued*

Foreign firms in contact with the Norwegian mechanical engineering firms

Firm	Location	Type of product sold	Dates
J. R. Dodge	Sheffield		1877
Elfving & Co.	Stockholm		1878–9
A. Field & Co.	Liverpool		1877–80
A. Fleming & Co.	Granton		1877
The Foreign and Colonial Power Storage Co. Ltd	London		1891
D. Foxwell & Sons	Manchester	machines	1892
Gesellschaft für Lindes Eismaschinen	Wiesbaden		1892
Gesellschaft N. Schmirgel	Frankfurt		1878–9
Th. Grauenhorst & Co.	Berlin		1877
R. H. Guiremand	Berlin		1877
E. Hagen & Co.	Hamburg (Norwegian)		1878
Hirst Brothers	Selby		1882
R. Hodd & Son	Britain		1882
Rob. Hudson	engineer and iron founder, nr. Leeds		1892
Hunt & Tavell	Britain		1880–4
Jeffrey & Blackstone	Stamford		1884–7
Kahler & Willecke	Hanover		1877–9
P. C. and W. Lord	Birmingham		1877–83
Markt & Co.	New York		1877

Name	Location	Product	Date
G. Maus & Co.	Remschied		1878–80
Motala Mek. Verk. Akt. Bolag	Motala, Sweden	machine parts	1873
Naxos Union	Frankfurt		1891
Newton & Nicholson	packing works, Tyne, South Shields	machine parts	1892–4
W. N. Nicholson & Son	Newark		1877–87
E. Page & Co.	Britain		1883
Picksley, Sims & Co.	Leigh		1877
Rheinhold & Co.	Hanover		1892
Reinholdt & Goern	Berlin		1877
Richmond & Chandler	Manchester		1877–80
Sachsische Stickmaschinenfabrik	Chemnitz		1877–9
Samuelson & Co.	Banbury		1877–87
Schaffer & Budenberg	Germany	machine parts	pre-1893
C. Scholermann	Berlin		1879
J. S. Schwalbe & Co.	Maschinenfabrik Germania, Chemnitz	machines, machine parts	1874
Sharp, Stewart & Co.	Locom, Works, Great B—— Street, Manchester	machines	1878
Siemens Bros. & Co.	Germany		1891
Steinmüller			1892
Wm. Summerscales & Co.	Britain		1883–5
Tangy Brothers	Birmingham		1878–83
J. Taylor & Sons Ltd	Nottingham		1892
Tysack Sons & Co.	Sheffield		1877

continued on page 280

Appendix 10.3 *continued*

Foreign firms in contact with the Norwegian mechanical engineering firms

Firm	Location	Type of product sold	Dates
A. Wernicke	Germany	machines	1892
Wickers Sons & Co.	Sheffield		1877–84
Mustad (1832)			
Barhus & Worth	Britain		1865
W. I. Brunton & Son	Britain		1896
Ch. Crawford	Britain		1864
Crownshaw Chapman & Co.	Britain		1896
G. Hallam	Havelock Steel & Wire Mill, Walkely Bank nr. Sheffield		1896
R. W. King	Bristol		
Kilbert Pickerson	Britain		1865
A. Lee & Sons	Britain		1896
Merry & Co.	Britain		1897
Parker & Sons	Britain		1868
T. W. Peterson & Co.	Birmingham	raw material	1873–95
Richardson	Britain		1896
Wm. Thompson & Co.	Birmingham	machines	1846
Thompson, Hatton & Co.	Britain		1863
White & Son	Britain		1896

F. Williams	Birmingham		1896
Wiingaard & Co. (Norwegian)	Birmingham	raw material	1866–7
Aker (1841)			
Altendorf & Wright	Birmingham	machine parts	1882–3
Jos. Baller & Co.	Britain	raw material	1857
H. Bessemer & Co.	Sheffield	machines	1884
Bird	Britain		1857
The Birmingham Patent Tube Co.	Smethwick	machine parts	1882–3
Bloomsfield Ironworks	Tipton	raw material	1882–4
Clarke & Chapman	Gateshead on Tyne	machines	1884–5
Cochran Carr	Britain		1854–57
Donner	Germany		1857
J. Denton & Co.	Leeds	raw material	1882
Fleischer & Holst (Norwegian)	Britain		1858
D. Foxwell & Son	1 North Parade, Parsonage, Manchester	machines	1884
Grüning & Co.	Germany		1857
Harfield & Co.	Mansion House Building, London, EC	machines	1883–84
Holbye, Mathiesen (Norwegian)	Newcastle	machine parts	1885
Hull Forge & Co.	Cannon Street, Hull	raw material	1883–4
Kenyon & Co.	Britain	machines	1883–5
The Leeds Forge Co. Ltd	Leeds	machine parts	1882

continued on page 282

Appendix 10.3 *continued*

Foreign firms in contact with the Norwegian mechanical engineering firms

Firm	Location	Type of product sold	Dates
M. Mail jnr.	Hudson Street, Tyne Dock, Newcastle–upon–Tyne	raw material	1883
Ths. March & Co.	Hull	machines	1883
Motala Mek. Verk.	Sweden	machine parts	1883
Nasmyth	Britain	machines	1854
Park Gate Iron	Rotherham	raw material	1882–5
Parr, Curtis, & Madley	Manchester	machines	1854
Pelly & Co.	Britain	raw material	1838–48
Robertson Brothers	Britain	machines	1883
G. Salter & Co.	West Bromwich	machines	1882
Sewell & Neck	London		1856
Smithwick Farbenworks (?)	Birmingham	machine parts	1882
Smithwick Tubeworks	nr. Birmingham	machine parts	1882–83
Ths. Turton & Sons	Sheffield	machine parts	1882–85
Whitworths & Co.	Manchester	machines	1851
Ziegler & Co.	Germany	raw material	1856
Jakobson (1845)			
Altendorf & Wright	Birmingham	raw material	1875–7
C. Aultman & Co.	USA	machine parts	1879

Barnes	Britain	machines	c.1902
J. Bedford & Son	Sheffield	machine parts	1875
Bentel & Margudant		machines	c.1902
Bernstrom & Co.	Stockholm	machines	1877
Bittrich & Simon	Chemnitz		1901
Burys & Co. Ltd	Sheffield		1868–9
Ch. Churchill & Co.	London		1875–6
Clayton Shuttleworth	Lincoln	machines	1864–1901
Cocoran Witt & Co.	Britain		1874–6
Wm. Cook & Son	Glasgow	raw material, machine parts	1874–9
Crosskill & Son	Britain		1875
J. Crowley & Co.	Britain		1875–6
Crowshaw Chapman & Co.	Sheffield		1876
J. and R. Dodge	Sheffield	machine parts	1868–77
Moses Eadon & Sons	Sheffield		1891–1907
A. Field & Co.	Liverpool	machines	1877–9
Flather	Britain	machines	c.1902
A. B. Fleming & Co.	Granton		1876
D. Foxwell	Manchester		1865–7
R. Garret & Sons			1874
Gerlach & Lie	Hanover		1896
Thos. Gibson	Birmingham		1864–7
Gilbert	Britain	machines	c.1902
Th. Grauenhorst & Co.	Berlin	machines	1872–7

continued on page 284

Appendix 10.3 *continued*

Foreign firms in contact with the Norwegian mechanical engineering firms

Firm	Location	Type of product sold	Dates
E. Hagen & Co.	Hamburg (Norwegian)	raw material	1877
Wm. Hall	Sheffield		1868–74
D. Heald & Co.	Christiania	raw material	1877–8
R. Hunt & Co.	Earles Colne		1888
Kupper & Co.	Berlin		1896–1901
Kohler & Willecke	Hanover	machine parts	1877–8
J. C. and W. Lord	Birmingham	machine parts	1877–9
Mahsden & Stidsberg	Sheffield		1866–7
Markt & Co.	New York		1877
G. Maus & Co.	Germany	machine parts	1878–90
H. R. J. Moser	Britain		1875–6
Naylah Vickers & Co.	Sheffield		1864–7
W. N. Nicholson & Son	Newark		1874–8
Orme	London		1875
Page & Co.	Bedford		1865
E. Parker & Sons	Sheffield		1864
Parry & Co.	Birmingham		1875–6
Pattello Brothers	Glasgow		1897
T. W. Petersen & Co.	Birmingham		1866–89
Picksley, Sims & Co.	Leigh		1875–6

Name	Category	Location	Date
Pratt & Whitney	machines	USA	c.1902
A. Ransome & Co.		London	1876
Wm. Reid & Co.		London	1897–8
Reinholdt & Goern	machine tools	Berlin	1876–7
Richmond & Chandler		Salford	1875–7
Samuelson & Co.	machines	Banbury	1877–8
M. Selig Jnr. & Co.		London	1875
E. Sonnenthal Jnr.		Berlin	1897–8
Smith Brothers & Co.		Britain	1871–4
Spong & Co.		London	1875
Tangy Brothers	machine tools	Birmingham	1879
Taylor & Wilson		Britain	1888–91
E. R. and F. Turner		Ipswich	1867–76
Wm. Tysack & Sons		Sheffield	1864–5
W. Tysack & Sons & Turner		Sheffield	1876
Vickers Sons & Co. Ltd.		Sheffield	1867–78
J. Warner & Sons		London	1861–74
Whitworth & Co.		Britain	1873
J. Wilder		Britain	1890

Christiania Nail Works (1853)

Name	Category	Location	Date
Bellamy Walker Hill & Co.	machine tools	Derby	1898–1905
Bennet & Sayer	raw material	West Hartlepool	1887
C. Berner			

continued on page 286

Appendix 10.3 continued

Foreign firms in contact with the Norwegian mechanical engineering firms

Firm	Location	Type of product sold	Dates
Downing & Co.	Sheffield	machine tools	1885–8
Fedden Brothers	Newcastle	raw material	1887–
Holst & Fleischer	Middelsborough	raw material	1887–
Iervelund(?) & Clapham	Middelsborough	raw material	1899–1901
A. G. Kidston & Co.	Glasgow	raw material	1897
J. Marx	Amsterdam	raw material	1894–8
Mitchell & Co.			
Neilson Brothers	Glasgow	raw material	1896–7
A. Wever & Co.	Hamburg	raw material	1897–8
Wm. Whitwell & Co.	Stockton	raw material	1897–8
William Brothers	Sheffield	raw material	1887–94
Kværner (1853)			
Brown & Knap	Birmingham	machines	1865
Nyland Works (1854)			
Allott, Thelwall & Co.	East Riding Iron Works, Hull	raw material	1865–8
Appleby Bros.	London		1884

Company	Location	Product	Date
Baerlin & Co.	Manchester	machine parts	1883
Balcke, Tellering & Co.	Benrath, Germany	machine parts	1885
A. Barber	Hamburg	machines	1861–82
Wm. Barrows & Son	Bloomfield Iron Works	raw material, machine parts	1857–69
Baselow & Co.	Middelsborough		1882
A. von Bergen	Darlington		1884–5
F. Berry & Sons	England	machine tools	1886
H. Bessemer & Co.	Sheffield	raw material	1884
Wm. Bird & Co	London	machine parts	1886
Birmingham Battery & Metal Co.	Birmingham	raw material	1884
Birmingham Patent Iron & Brass Tube Co.	Birmingham	machine parts	1859–82
Bodin & Co. (Norwegian)	Glasgow	(agent)	1883–85
Bolderman Borris & Co.	Newcastle	raw material	1855–65
E. Bremeyer	Hannoversche Gummiwaaren Fabrik, Germany		1884
Britannia & West Marsh Iron Works	Middelsborough	raw material	1883
Broughton Copper Co.	Manchester	raw material	1865–8
J. Brown	Atlas Works, Sheffield	machine parts	1866
J. Buchton & Co.	Well House Foundry, Leeds	machine tools	1859–84
Bullivant & Co.	London	raw material	1885
Caledonia Tube Co.	Coatbridge	machine parts	1886–7
Campbell & Hunters	Dolphin Foundry, Hunslet, Leeds	machine tools	1883–4

continued on page 288

Appendix 10.3 continued

Foreign firms in contact with the Norwegian mechanical engineering firms

Firm	Location	Type of product sold	Dates
Carrick & Brockbank	Manchester		1865–6
Clarence Iron Works	Leeds	machine parts	1882
Clarke, Chapman & Gurney	Victoria Works, Gateshead on Tyne	machines	1882–4
Wm. Collin & Co.	Manchester		1865–6
W. Collier & Co.	Engineers, Salford	machine tools	1864–6
Danday Mailliard	France	machines	1882
Darlington Forge Co.	Darlington		1883
W. Davies	England	machine parts	1883
Day, Summer & Co.	Engineers & Shipbuilders, Southampton	machine tools	1882
John Denton	Britain	raw material (agent)	1861–6
Denny & Bros.	Iron & Steel Shipbuilders, Dunbarton		1884
Dorman, Long & Co.	Iron Manufacturer, Middelsborough	raw material	1883
Duncan Bros.	Engineers, London	machines	1884
Dunkerley & Co.	Britain		1886
Eastbrook & Allcard	Sheffield		1869–90
Everret Allen & Sons	Birmingham	machine parts	1882–3

J. Fenn & Co.	London		1856–66
Forrest & Barr	Engineers, Glasgow		1867
Fox, Head & Co.	Manchester	machine parts	1865–84
D. Foxwell & Sons	Export Agents, Manchester		1883–5
Garforth	Dickinfield Foundry, Ashton	machine tools	1862
Gibbins & Co.	Birmingham		1883
J. Glasgow	Manchester	machine parts	1865–6
Gutehoffnungshütte(?)	Germany		1885
J. & H. Gwynne	Engineers, London	machines	1885–6
Hall & Co.	Hull	machine parts, raw material	1862
Harfield & Co.	London		1879–83
Hartmann & Mallet	London	(finance)	1862
Harpers Ltd.	Aberdeen	machine tools	1885
A. Harris & Co.	Middelsborough	raw material	1866–7
W. B. Harrison	Hull		1867
Hawkes, Crawshaw & Sons	Newcastle	machines	1882–4
Head & Co.	Newport Rolling Mills, Middelsborough	machine parts, raw material	1865
Hopkins, Gilkes & Co.	Tees Side Iron and Engine Works, Middelsborough	raw material	1865–8
J. Hulse	Ordsal Works, Salford	machine tools	1857–65
R. Irvine	West Hartlepool		1866–7
Johnsen & Jorgensen (Norwegian)	Agents, London	(agent)	1884

continued on page 290

Appendix 10.3 continued

Foreign firms in contact with the Norwegian mechanical engineering firms

Firm	Location	Type of product sold	Dates
Jonas & Colver (Swedish and British)	Continental Steel Works, Sheffield		1883–4
Kenyon & Co.	Birmingham	raw material	1882–4
Kittel, Downing, & Wilson (Swedish and British?)	Scandinavian Works, Sheffield		1884
S. Lawson & Sons	Britain		1860–2
Llewellins Patent Machine Co.	Bristol and Glasgow	machines	1882–3
Lloyd & Lloyd	Albion Tube Works, Birmingham	machine parts	1865–84
Lobnitz & Co.	Engineers and Shipbuilders, Renfrew		1884
London Joint Stock Bank	London	(finance)	1866
Low Moor Iron Works	Bradford	machine parts	1860–8
Maschinenfabrik Germania	Chemnitz	machines	1887
McDonald & Co.	Liverpool		1885
Moreaux aîné	France		1886
Morgan Crucible Co.	London	raw material	1882
Morley & Co.	Hull	(finance, freight)	1860–8
Morrsion & Co.	Britain	machine tools	1861
Motala mek Verksted	Sweden	raw material	1882
Napier Bros.	Glasgow	machine tools	1886

Naxos Union	Germany	machines	1884
Neild & Co.	Dallam Iron Works, Warrington	raw material	1862
Palmers Shipbuilding & Iron Co.	Jarrow		1882
Park Gate Iron Co.	Rotherham	raw material	1868–84
E. Parker & Sons	Sheffield	machine parts	1862–6
Paw & Fawcus	North Shield Iron Works	machine parts	1866–8
Pearson & Co.	Glasgow	machine parts	1883
Pelly & Co.	London	(finance)	1861
Pile, Spence & Co.	West Hartlepool		1866
Powis, James & Co.	Saw Mill Engineers, Victoria Works, London	machine tools	1867
Quitzow, Schlesinger & Co.	London and Low Moor Iron Works, Bradford	machine parts	1861–4
Ransome & Co.	London	machine tools	1865–7
Wm. Reid & Co.	London		1884–7
J. Rennie & Son		machine tools	1892
Ths. Robinson & Son	Rochdale	machines	1867–8
Rolfe & Green	Hull	machine parts	1860–2
H. Ronnebeck	Middelsborough		1882
Schaffer & Budenberg	Germany	machines	1885
Schaffer & Walcker	Berlin, Dresden		1887
Schlesinger & Wechmar	Bradford		1879
Schonfeldt & Wolfers	Hamburg		1887
Scorer, Fawcus & Co.	North Shields		1883
Scriven & Co.	Tool Makers, Leeds Old Foundry	machine tools	1884

continued on page 292

Appendix 10.3 continued

Foreign firms in contact with the Norwegian mechanical engineering firms

Firm	Location	Type of product sold	Dates
Seebohm & Dieckstahl	Dannemora Steel Works, Sheffield		1885
W. Sergant	England	machines	1862
Sewells & Neck	London	(finance)	1855–60
Shakers & Lowe	Newcastle	raw material	1883
Smith, Peacock & Tannet	Victoria Foundry, Leeds	machine tools	1856–66
Steinmetz & Knecht	Kassel		1887
Stones Settle & Wilkinson	Plumbers, Brass Founders, Hull		1867–8
Thompson & Woods	Spring Garden Engine Works, Newcastle	machines	1856–63
Thwaites Bros.	Vulcan Iron Works, Bradford	machines	1882–7
Torrey & Co.	New York, Denmark, Sweden	raw material	1885
Tube & Co.	Birmingham	machine parts	1856–9
Turton & Sons	Sheffield	raw material	1882–5
Walen	Sheffield	machine parts	1883
J. White	Glasgow		1887
M. Whitwell & Co	Stockton		1882–3
Whitworth & Co.	Manchester	machine tools	1857–84

Wilson & Son	Hull	machines	1867
J. Wood	Britain	machines	1885
Worsam & Co.	Britain	machines	1856
Kampen (1865)			
Allgemeine Elektrizität Gesellschaft	Berlin		1898
J. Bedford & Sons	Lions Work, Sheffield	machine parts	1874–6
Blake & Knowles	Christiania	machines	1898
Erie City Iron Works (est. 1840)			1898
Gerlach & ?	Hanover		1898
Globe Elevator Works	Stockton on Tees		1898
Huldschinsky	Gleivitz		1898
Krigar & Ihssen	Hanover	machines	
Lange & Gehrckene	Altona		1898
Marsh Brothers & Co.	Ponds Steel Works, Sheffield	raw material	1898
Naxos Union	Frankfurt	machines	1898
Picard Frères	Alsace		1898
Pickerings Ltd		machines	1898
C. Scholermann			1897–8
Schuchard & Scütte	Berlin	machine tools	1898
E. Sonnenthal Jr.	Berlin	machine tools	1898
O. Thost	Zurich		1899
H. Woltersdorf	Stettin		1898

Sources: firms' archives held in Norway and Britain, account books, correspondence, notes, etc. Exact references are available from the author.

11

The Introduction of the Bessemer Process in Sweden

Claus Wohlert

Many of the ideas which brought about innovations in Sweden during the nineteenth century were not Swedish. This holds true, for example, in telecommunications, in the use of centrifugal force principles in the separation industry, and in the Bessemer steel production method. It must be emphasised, however, that foreign knowledge was not only applied but also further developed in Sweden. This essay will deal with the introduction and application of the Bessemer process – the method which made it possible to mass-produce steel and which therefore proved to be of pioneering importance for the steel industry.

It is well known that the introduction of new production methods can bring with it a string of readjustments to earlier production processes. The shift from handicraft to machinery is a good example of this. The handicraft tools and machines which workers use are similarly subject to change, just as the craftsman functions in themselves change. Mechanisation or rationalisation in production can manifest itself as a replacement for an entire production line, for certain parts thereof, or can aim at small changes without influencing other units of production in any appreciable way. An example of the latter was the shift from German forging to Lancashire forging. A readjustment of this kind, at reasonable costs and with only marginal changes in production methods, occurred within the traditional framework of wrought iron production. The different methods of wrought iron production – German forging, Walloon forging, Franche-Comte (Saint-Cyr) forging, puddling, and Lancashire forging – do not deviate appreciably from one

The author gratefully acknowledges financial support from the Swedish Council for Research in the Humanities and Sciences in the writing of this paper.

another. On the other hand, the main ingot steel processes – the Bessemer process, the Siemens–Martin process, the Thomas process, and the electro-steel process – signify major, pioneering changes in production methods and in the course of production in steel manufacturing, by way of different refining processes for different end-products.

The innovator of the Bessemer process was the British foundryman, Sir Henry Bessemer (1813–98), who was able to patent this new method with the Commissioners of Patents in London on 14 March 1856. The patent was granted in Sweden on 1 July in the same year. The first Swedish attempt to apply this new steel-refining method took place shortly after, in Dormsjö, which at the time belonged to the Garpenberg *bruk*.[1] However, it was thanks to Göran Fredrik Göransson, a merchant from Gävle, and later the founder of Sandviken's Ironworks, that the Bessemer experiments seriously began in Sweden. This took place in 1857 at the Edsken blast-furnace and Högbo *bruk*, just outside the community of Sandviken, which was established in the 1860s. After several troublesome reconstructions, success was achieved with these experiments during the early summer of 1858 through a combination of the know-how of the local workforce and techniques and scientific knowledge borrowed from outside the area. The Bessemer method could thus be practically applied.

This concrete example is well suited to illustrate how a technical idea takes practical form. By working from the assumption that different types of skills were available at the ironworks where the experiments were conducted, this essay will try to map the connections between the local work force, technique and science.[2]

1. The *bruk* was a rural-based iron works, 'thus a typically mixed enterprise, combining industrial and non-industrial pursuits in a sometimes perplexing but, in principle, quite meaningful whole . . . A typical *bruk* was not only a centre of industrial activity but also a combination of forest rights and land holdings': see Karl-Gustaf Hildebrand, *Den klassiska bruksekonomins upplösning*, Uppsala, 1974, which is mentioned in my article 'Concentration Tendencies in Swedish Industry before World War I', in Hans Pohl and Wilhelm Trene (eds), *The Concentration Process in the Entrepreneurial Economy since the Late 19th Century*, *Zeitschrift für Unternehmensgeschichte*, Beiheft no.55, Stuttgart, 1988, pp.14–47, at p.16.
2. Some concept definitions: The concept of *technology* refers to knowledge of technical relations and familiarity with how these can be exploited to fill specific needs, especially in production. The concept thereby includes the doctrine of technique. *Technique* means the way in which technology is in fact applied, thus the term includes work instruments, machines and other equipment which are used by workers under the instruction of entrepreneurs or technicians. *Technicians/engineers* refer to such individuals who have gone through a practical-theoretical education at one of the country's schools of mining. *Worker* refers to one who is employed at the

1 The Significance of the Work Force

The Swedish economic historian Professor Torsten Gårdlund has, in several contexts, called attention to the significance of Swedish craftsmanship in relation to industrial development. He maintains that the Bergslag's (the well-known Swedish mining district) iron manufacturing was favoured by 'rich funds of handed-down labour skills and great technical capability', which essentially eased the adaptation to the new technique. Gårdlund is not sure how long and to what extent the industry took advantage of these assets. However, there is no doubt that the rapidly expanding industry immediately absorbed a workforce already trained. The spread of the factory system, however, was in many cases faster than the spread of worker skills. The increased demand for skilled labour led to labour shortages, above all in the cities' engineering workshops.

Within the period of handicraft manufacture, the master craftsmen had a monopoly of production until 1846. However, the abolition of the craft guilds in 1846 did not put an end to master craftsman examinations; instead they remained for some length of time in a number of trades, for example, among tinsmiths and blacksmiths. After the passing of the act giving freedom to pursue trades in 1864, master craftsmen's examinations were no longer formally demanded. In spite of this liberalising measure and in spite of the declining importance of handicraft tools, there was, within most trades, a strong tradition in the form of inherited worker skills, and this advantageously served the introduction of the new technique.

Unfortunately there are very few specific studies on the importance of labour skills with regard to the transfer of knowledge.[3] In

---

ironworks. Particularly included in the *entrepreneurs'* function is their authority to decide on production factors, such as investment and the hiring of personnel. The entrepreneur, whose role in our context is played by G. F. Göransson, is the owner of the means of production. The concept of *innovation* – meaning renewal or change – is defined as a technical invention which can be practically applied.

3. See e.g. Klaus Wohlert, *Framväxten av svenska multinationella företag: En fallstudie mot bakgrund av direktinvesteringsteorier. Alfa-Laval och separatorindustrin 1876–1914* (The emergence of Swedish multinational corporations: case studies in the light of direct investment theories: Alfa-Laval and the international separator industry 1876–1914), Stockholm, 1981 (English summary). In a more general context see e.g. Professor Lennart Jörberg, who pointed out that 'Sweden's industrial development was to a high degree a process of adapting to events outside the country.' Others have stressed the importance of craftsmanship within the proto-industrial period; see Maths Isacson and Lars Magnusson, *Proto-industrialisation in Scandinavia: Craft Skills in the Industrial Revolution*, Oxford, 1987; L. Jörberg, 'The Nordic Countries

1867 one of the Jernkontoret (the Swedish Ironmasters' Association) representatives, after studying the iron industry in several different countries, wrote that the 'wise use of the advantages our country owns in first rate iron ore . . . the purest of all fuels, or in charcoal, cheap motive power in our waterfalls, in general low work wages and an intelligent population' should constitute good prerequisites for profit and an ability to compete against other countries. In the same spirit the Swedish engineer Johan (Janne) Albert Leffler (1870–1929) maintained that the superiority of Swedish iron, among other things, depended on the methods of iron manufacturing, which was carried out by 'an excellent, responsibly conscious stock of workers with traditions from time immemorial'. To this can be added C. Sahlin's (the managing director of Mining Engineering) curious anecdote of how he guided a foreign iron manufacturer around; Sahlin first praised 'the age-old inherited skilfulness' of the workforce, whereupon the foreign visitor cried out: 'But this is surely a chemist's shop, not an ironworks! Why, you treat everything as exactly, as if it had to do with the preparation of medicine, not iron and steel!'

It is common knowledge that industrial development in countries such as Britain, France, Germany and the United States was further advanced than in Sweden. One can therefore assume that labour know-how in these countries was at a higher level. However, the discrepancy between Swedish and foreign industrial development was not greater than could be bridged. The short-lived efforts of two English technicians – which will be dealt with in more detail in the following section – were the only contribution of foreign labour know-how during the testing stage of our case study.

When it comes to the Swedish workers' efforts to achieve a higher level of steel refining, E. Westman, director of the Jernkontoret, in 1871 maintained that Swedish 'workers are not less intelligent; however, on the other hand they are less costly than the English', and that the cost advantage should be exploited. He also considered the advantage of being able to judge the degree of hardness of the material, which was important for further refinement.

The rising work wages in England would, according to a stipendiary's travel account from 1864, have produced an 'obstructive effect on English iron manufacturing's free development', and workers' dissatisfaction expressed itself in strikes, which in turn

1850–1914,' in C. Cipolla (ed.), *The Fontana Economic History of Europe*, vol.4.2, London, 1975, p.439 (cited in Isacson and Magnusson above, p.4). For valuable suggestions I am grateful to Dr Alf Johansson, Uppsala University.

caused the English producers to mechanise production at an increased pace. For Sweden this had the unfavourable consequence that prices could be further lowered in England.

The long tradition of skilled labour certainly set its mark both in the recruitment of new personnel and in their training in the smith's trade. The smith had to pay for his ignorance as he 'hammered away' at the iron. Since the head smith and his assistants received their wages after the completion of the forged iron, it was in his interest that the group produced iron of first-rate quality. The transformation of cast steel into bar iron was a complicated process, and demanded, among other things, six to eight heatings during forging with the steam hammer. Later it was wrought under the bar iron hammer with borax and English forging sand, and for smaller dimensions placed under a rapidly striking hammer. Great skill was required, and through practical training under the supervision of the smith, the newly recruited workers gradually achieved proficiency. It is well known, that skills were frequently passed on from father to son, and it was common to recruit workers from the same industrial communities.

In this particular case study we note that the Sandviken iron manufacturing community was not established until the 1860s, and therefore lacked such traditions. Seventy-five per cent of the smiths and rollers came 'from outside'. On the other hand, the Högbo works had old manufacturing traditions; similar to Edsken, the Högbo works had skilled labourers at its disposal, skilled in the traditional and proven method and procedures of production.

In Högbo there was a head manager, a works inspector and two bookkeepers. The works manufactured bar iron and employed a total of twelve qualified craftsmen: four master smiths, one at every hearth, and each one had his journeyman and an assistant smith. In addition each smith kept a salaried assistant, and a number of boy apprentices and coal boys (whom he paid). There was also a temporary workforce, the so-called *dagakarlar* (day-labourers). Information beyond this is missing from the ironworks' records.

The Högbo *bruk*'s blast furnace was in Edsken, which was located approximately 35 km from the steel refinery site. After Göransson's purchase of the Edsken blast-furnace in 1856, the entire works was rebuilt and this reorganisation was completed at the end of 1857. During 1858 the workers employed consisted of one furnace master, two smelting-house assistants, one foundryman, one machinist, and twelve day-labourers.

Communication between the workforce at the Edsken blast-furnace and the officials at the Jernkontoret in Stockholm functioned in an exemplary manner. In one of the 1858 reports to the Jernkontor 'about the application of the Bessemer steel process at Edsken's blast-furnace', Andreas Grill, the director of the Jernkontor and also an active ironworks owner in Mariedam, reported, in eulogistic terms, how a worker had delivered 'the most favourable testimony' about a test sample. According to Grill, who had spoken with 'many skilled workers . . . the steel fulfils everything that one can want from the best goods.' The steel's degree of hardness was indicated on a five-degree scale and this 'numerical classification, which at Edsken was conducted by a skilful jobbing smith', later became the standard indicator at many other works.

Thus, the judgements of the workers were highly valued, and they were given tasks of responsibility. However, at the same time new methods brought with them labour and technical problems which were difficult to overcome. This was for example, the case also in the later-introduced Siemens method of refining. At some ironworks the process happened so quickly that the workers could not keep up or were unable to establish the correct point at which to terminate the process. The greatest difficulty, as expressed in 1866 by the Jernkontor's director, L. Rinman, 'remains in the workers' training; for however skilful and willing they may be, the method requires first and foremost some familiarity with the handling of cast steel, and then familiarity with the management of air furnaces', and the ability to follow the process 'which earlier had not been done by ordinary bar iron smiths'.

These problems were touched upon by Göransson in a letter to the Jernkontor's director, C. O. Troilius, in which he also maintained that in the new method of steel production, workers displayed a certain 'unfamiliarity in handling cast steel when flattening and heating', and that there existed a lack of suitable furnaces. He added that hardness tests, before flattening the steel, had not successfully been developed in Britain either. However, gradually the workers became increasingly familiar with the treatment of the material. For example, on Christmas Eve 1858, Göransson informed Troilius that by looking at the flame 'each and every one of the workers employed there could tell you when the [steel] is ready.' But he still maintained that the lack of experience was a disadvantage. It was the smith's task to achieve this quality. Even when numerous attempts finally, on 18 July 1858, led to full industrial production, Göransson expressed certain hesitations. 'I am not certain', he wrote to his three sons studying in Lausanne, 'until I have the steel examined by a fully experienced expert.'

Summing up, we note that our findings, based on fragmentary

source material, do not permit us to draw definite conclusions about whether labour skills were a necessary prerequisite for the introduction of the new technique. In that respect, this study has not come much further than Torsten Gårdlund's findings. On the other hand, it has sought to show that workers' skill nevertheless contributed significantly to the successful application of the new technique.

2 The Significance of Technicians

Edward Scholey, an English instructor from Ecclesfield in Yorkshire, was hired to counteract the lack of practical experience with the Bessemer steel process at the Högbo *bruk*. Scholey led the steel flattening work and was in charge of the new industrial works during March 1858. He did not stay long in Högbo. According to the Jernkontoret, he was specifically engaged to follow 'all the precautionary measures concerning cast steel . . .' However, since the Bessemer process had not been tested out in England either, it is unlikely that Scholey advanced steel flattening in Högbo.

Furthermore during Edsken's reorganisation to the Bessemer technique, the English machinist Price from Manchester spent forty-eight days in Edsken assembling the steam engines. Names of other foreign experts have not been found in the ironworks' records or in other source material.

The Swedish engineer Carl Johan Laurens Leffler (1825–1908), who had worked with Bessemer, had acquired more experience in this area than any other technician. Leffler was regarded as well suited to convert Bessemer's idea to Göransson's innovation. He is named in several letters, was the one who commissioned orders for complementary equipment which was ordered from Britain, and initially led the experiments. It is difficult to say to what extent the method's successful application can be ascribed to Leffler's efforts. Even if Göransson, as is apparent from a letter he wrote twenty years later, did not have a high opinion of 'the Engineer [Leffler], not knowing anything more than myself', Leffler's contribution during the initial phase of applying the Bessemer process can nevertheless clearly be demonstrated.

Another Swedish technician, a so-called steelsmith by the name of Hans Nordgren, was recruited at the steelworks in Högbo. Furthermore the construction work at Edsken was led by Swedish building contractors from Leksand, an iron plate-worker from Kloster's *bruk*, who was responsible for blast equipment, and

301

blast-master Per Eriksson from Bommersbo, who, with his staff, attended to the rebuilding of the smelting-house.

Summing up, we see that foreign participation in the introduction of the new technology was principally in the installation of British machinery. The foreign contribution was confined to purely mechanical proficiency, which was of small significance, especially when compared to the contributions made by Swedish technicians.

3 The Significance of Institutionalised Knowledge

As already mentioned, the most significant institution for Swedish iron manufacturing was the Jernkontor in Stockholm, an organisation which was 50 per cent state-founded, and whose statutes were approved by the estates of the realm as early as 1747. Its activity was directed towards the science of mining research: mining, blast-furnace operation, hammer forging etc. Information about foreign developments was gathered during study visits and conference participation, both of which were funded by the Jernkontor. Since 1817 the Jernkontor had published the frequently cited *Annalerna*, one of the world's oldest journals about the science of mining, which regularly reported on recent research and experiments, discussions and developments in the above-named areas.

During the period with which we are concerned, a number of high-ranking, well-trained officials were employed at the Jernkontor. They contributed innovations and assisted with the introduction of new methods, for example the introduction in the 1830s of the Lancashire method, the open-hearth process during the 1860s, pig iron production through the use of electrical processes, and much more. They were also active in the introduction of the Bessemer process in Sweden.

One prerequisite for continued experimental activity at Edsken was the loan of 50,000rd, which was granted by the Jernkontor. The loan application was accepted on the conditions that, *inter alia*, the Jernkontor exercised the technical supervision and, further, that other ironworks owners would be able both to participate in the experimental work and to conduct tests on their own iron ore. An important consequence was that the Jernkontor's officials were almost always present at the Bessemer experiments. We shall soon return to the significance of these experts in this context.

Two other important institutions should be mentioned: first, the Falu Bergsskola (Falu School of Mining) established in 1822; second, the Filipstadsskola (Filipstad's School), started in 1850, which, according to Torsten Gårdlund, was an 'educational institution for ironworks managers and servants'. The less technical, so-called elementary schools, were founded in Norrköping, Boraas and Örebro, and were often connected with Sunday and evening classes. On a lower educational level various arts and crafts schools and associations were founded during the last half of the nineteenth century. Sometimes staff such as bookkeepers, cash managers, correspondence clerks and others attended these educational institutions, while factory workers, warehouse clerks and the like did not receive such training.

It would seem that agencies, organisations, associations or other groups did not demand technicians trained in economics before the beginning of the twentieth century. Thus it should be remembered that the field engineers and scientists dispatched by the Jernkontor were not trained in economics. They primarily advised on technical questions. Frequently, when these *tecnici* promoted the application of technically interesting, and for Swedish iron manufacturing potentially fruitful, innovation, these engineers even supported economic subsidies. In the case of the exploitation of the Bessemer method, Göransson presented the Jernkontor with an estimate and proposal on the use of the contingent subsidy. Even though the Jernkontor's executive committee was composed for the most part of experienced ironworks owners well acquainted with the problems of iron manufacturing, the Jernkontor did not conduct any economic evaluation of Göransson's solicitation, nor did it request any alternative cost estimates.

Among the Swedish technicians, we find the mining assessor Sieurin, from the Falu Bergskola. He made the designs for the arrangement of steam boilers, the blast machines, the installation of furnaces and also the required buildings. Sieurin was, according to the author Per Carlberg, 'a highly qualified man with elegant academic degrees and thereafter studies at the Falu Bergskola'. During the tests of 1858, from June to November, Sieurin travelled to London, Hull, Birmingham, Manchester, Newcastle and Liverpool, where he studied British ironworks. Permission to enter various work sites was granted to Sieurin only after payment of a costly fee. This was totally financed by the Jernkontor.

The building contractor C. P. Lindberg, also a graduate of the Falu Bergskola, led the construction work for more than three months, until 1 November 1857. However, as an apprentice on the Jernkontor's staff, he returned to Edsken in February 1858 and remained on duty there, according to the register, until August 1859. Lindberg led the work at the steel furnaces. However, in the

303

middle of 1859 he temporarily left Edsken to build up the Bessemer plant at his uncle's works at Carlsdal *bruk*.

A certain bookkeeper, Lundvik, was at first in charge of the office duties at Högbo *bruk*, and when the steel manufacturing experiments at Edsken started, he took part in steel blowing. Lundvik 'learned how to do coal tests with Victor Eggertz' at Falun. Eggertz was a teacher and a principal at the Falu Bergskola, and his decisive contribution to steel manufacturing through developing a method of determining carbon content is generally acknowledged. The problem, up until then, had been to find reliable methods and tests to determine hardness other than by the naked eye. Eggertz's method, permitted technical determination of the percentage of carbon in steel. Even the Jernkontor's director, Andreas Grill, appreciated Edsken's experimental work, and in optimistic terms he expressed the view that 'the entire method's usefulness rests . . . on the ability to reduce the pig iron's carbon content with certainty to precisely that degree at which the product becomes steel, that is to say, malleable and forgeable.' Eggertz applied himself also to other areas with devotion and commitment, including problems in technological education which are not treated here; but his lively interest in technological-scientific training and other educational demands, which to the highest degree affected Swedish workshop industry, deserves to be mentioned. His contribution as the country's foremost mining chemist and scientific metallurgist, combined with his insight into foreign developments, must have played a significant role in the qualitative improvements in steel production at the Edsken works.

Among the Jernkontor's officials, the director, Grill, and his two assistants, P. A. Ahlberg and N. Mitander, stood at the head of the technological management at the Edsken and Högbo *bruk*. Before his employment at the Jernkontor, Ahlberg was active as a master smith at a string of Swedish works, leading the readjustments from Saint-Cyr to Franche-Comté forging. Mitander was similarly active at several works and furnaces, and already during 1857 he supplied Grill with drawings for the erecting of blast-furnace buildings, started up blast-furnaces, and so on. Earlier he was engaged in blast-mastery, forging and bar iron rolling. Ahlberg and Mitander's work with the ongoing Bessemer experiments was published in *Jernkontoret's Annaler*, and through it their work became available to every interested ironworks in the country.

Andreas Grill joined the Jernkontor as a stipendiary in 1849 and at the same time acted as the administrator for Mariedam's and Godegård's works, which were recorded as having four hearths and one steel furnace. All changes and improvements in relation to the Bessemer experiments at Edsken were, according to the Jernkontor's recommendation, to be forwarded to them by Grill for subsequent publication, thereby 'preventing an extension of Bessemer's patent through additions'. Ahlberg and Mitander assisted Grill in his function as technical adviser. When Ahlberg left Edsken he was replaced by Mitander, who later was succeeded by C. P. Lindberg, and he in turn was replaced by mining assessor L. E. Boman. Consequently, the Bessemer experiments were constantly attended by Jernkontor's personnel, and also the contact with the Falu Bergskola was important for the development of the new method.

In 1859 Grill noted that the new *'method liberates the making of steel from dependency on the specific hand skills of workers and lays the control of the same almost entirely within the area of intelligence'* (my emphasis). The Bessemer innovation marked the beginning of a new phase of development from individual handicraft skills to organised mass-production by large-scale industrial operations, and consequently made the future production of steel less dependent on the traditionally acquired labour know-how of the workforce.

To sum up, we can point out that the British innovation could first be used only when the entrepreneurs had brought in the Jernkontor's technological know-how. The cost of exploiting the innovation was not the primary concern of the Jernkontor, but rather the estimated future good which would later bring benefit to the entire Swedish iron industry. Published accounts of practical applications of the Bessemer method in journals such as the Jernkontor's *Annaler* and others, also played an important role in spreading the Bessemer method to other Swedish iron works.

4 Conclusions

With regard to questions concerning the transfer and application of new techniques, the following conclusions, in the form of hypotheses, may be drawn from this study:

1. In the introduction of British technology, the previously acquired skills of Swedish workers constituted a real asset and simplified the transfer of the technology.
2. With new production methods and increased product refinement, the workers also accumulated experience which allowed for the continuous application of the new technique.

3. With the introduction of English technology, Swedish technicians and engineers were, through theoretical training and practical work, capable of adapting the new technology.
4. The prerequisite for their technological know-how was the systematised combination of high quality theoretical schooling with practical work in the Swedish mineral and iron industry.
5. Swedish entrepreneurs functioned as coordinators through whom existing production techniques – as practised by skilled workers – were connected to the Jernkontor's knowledge of foreign ideas.
6. To the extent that technological know-how could be gathered and organised by the Jernkontor during the early nineteenth century, a technological bridgehead was created which made the transfer of new technology possible. The institutionalisation of knowledge benefited Swedish enterprises well into the twentieth century, before research and development had begun to be carried out *within* the enterprises themselves.

References

The source material consists mainly of work reports published by the Jernkontor's officials. Furthermore, certain information and dates have been obtained from *mantalslängder* (population censuses), registers of taxpayers archived by the office of public records in Stockholm, and from the directors' records of the Sandviken corporation. The experimental course of events for the Bessemer process itself is treated in P. Carlberg's 'Bessemermetoden's genombrott vid Högbo och Edsken', *Med hammare och fackla*, vol.22 (Stockholm), 1962. On Sandviken's ironworks, see G. Hedin (ed.), *Ett svenskt jernverk: Sandviken 1862–1937*, Uppsala, 1937.

This essay is based on my study titled 'Svenskt yrkeskunnande and teknologi under 1800-talet' (Swedish know-how and technology during the nineteenth century. 'Prerequisites for the transfer of knowledge – a case study', *Historisk Tidskrift*, vol.4, 1979, pp.398–421. For complete source and text references, the reader may refer to the above-named essay.

12

Laggards as Leaders: Some Reflections on Technological Diffusion in Norwegian Shipping, 1870–1940

Håkon With Andersen

Between the 1880s and 1940 the Norwegian merchant marine underwent profound changes. With regard to technology, two shifts have traditionally been regarded as particularly important: the transition from sail to steam and the somewhat later change from steam to motor. These major changes transformed the fleet twice in the space of sixty years, each time making the former technology more or less obsolete. Furthermore, close to half the fleet (at the 1914 level) was lost during the First World War. In other words this was a period of turmoil and deep structural change. The significance of these transitions was not simply a matter for Norway's internal economy: throughout the period the Norwegian fleet accounted for between 4 and 7 per cent of the total world fleet, making it one of the largest national fleets in the world.

The purpose of this article is to analyse these changes in the Norwegian fleet as a process of technological diffusion. Neither the steamship nor the motor ship was invented in Norway; both technologies were imported. It looks from the outset, therefore, as if we have a case which lends itself to the use of conventional diffusion theories. This is the main point in what follows: not so much to contribute to maritime history as it is usually understood, as to use the specific example of technological transitions in the Norwegian fleet to discuss and evaluate the concepts and the analytical tradition of technological diffusion. This follows partly from the topic of this book and partly from the fact that – from a maritime history point of view – these changes have been described in a broad range of texts, from amateur writings to professional monographs. The splitting up of maritime history into subdisciplines has, however, not been particularly fruitful to the subject in question.

This article deals with the diffusion process primarily on an aggregate level, since it is on this level that different approaches can be compared and theoretical problems discussed. It goes without saying that it is also on this level that the most general conclusions can be attempted. This is not to deny the importance of microstudies. On the contrary, the present approach builds on several detailed studies of maritime history, but as long as our aim is to deal with diffusion processes in general it is only natural to investigate the problems through an aggregate diffusion process approach.[1]

Who were the suppliers of technology, and where do we have to look to identify a diffusion process? First of all it is necessary to clarify the focus of this study. We concentrate here on the shipowning companies. It is their decisions that we are looking for; how did they decide which technology to choose? For the sake of simplicity we will consider only two kinds of supply: firstly, supply by contracting for new ships from shipyards, and secondly, by purchasing second-hand ships from other ship-owning companies. The buying and selling of ships could occur through different channels in our period, one in particular becoming of increasing importance, namely the system of ship brokerage. However, we assume that the technology comes from either the yards or other ship-owners, perhaps mediated through a system of brokers, at least up until the end of the period. In other words the emphasis here is on shipping and the operation of a deep-sea fleet, not on the question of yard and shipbuilding technology. True, the question of how the yards obtained their product technology is important, but we shall put this question aside in this article.[2]

The structure of the analysis is as follows. We begin with a short review and critique of the two important transitions in the Norwe-

---

1. A basic source of information on Norwegian shipping history is still the monumental work edited by Jacob S. Worm-Müller, *Den norske sjøfarts historie*, vols1–3, Kristiania/Oslo, 1923–1951. An overview in English is to be found in H. W. Andersen and J. P. Collett, *Anchor and Balance. Det norske Veritas 1864–1989*, Oslo, 1989.

2. This problem relates to quite another set of literature. Good lists of references are to be found in Håkon with Andersen, *Fra det britiske til det amerikanske produksjonsideal, forandringer i teknologi og arbeid i norsk skipsbyggingsindustri*, Trondheim, 1986 (new forthcoming); Kent Olsson, *Från pansarbåtsvarv till tankfartygsvarv. De svenska storvarvens utveckling til exportindustri 1880–1936*, Göteborg, 1983. S. Pollard and P. Robertson, *The British Shipbuilding Industry 1870–1914*, London, 1979.

gian fleet (from sail to steam and from steam to motor). We then make an international comparison and introduce an apparent paradox: the shift, in the Norwegian fleet, from extreme backwardness to an avant-garde fleet in a time span of twenty years. From a position as a technological laggard, slow in adopting steam technology, the Norwegian marine became an advanced motor-powered fleet in a very short period. In the rest of the article we shall discuss alternative perspectives which can explain this seemingly paradoxical process, and thereby perhaps overcome some shortcomings in traditional diffusion theory.

I Technology as Artefacts or Objects – an Approach

Let us turn to a simple description of the process of change through the sixty-year period, namely figures 12.1 and 12.2. See pp.313, 314. The relative shares of the different technologies are shown as shares of the total Norwegian fleet. As we would expect from traditional diffusion theory, we find two nicely shaped S-shaped curves, one for steam and one for motor, as well as falling S-curves for the outdated technologies, sail and steam.[3]

This traditional picture from diffusion theory is usually interpreted in terms of the individual behaviour of the actors in question. Some early entrepreneurs start the change, and adopters follow suit in an exponential way, making the breakthrough phase evident to everybody. Then follows rapid growth, during which the technology diffuses through the economy as more or less 'conservative' actors take up the new system. At last, during a phase of slowing adoption, comes the time of the laggards.[4]

Even when this approach addresses the motivation and rationality of the individual actors it has a strong bias towards technological determinism. Technological innovations are viewed as something not in need of explanation, something developed

---

3. The sources for the Norwegian fleet are derived from Det norske Veritas yearly statistics, most easily accessible through *Det norske Veritas 1864–1914*, Christiania, 1914. See also Andersen and Collett, *Anchor and Balance*, p.480.
4. One standard reference in this field is E. M. Rogers, *Diffusion of Innovations*, 3rd edn, New York, 1983. N. Rosenberg is perhaps the best-known opponent of this 'diffusion approach', followed by C. Freeman et al. See N. Rosenberg, *Perspectives on Technology*, Cambridge, 1976, and *Inside the Black Box*, Cambridge, 1982; C. Freeman, *The Economics of Industrial Innovation*, 2nd edn, Cambridge, Mass., 1982; G. Dosi, C. Freeman, R. Nelson, G. Silverberg and L. Soete, *Technical Change and Economic Theory*, London, 1988.

outside the society but with a large impact on it. The theory of diffusion assumes a given, completed innovation and then considers its spread: so technology is regarded as an object moving through and spreading throughout a system as if it were a virus or some pollutant. While the technology is spreading, it is considered to be basically homogeneous in form. The direction of the process, the progress, is unidirectional, determined by the technological development. Hence this approach also embodies some kind of linear, progressive concept of time.[5]

We cannot here argue that this approach is wrong when applied to particular problems. We can, however, note some of the hidden assumptions behind it. It involves a tacit evaluation of the kind of processes involved. It is structural in the sense that it focuses strongly on the mental resistance of the population in question. In fact, it is commonly argued, it is this resistance that determines the shape of the S-curve. An alternative and equally legitimate approach would be to examine the rationality behind the pattern – opening up the possibility that not switching to the new technology is based on rational choice.

Perhaps the most important objection to the approach is, however, the dichotomy involved and the implicit concept of the changing technology. The model embodies a concept of two very different technologies with very different characteristics. In the case of sail, steam and motor this is obviously relevant to some extent. It is difficult to see the small-step process making a sailing-ship gradually change to a steamship, and a steam engine growing to become an internal combustion engine. However, there are two problems connected with this dichotomy. One concerns the actual technologies, and one relates to other changes surrounding the technologies.

The technology of steam propulsion was not one technology, but involved a whole array of improvements, changes and prerequisites. Changing boilers and development of more efficient and reliable machinery continued throughout the period. Even the fuel changed, from coal to oil at the end of the interval. The concept of one change is thus rivalled by a process of incremental improve-

---

5. In the history of technology this is discussed in J. M. Staudenmaier, *Technology's Storytellers: Reweaving the Human Fabric*, Cambridge, Mass., 1985. See also H. W. Andersen, 'Manna fra himmelen. Om teknologihistorie og teknologideterministisk historie', *Historisk Tidsskrift* (norsk), no.2, 1987. For a social science account of technological determinism see D. MacKenzie and J. Wajcman, *The Social Shaping of Technology*, Milton Keynes, 1985. See also L. Winner, *Autonomous Technology*, Cambridge, Mass., 1977.

ments – within some kind of framework of steam technology. Although it is an unquestionable fact that a major change took place, many small-scale developments and improvements occurred within the 'technological paradigm' established by steam engines. This point is, moreover, not confined to steam. It holds equally for the competing technologies, sail and motor. Sailing-ships were improved and made more efficient from at least the start of the period up to the turn of the century, making the rig more economic in use (with lower levels of manpower needed) and the ship easier to manoeuvre. Much the same holds for motorships, particularly when it comes to the reliability and economic efficiency of the different motor types.[6]

Hence we are left with a picture of three more or less paradigmatic shifts, opening large possibilities for competing developments inside the paradigms. The paradigm shifts were dramatic and it is possible to call them sail, steam and motor respectively, even if we have effects of the competing technologies inside each paradigm. The fact that outdated technologies 'mobilise' to defend their position by developing technological improvements is a well-known phenomenon. Frequently the result was an extended life-period for the technology in question.[7] The crucial question is whether it is possible to bring the technology into a new position where the potential for improvement is considerable. The long time-periods for the changes we are discussing seem to indicate that the competition between the technologies was fierce and that all of them developed under the pressure of potential obsolescence. The picture given above of early adopters and laggards thus shifts to one in which the different technologies compete and the adopters of the technologies are not laggards but competing in different arenas and with different aims.[8]

When we look at the fleet, there were other changes as well, not necessarily connected with the shifts in propulsion. The most commonly recognised change in the period is the use of new building materials, and hence new building techniques. The shift in

---

6. For motor technology see Andersen, *Fra det britiske* and J. Hetmar, 'The Role of the Diesel Engine in Shipbuilding', in F. M. Walker and A. Slaven (eds), *European Shipbuilding: One Hundred Years of Change*, London, 1983.

7. Later the 'fight' between diesel engine and steam turbine is a good example. See Andersen, *Fra det britiske*, chap. 12.

8. A good example of an econometric approach to this is C. K. Harley, 'The Shift from Sailing Ships to Steamships, 1850–1890, a Study in Technological Change and its Diffusion', in D. N. McCloskey (ed.), *Essays on a Mature Economy: Britain after 1840*, Princeton, 1971.

particular from wooden ships to composite ships (combining wood and iron), to iron and finally steel hulls was a development that affected both steam and sailing ships. This development was not without dramatic consequences for the steam–sail competition, but it was not a phenomenon inherent only to steam.

The results displayed in figure 12.1 provide an overview of a paradigmatic change with particular focus on technology. There were, however, also other important changes that were a part of this without being technical in their nature. One obviously relevant factor is the type of trade and the socio-economic setting of the different ship-owning companies. The Norwegian maritime tradition, from the second half of the nineteenth century, was based on small companies competing worldwide in the tramp markets. It is significant that in Norway we usually talk about the 'ship-owner', not the ship-owning or shipping company. The Norwegian language also has a separate word for this: *reder*. We shall return to this later on in the context of another model of diffusion.

A final word on the diffusion pattern displayed in Figure 12.1 concerns the measurement of change. Accepting the paradigm as a real and useful concept, we still cannot say that we have found a good way of measuring the technological change involved. In Figure 12.1 we have used the tonnage of the ship as a criterion for calculating the relative shares. But it is not evident that this is relevant to the change in propulsion. It is, however, the simplest way of calculating shares, and is also the most traditional way of doing this. The really important argument for this way of calculating, however, is that it is the best way of looking at the economic importance of the different changes. Which share of the capacity is propelled by what technology? In doing this we twist the problem to become one of economic importance, not of the technology *per se*. Hence we add a new premise for the diffusion process, which perhaps is useful and certainly is relevant, but nevertheless changes the concepts in question. Thus the attempt to demonstrate technological change by examining the artefacts themselves is undermined even in an approach focusing particularly on the objects.

There is an even more problematic issue which is tacit in the data on which Figure 12.1 is based. This is that the productivity of the technologies does not correspond with tonnage. Steam technology is often regarded as somewhere between two and three times as efficient as sail even if the tonnage is the same. It depends, however, heavily on the particular trade and on time. In the figure this has been corrected to some degree by using net registered tons for sailing-ships and gross registered tons for steam and motor. We

Figure 12.1 Relative shares of the Norwegian merchant fleet, separated by propulsion. Only ships larger than 100 tons. Sailing-ships in NRT, others in GRT

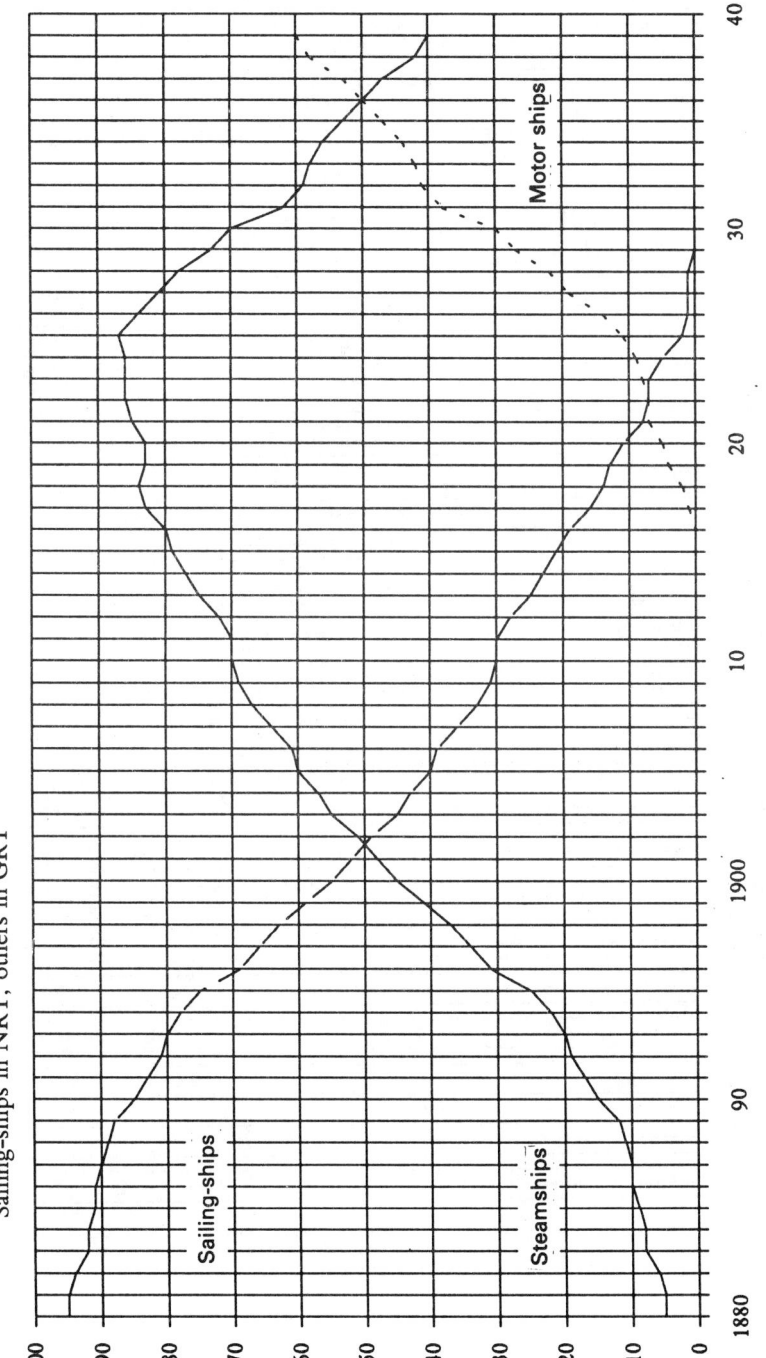

Source: Veritas 1864–1914, Veritas 1864–1939, annual reports and information from Det norske Veritas Classification A/S.

Figure 12.2 Tonnage of the Norwegian merchant fleet, separated by propulsion. Only ships larger than 100 tons. Sailing-ships in NRT, others in GRT.

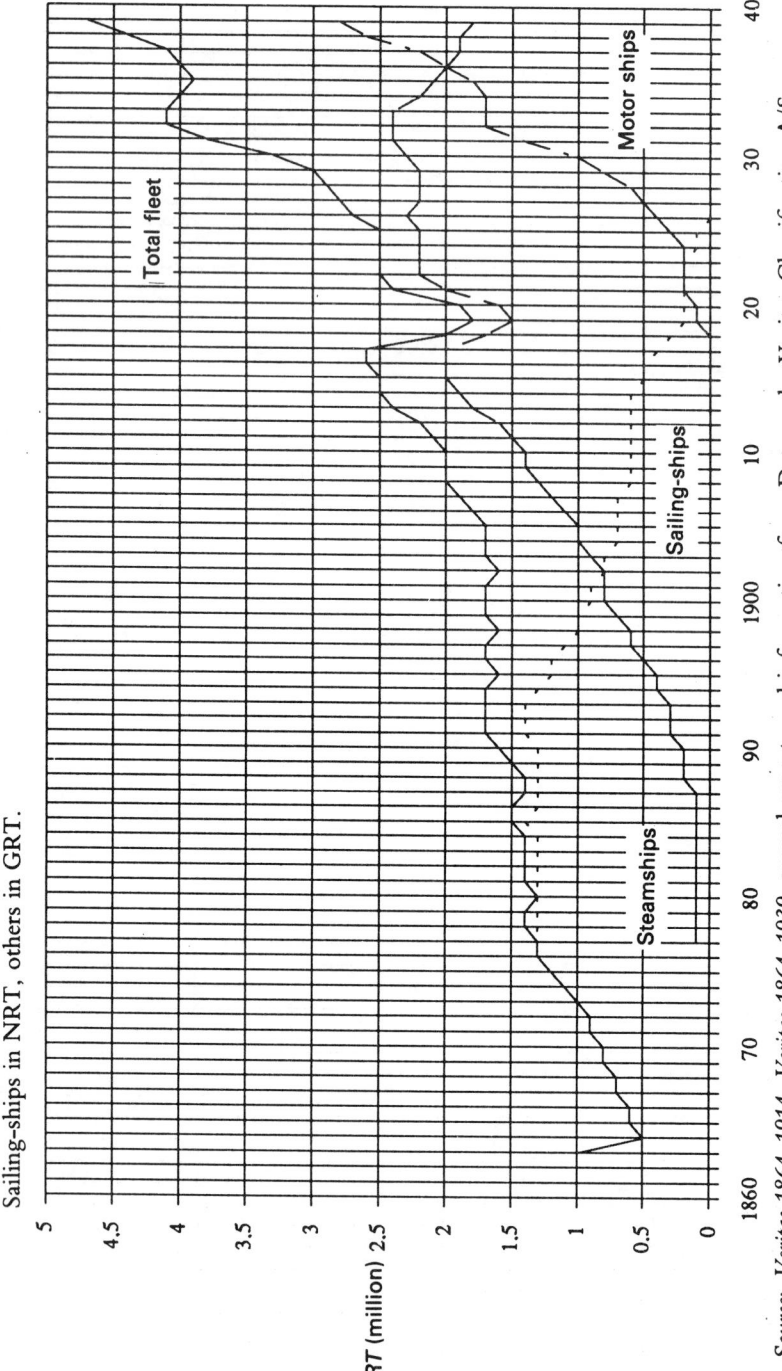

Source: Veritas 1864–1914, Veritas 1864–1939, annual reports and information from Det norske Veritas Classification A/S.

have also excluded ships of less than 100 tons GRT or NRT to remove the typical local ships and fishing boats.

An approach based on the amount of machinery installed would perhaps be more relevant if we wanted to measure the number of objects. Here, however, we meet the problem of economies of scale. In the 1870s Norwegian steamships were rather small compared to the deep-sea part of the sailing-ship fleet. As steam technology was developed, steamers came to be the largest ships within the overall fleet. However, as sail technology evolved in the 1890s sailing-ships also became large as steel replaced wood as building material. Much the same holds for motor ships. In the 1910s these ships were often small, but in the late 1920 and 1930s the average size of these ships grew much more rapidly than steamers and ended in 1940 more than three times as big as the average steamer.[9]

A picture of this development based on objects, rather than some crude measure of economic importance, would thus show a much quicker start for steamers and motor, but a slower diffusion rate, particularly for motor ships. However, measuring numbers depends very much on the composition of the fleet, particularly on the number of smaller vessels, above the lowest counting level (100 tons). One sailing-ship of 101 NRT would count the same as one large tanker of 9,000 GRT. The idea of counting objects ('technologies') is problematic as long as the scale is so different. The problem is not only that the economic importance of the ship-types differs greatly, but also that it is fundamentally difficult to compare a large diesel engine and the modest rig of a small sailing-ship. They are basically two different things.

2   The 'Object' Approach in International Perspective.

Beyond the description of the phenomenon it is impossible to say anything about the pace of the diffusion process without comparing it with other processes in other countries. It is commonly accepted that the Norwegian transition to steam was slow, and that the transition to motor was rather quick. We also know that the American and the Canadian fleets also were rather slow in the transition to steam, while Britain and the main Continental powers

---

9. Statistics from Det norske Veritas.

were rapid.[10] Instead of comparing the Norwegian fleet with the different countries, we shall take another view and shorten the discussion to a comparison of the world average. In Figure 12.3 we have shown the Norwegian fleet's share of the world fleet, and those of the Norwegian sailing-ship fleet, steamship fleet and motor ship fleet as proportions of the world's sail, steam and motor fleets respectively.[11]

This, of course, is based on an acceptance of the approach we have discussed above and the mixture of both an economic impact study and a technological object approach. We shall, however, use it to raise some new question about the concepts we have already introduced.

Figure 12.3 displays a peculiar phenomenon, at least if we stick to the traditional approach of quick adopters and laggards. From the 1890s up to the First World War the Norwegian share of the world's sailing fleet increased from double to three times the share of the total Norwegian fleet. In other words, a sign of more or less extreme laggards. After the war, however, the opposite was the case: the Norwegian proportion of motor ships in its fleet was approximately three time the world average. Indeed, a very early adopter! In general, steamships were under-represented in the Norwegian fleet throughout the period.

It is perhaps a little problematic to use the figures around the First World War as long as the total tonnage of both sailing-ships and motor ships in the world fleet was rather low. But this aside, in the other periods when the figures are reliable we must conclude that an extremely backward fleet was transformed to an avant-garde fleet in a time span of very few years. In the terms of technological determinism and a linear, progressive attitude to history this might be regarded as the triumph of reason and clever ship-owners. However, this Whiggish approach to history is not acceptable to the serious student of the history of technology, and we want in what follows to look behind these figures.

It has often been argued that economic backwardness can be turned to an advantage because one does not have to repeat the mistakes made by forerunners.[12] In this case, we could argue that the reason for the swift transfer to motor technology has its roots in

---

10. Statistics from Lloyds and Det norske Veritas.
11. Statistics from Lloyd's and Det norske Veritas. In particular Veritas (1914) and *Det norske Veritas. Beretning til 75 årsjubileet*, Oslo, 1939.
12. The tradition of this view goes back to A. Gerschenkron, *Economic backwardness in Historical Perspective*, Cambridge, Mass., 1966.

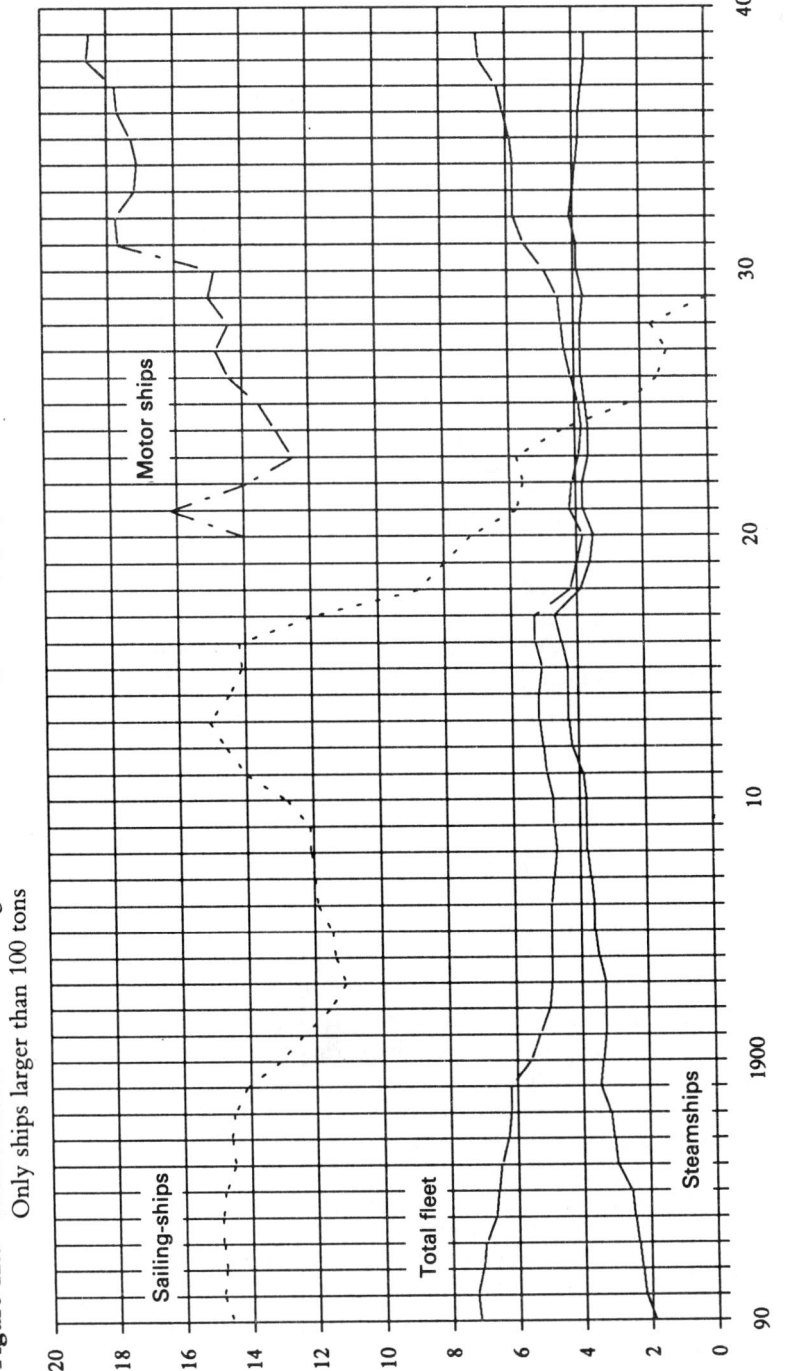

Figure 12.3 Relative shares of the Norwegian merchant fleet, separated by propulsion, related to the total world fleet. Only ships larger than 100 tons

Source: Det norske Veritas. Beretning til 75 årsjubileet 1939, Oslo, 1939

the weak steamship tradition, which again is explained by the strong sailing-ship tradition. However, such an approach under-estimates a well-known problem for the process of technological change: the binding and stiffness of the capital already invested.[13] Investments made on land are often an irreversible act in the sense that it is difficult to sell the capital goods and start again with a new technology. At sea, however, the buying and selling of second-hand tonnage is quite common, and has always been. Hence the traditional argument for the advantage of being backward is not linked to the investment in technologies, but more to other el-ements: new ways of organising the fleet, the possibilities for quickly entering new trades and so on. Thus another tacitly ac-cepted premise of the 'object' approach does not hold: the irreversi-bility of earlier investments. In shipping the investment cycles are different because of the possibilities of selling capital goods on the international market. The commonly accepted putty–clay approach turns into more of a putty–putty situation for shipowners.[14]

The nature of the capital stock in shipping should not therefore have prevented the diffusion of steam. Nor are there any indica-tions that the organisation, financial structure or size of firms in the Norwegian industry were obstacles to diffusion. We are left with the problem of the laggards who became leaders.

3 An Economic Approach.

The object approach and the language of technological determin-ism (laggards and early adopters) have their obvious limitations. As we have seen, when we start counting the technologies, economic impact creeps into the concepts. Why then not try to use a purely economic approach to the process of diffusion, translating anything of importance into economic concepts and values? Perhaps it is possible to compare steam-engines, rigs of sailing ships and motor-ships, and construct a common denominator – a measure of value?

At hand we find the neo-classical approach of production func-tions whose main objective is to compare different input–output

13. Typical in this tradition is a study by W. E. G. Salter, *Productivity and Technical Change*, 2nd edn, Cambridge, Mass., 1966.
14. The concept of putty–putty models and putty–clay is taken from L. Johansen and E. S. Phelps. L. Johansen, 'Substitution vs. Fixed Production Coefficients in the Theory of Economic Growth, a Synthesis', *Econometrica*, vol.27, 1959, pp.157–75; E. S. Phelps, 'Substitution, Fixed Productions, Growth and Distribution', *American Economic Review*, vol.51, 1961, pp.638–43.

functions on the basis of three different input concepts: labour, capital and the rest – technology. The most important analysis done on the Norwegian fleet along these lines is a doctoral thesis by Ole Gjølberg.[15] His work is restricted to the first transformation, from sail to steam, so we limit ourselves to this first change.

Gjølberg's work is interesting, because he sets up two models of behavior to explain the diffusion of new technology. One is based on what I have called early adopters and laggards; Gjølberg's approach here is built on the psychological behaviour of the owners. The other is based on rational and maximising behaviour of the firm. Gjølberg concludes that the shift was based on a continuous adoption process based on prices, costs and productivity of the different technologies in question. He arrives at this after having estimated production functions, including having made the traditional marginalist arguments. In his eyes, steam was accepted from the start in the Norwegian fleet while the implementation rate was dependent on real economic considerations.[16]

Arguments can of course be made against Gjølberg's approach and his conclusions. One important critique has been based on the estimate of wages, which has been criticised for being too low.[17] Another is of course the assumptions of the model, where the marginal substitution rate and the marginal economic approach in general are taken for granted, and result in a conclusion that the owners behaved in the way that they were supposed to act. Gjølberg is, of course aware of this. Nevertheless his conclusion of rational behaviour is probably correct. There are, however, two important modifications. In history, that is in real life, we never find static equilibrium, and there is no reason to believe that any real situation can be simulated through the theoretical assumption of such an equilibrium. What Gjølberg misses, and hence underestimates, is also the different freight options open to the owners. Two factors, namely dynamic instability and knowledge of market alternatives, limit flexibility and put serious constraints on what can be conceived as rational behaviour.

The critique of this type of approach is not only based on the

---

15. Ole Gjølberg, *Økonomi, teknologi og historie. Analyse av skipsfart og økonomi 1866–1913*, Bergen, 1979 (mimeo).
16. Ibid., p.254.
17. L. R. Fisher, H. W. Nordvik, 'From Namsos to Halden: Myths and Realities in the History of Norwegian Seamen's Wages, 1850–1914', *Scandinavian Economic History Review*, 1986, pp.41–64. They criticise Gjølberg for his estimates because he based his analysis on wage levels in Norwegian ports, not on international ones. Besides the deflation of the wages has been critically revised in Fisher and Nordvik.

assumptions made about behaviour and the transformation of different technologies into a common value concept, but is also a question of the limits to rationality. It is possible to argue that the owners were locally rational with regard to the tradition they were acting within, but were not necessarily rational in any global meaning of the word. In other words it is possible to argue for a type of bounded rationality, where economic considerations are important within the boundaries, but not outside.[18]

This leads us to an important question: what were the boundaries and how were they perceived? Gjølberg is at least halfway right: technology was not the limit of this bounded rationality, at least not alone. The result of action taken by ship-owners, however, made it look as if technology was the problem.

Gjølberg has an interesting observation in his material on this. If we accept his neo-classical model and his empirical data series, he shows a switch in the bias of technological change in the 1890s: up to 1890 and after 1900 he finds a neutral technological change. However, in the 1890s he finds labour-saving change. This change does not necessarily have anything to do with technological shifts, but is more an answer to increased competition and rationalisation of both sailing-ships and steamships. However, during the 1890s average size in the Norwegian steamship fleet increased, from 600 GRT to close to 1,000 GRT (that is, of ships above 100 GRT). From 1900 to the First World War steamers grew only very slowly in size. The sailing fleet, however, had a fairly stable average size of around 500 NRT.[19] Hence Gjølberg's conclusion of labour-saving technological change may be based on two different phenomena: increased investment in larger steamships, including labour-saving effects due to economies of scale, and rationalisation on board sailing-ships – making it possible to sail the same ships with a smaller crew. In addition Gjølberg has also calculated a stagnating real wage. However, it is not crucial to the theoretical argument that the real wage was stable or falling – the point in question was the tendency to cut costs, of whatever kind possible.[20]

This leads us back to the question of bounded rationality and the traditions within which the ship-owners worked. This tradition

---

18. See H. Simon, *Administrative Behavior*, 3rd edn, New York, 1976; R. R. Nelson and S. G. Winter, *An Evolutionary Theory of Economic Change*, Cambridge Mass., 1982; G. Dosi, 'Technological Paradigms and Technological Trajectories', *Research Policy*, vol.11, 1982, pp.147–62.

19. *Det norske Veritas, 1864–1914*, pls 7 and 8.

20. This may be somewhat underestimated by Gjølberg's opponents, L. Fisher and H. Nordvik, *From Namsos to Halden*.

was not marked with different technological trajectories.[21] On the contrary the limitations must have been of other kinds. If this is correct we can go a step further to look at the introduction of the motor ship. If the constraints were not technical in nature then the paradox of figure 12.3 is no longer a paradox, but simply a measure of something that had some basis other than a shift from technological conservatism to technological radicalism.

4 The Sources of Ship Technology.

One way of studying the hypothesis of technological conservatism is to take a look at the flow of technology in shipping. As we have argued above, the ship-owners themselves usually had rather small companies with little technical expertise in-house. In fact the companies often were rather loose organisations, including for example part-ownership companies and single-ship companies. Typical is our earlier remark concerning terminology, namely the ship-owner as *reder* in Norwegian. Hence it was tacitly understood that the ship-owner (*reder*) was the same as the company and that his obligation was to run a ship and not to have a staff of experts in-house.

The technology had then to be transferred to the owner from outside experts. This was done in several ways. Other owners, shipyards and brokers were the most important ways of acquiring ships, new or old. Here it is again very important to make the distinction between ship-owners and shipyards. The Norwegian shipyards had to acquire new technologies in quite other ways than the ship-owners. The active *building* process presupposed quite other learning processes than the *use* of their different products.

Looking at the social and geographical context of Norwegian yards we can see the problems of switching from sail to steam. The sailing-ship yards were situated along the south coast of Norway, integrated into the local communities in numerous ways. These communities also of course included owners, who in the days of sail were also predominantly situated along the south coast, and sailors. However these yards had severe problems shifting to iron and steel steamships. In fact very few of them managed to do so. Instead another tradition took over, that of the old mechanical

---

21. The concepts of technological trajectories have been stressed by economists criticising the marginalist argument, as for instance Nelson and Winter, *An Evolutionary Theory*, and Dosi, 'Technological Paradigms'.

engineering companies in the larger cities, situated elsewhere. However only one of these yards attempted to, or succeeded in producing motor ships before the Second World War, namely the Aker company in Oslo.[22]

If the yards had problems with the changing technologies, however, it was much easier for the owners. A ship may be regarded as a package of technology, bought ready for operation. The learning process of operations is significantly easier than for the production of the same package. Besides, operations in international waters also made possible the use of the international labour market for skilled seamen. Ships also had the particular feature, noted above, of being an investment which was easy to trade after the initial outlay had been made.

In fact, Norwegian owners had a very long tradition of buying and selling ships internationally. From the 1860s up to around 1880 most of the new Norwegian sailing ships were built in the south-coast yards. From then on, however, a steady and large influx of second-hand sailing-ships took over. They were bought from all over the world, wherever ships were sold. The transition to steam rendered old sailing-ships obsolete to foreign and better-off owners. These ships found their way to the Norwegian fleet. They were cheap, but not always in good condition. The Norwegian fleet changed slowly to a fleet with one of the worst wrecking statistics in the world around the turn of the century. The figure was on average two or three times as high as the world average in the ten years following 1900. It is from this period also that the fleet got a name for being backward, and the owners for being technological laggards.

New steamship technology entered the fleet from two main sources: national yards and British yards, with about half from each country up to the First World War.[23] The important point was, however, that this development was rather slow, compared with other national fleets. British shipbuilders became, of course, the major suppliers not only to the Norwegian fleet, but to most other countries as well.[24] Subsequently a second-hand steamer market developed around the world.[25]

22. Håkon W. Andersen, 'Norsk skipsbygging gjennom hundre år', in E. Lange (ed.), *Teknologi i virksomhet. Verkstedindustrien i Norge etter 1840*, Oslo, 1989.

23. Statistics from Central Bureau of Statistics in Norway, shipping statistics, several years.

24. See Pollard and Robertson, *The British Shipbuilding Industry*, and Lloyd's statistics, several years.

25. This became a major problem for the classification company in Norway. See Andersen and Collett, *Anchor and Balance*.

The source of motor ship technology was, however, other countries. Of crucial importance here were the Swedish yards, in addition to Danish and German. In the early phase, during the interwar period, the second-hand market for this technology was undeveloped, and hence the technology was mainly acquired from new buildings at yards in these countries. As mentioned, only one Norwegian yard built large diesel motors in the period, as a licensee of the Burmeister and Wain yard in Copenhagen.

This picture of the typical Norwegian ship-owner is one of a man who could choose between technologies without too severe constraints on the choice of technology seen from the supply side. The international market for ships was well known to him, as we might expect, since his business was to a large extent based on international trade. Typically, only a small fraction of his business concerned Norway and Norwegian goods. Most of it was cross-trade between foreign countries.

From this short overview of the availability of new technologies we must conclude that the typical Norwegian owner did have a real choice of alternative technologies and that the different options were open to him most of the time. In shipping the diffusion process or, better, the potential diffusion process was only to a very small extent hindered by lacking technology. For shipyards, however, the problems were quite different and much worse. This important distinction, between owners and yards, is crucial to the understanding of the diffusion process, as the two activities, shipping and shipbuilding, display quite different characteristics both as social and as economic activities.

5 Traditions and Constraints in Shipping.

Even if the owners were free to choose technology, their choices were not without constraints. Technology was only a part of their business strategy or business tradition. The third way into the problem of technology transfer is as typical as it is traditional: is the problem of diffusion put in a fruitful way? In fact this is a question often raised regarding technological innovation and diffusion. Is it possible to take out some part of the human experience and study it in its own context? Obviously the answer is yes (history would have been impossible without it), but with important modifications. What we actually do when we study the transfer of ship technology is to isolate one part of an experience that to the ship-owner presents itself as a whole. Traditional methodology in history makes this problematic. In fact we devalue what is the strong side of historical studies: to understand a particular phenomenon as a whole, regardless of the type of variables or field of human experience that can contribute to the understanding.

To the owner, technology is only one part of his activity, maybe not even the most important one. Economy, technology, cultural tradition, financial strength, social facts and phenomena, relationships, information on trades and so on add up to the system in which the owner is engaged and the background against which he makes his decisions. Technology is integrated into this web of knowledge, phenomena and facts in a very organic way, not as some external force with an even more external impression of progress.

This approach, it can be argued, is typical for historians.[26] Thomas Hughes, for instance uses the metaphor of technological systems where technology, economy, culture and politics are woven in a 'seamless web'.[27] However, social scientists have likewise taken up the idea and tried to develop new concepts and theories around it. It is not so difficult to argue a kind of convergence between historical and sociological studies of technology. However, though the concepts may vary considerably, the ideas in this new sociology of technology are not as diverse as the concepts may indicate. Concepts like actor-network building of heterogeneous associations, stabilisation and closure of technological solutions, socially shaped technology and heterogeneous engineering are only some examples of this very interesting development which probably will also influence the historical studies of technological diffusion.[28] Another slogan in science and technology studies is set by B. Latour: 'Follow the scientists'.[29] In our context it should be rephrased as 'study the owner', regardless of what field of society's web he is currently working within: technology, financing, culture, politics and so on.

Let us try to follow this path of inquiry to see if it can offer a more satisfying explanation of the seemingly contradictory pattern: the owners as laggards in the transition from sail to steam and after

26. See D. Noble, *Forces of Production*, New York, 1984. Most systematic perhaps is T. Hughes, 'The Evolution of Large Technological Systems', in W. Bijker, T. Hughes and T. Pinch, *The Social Construction of Technological Systems*, Cambridge Mass., 1987, pp.51–82.
27. T. Hughes, 'The Seamless Web: Technology, Science, Etcetera, Etcetera', *Social Studies of Science*, vol.16, no.2, 1986.
28. An introduction is to be found in Bijker, Hughes and Pinch, *Technological Systems*. See also M. Callon, J. Law and A. Rip, *Mapping the Dynamics of Science and Technology*, London, 1986.
29. B. Latour, *Science in Action*, Milton Keynes, 1987.

a short time, the avant-garde of the new motor ship technology. To follow the owners is not an easy task, particularly not on an aggregated level. Such a study can be based only on micro-studies of the owners and their business strategies, their traditions and their hopes. This alone is, however, not enough. We also have to compare their attitudes and strategies with those of shipping companies in other seafaring nations. That is we must study Norwegian owners as members not of Norwegian society but as members of the international shipping community. Then at last it may be possible to draw some more stable conclusions on the particularities of Norwegian shipping.

Much has been done in studying Norwegian owners' behaviour. However, the literature has two main drawbacks. Firstly, it has been occupied with the large and usually successful companies, that is companies which have succeeded in the different technological transitions. For our purpose the selection of histories of shipping companies should be an average. Secondly, shipping history in Norway has always been a part of Norwegian history, not part of the world's shipping community's history. This can be understood only against the background of the particularities of Norwegian history. As a small country on the periphery of Europe with one of the world's largest fleets, shipping was a way to success in Norway. Nobody bothered how fortunes were made in a large and unfriendly world, as long as status and recognition inside Norway were obtained.

The first decades of the century put some owners at the top of society's ranking list. Owners became prime ministers and held other ministries. Hence the story of shipping in a way became partly mythology inside Norway, stressing the pioneering efforts of new trades and the heroic efforts of owners, captains and seamen as well. This legacy also put its mark on many of those who wrote the history of Norwegian shipping.

Even though the situation is cumbersome and our knowledge about the owners' behaviour and strategies is far from sufficient, we may at least draw some tentative conclusions, or perhaps, even better, hypothesise about their behaviour as members of the international shipping community.

6 Follow the Owners!

The structure of the sailing-ship industry in Norway imposed some crucial limitations on the transition to steam. Let us start with the structure of ownership. Even though different arrangements were made, it would be fair to say that the typical form (in the sense of an ideal type) of ownership in sail was the part-owning arrangement. That is participants bought one or more shares in the ship, they were responsible for this share, but not for anything more. This system worked very well on the south coast of Norway, because it made it possible to concentrate capital on ships. In a district with scarce access to liquid capital, a system of part-ownership made it possible to concentrate the available capital, whether in the form of money, raw material, labour or other services. The system of shipping and shipbuilding established on the south coast became a cultural form, involving the small communities in a number of ways: as shipbuilders, as sailors, as owners and as part-time farmers and fishermen. But this served the purpose only up to a certain point. The capital was concentrated in one ship, but the command over larger capital was problematic.

When a part-ownership was dissolved, the capital was spread among the part-owners. It made it difficult to build up larger capital under single control. In the early period this was not much of a problem. Shipping was a growing industry, and the particular type of trade chosen made the system capable of reproducing itself. Tramp-freight became the typical kind of trade for the Norwegian owners.

As the price of transportation based on steamships fell and competition hardened during the last decades of the nineteenth century two solutions were open to the south-coast owners: one was to enter the more capital-intensive steam freight service, the other to continue to reduce costs and stay in the business of sailing-ships. The first option was hindered by the problems of concentrating capital, changing trades and price competition; the other was more compatible with the social and economic setting of the south-coast owners.

To continue to compete with help of the old technology was not easy. However, as long as tramp-freight involved cargoes with a rather low price per ton, such as timber, coal and other goods, the time spent in transportation was not very important. The capital requirement was reduced by buying second-hand tonnage abroad, or continuing to sail the old sailing-ships. The logic of competition became to cut whatever cost possible. Ships were rigged down in order to be sailed with smaller crews, wages were cut as much as possible, and insurance, maintenance and classification expenses also cut back. At the same time the slow-down of world trade reinforced the cost-cutting line and made the competition even more fierce.

Of course this was a strategy that could earn the owners income only for a limited number of years. There were owners who tried to change to steam, and managed. But overall, the days of the south-coast owners were numbered for the time being. Instead the merchants in Bergen became the dominant owners of the early steamship fleet. Bergen gave much easier access to the restricted available capital, and the organisational form of limited liability companies made for better control and command over the accumulated capital. In the first decade of the twentieth century, however, the crucial point of steamship technology was linked to a new arrangement of the trades, that of the liner companies. Contemporary observers argued that the future of Norwegian shipping was dependent upon a change away from the small tramp-based companies towards larger multi-ship liner companies. This was the more so as internationally the more advanced western societies had for a long time developed this kind of trade. The important fact was that the quasi-monopoly established by the liner conferences expelled the small Norwegian owners from some kinds of trade and made it difficult for them to enter any trade where they dominated.

Some owners, mainly in Bergen and Oslo, succeeded in building large companies, but their part of the fleet remained rather modest. In 1911 9.2 per cent of the fleet consisted of liners, but only a small 0.2 per cent was working exclusively abroad.[30] Steam technology remained problematic for the bulk of Norwegian owners as the old system of capital accumulation and concentration was not sufficient for the new technology, and at the same time their tradition of tramp trade was threatened.

What then about the new technology, the motor ship? How is this to be explained against the background of the typical tradition of Norwegian shipping, tramp shipping in rather marginal fields with a rather weak company structure? To understand the development of the motor fleet is to understand the particularities of the international oil trade and the new shipbuilding nation, Sweden.

Traditionally, oil transportation had been in the hands of the larger oil companies. In the second part of the 1920s, however, as the demand for oil rose steadily, the companies changed their policy. From then on the marginal transportation capacity was hired from independent owners. The breakthrough for Norwegian owners came with the selling of Shell's old Anglo-Saxon fleet of

---

30. Worm-Müller, *Den norske sjøfarts historie*, vol.2:3, p.408.

steam tankers from 1926 onwards.[31] With the sale came also contracts for carrying oil for the oil companies for several years. The strategy of the oil transportation company was thus to hire marginal capacity and let independent owners bear the risk of reductions in the volume of oil trade. This set-up was not particularly appealing to the larger shipping companies, but for the smaller ones it offered a possibility. Even for the former south-coast owners it became a possibility to be back in trade. Being rather marginal in international shipping overall, taking large risks was not new to the smaller Norwegian owners. Buying the old and outdated Anglo-Saxon fleet with its very close connection to Shell, the owners took a risk, not a too big risk, however, for individuals with very little to loose.

It was typical of the trade that at the same time as these new and small owners entered the oil trade on a larger scale, some of the larger shipping companies in Norway withdrew from the same trade, probably evaluating the conditions as too insecure.[32] The oil trade was in a way a continuation of the old tramp tradition, at least if we compare it with the liner tradition. In a way at least part of the transportation contracts involved several years of work for the oil company. In this respect the contract terms differed from the traditional tramp contract. In oil, the ship was not only chartered for one trip or a shorter series, but for several years. I have chosen to consider this as a prolongation of the tramp tradition since it did not involve a large professional organisation at the ship company's head office. The oil trade instead invited a decrease in this organisation. Subcontracting for the larger oil companies removed even the work with different tramp contracts from the owners' organisation. The residue was linked to personnel administration, supplies etc, much of it taken care of on board the ship itself. The tradition from the days of the sailing-ship with strong captains and weak organisational structures on shore prevailed owing to the oil companies. But in the modern days of the 1930s the captain's work was confined to operating and maintaining the ship, and not to negotiating shipment contracts any more.

This could all have been a sad and tragic story of the fall of Norwegian shipping if two important features had not occurred, one independent of the owners and the second in symbiosis with

31. L. Nørgård, *Tankfartens etablerings- og introduksjonsperiode i norsk skipsfart 1912–1913 og 1927–1930*, Bergen, 1961. See also Andersen and Collett, *Anchor and Balance*, chap. 3.

32. The Wilhelmsens company was one that withdrew. See Nørgård, *Tankfartens* and Kr. A. Olsen, *Wilh. Wilhelmsen i 100 år*, Oslo, 1962.

them. The first was the fact that oil demand increased in the world during the whole interwar period. This was partly due to an increase in energy consumption, partly to the substitution of oil for coal. This stable increase in demand, combined with an increasing distance of transportation, made the marginal fleet essential to the oil companies, which were trying to require sufficient transportation capacity. Hence what could have been a traditional and very dangerous situation turned to a very favourable one. The tragedy was not to be seen until the collapse of the oil market in the mid-1970s, when the Norwegian tanker fleet was hit perhaps harder than any other nation's fleet.

The oil trade became a flourishing trade for the Norwegian owners. It was a trade that made it possible to pursue economies of scale to a hitherto unknown level. A more or less fluid cargo, carried over very long distances, between a few specialised ports over several years opened up possibilities for a specialisation of ships in just this trade. Motor ships early became the most economic alternative for this type of trade (large ships, long distances), while steam was still dominant among the more flexible tramp carriers and liners.

Just as the Norwegians entered the oil trade as subcontractors for the large oil companies, and thus as outsiders, the Swedes entered modern shipbuilding of this type of tankers. The link between Norwegian owners and Swedish yards grew strong; both had in a way hit 'black gold' as outsiders. One reason for the fruitful symbiosis was that the Swedes specialised in welded motor tankers built particularly for the growing number of Norwegian owners.

There was however one more link in this system, relating to the capital question. With freight contracts with large multinationals for up to eight years it was possible for the Swedish banks in collaboration with Swedish yards to offer large credits and loans to the always capital-hungry Norwegian owners. This system worked through bills of exchange. At what future date these bills were to be redeemed, and at what rate of interest, was regulated by the shipbuilding contract. The yard could then cash the bills at an earlier date and the bank carried the credit.[33] There were several reasons for the banks to accept this system: the contract with the oil company was a good security, as were the guarantees of the large Swedish yards.

33. Jan Bohlin, *Svensk varvsindustri 1920–1975, Lönsamhet, finansiering och arbetsmarknad*, Meddelande från ekonomiskhistoriska institutionen ved Göteborgs Universitet, no.59, Göteborg, 1989, pp.105–31.

In this way a modernisation based on large motor ships was a continuation of the old tramp tradition with small ship-owning companies. What appeared as a paradox, the backward owners who became the most progressive can instead be analysed as both a continuation of a tradition (size of companies, trades, and problems with capital concentration) and as intentional choices based on the limitation set by the same tradition (taking the risk of being dependent on just one oil company and one trade). Developments outside the Norwegian owners' control secured the success of the owners' choice (the increasing demand), just as the owners' demand for ships was met by the much stronger Swedish yards and bank system. Hence the success of the Norwegian owners and the Swedish yards represented two sides of the same international development.

We have come a long way now from the start, which stressed the dichotomy of the technology in question and raised the problem of the laggards that became the modernists. We looked briefly at the rational choice approach as a base for neo-classical economics and ended with a combination of culture, traditions and intentions, a kind of bounded rationality where the boundaries are explicitly set, both with regard to the culture and to the economic and structural facts. In this we have subsumed technology under a larger understanding of the situation and intentions of the owners, where technology was only a part. What appeared as a contradiction evaporated and became understandable as soon as we stuck to the old tradition of historical analysis: never to extract pieces of reality out of their proper context, whatever the proper context may be. History, even the history of technology, is concerned with the history of human beings. It is here we find the whole, or the proper context.

Even though the empirical result of this study is tentative and stresses the theoretical sides of technology transfer and diffusion in perhaps an unusual way, it is possible to draw some conclusions of how technological transfer and diffusion should be studied. The danger is that of being seduced by the genealogical approach so easy to accept. The artefacts are charming, so is the model with a country of origin and the spreading of technology. But the origin is only one part of the story. As we switch to a more dialectic approach it is not only the origins that account for the diffusion, but the whole process of diffusion has to be studied as a field in its own right, stressing the human activity that is involved. It can never be explained only from the 'origin' of the technology.

The history of technology, economic history or the history of technological diffusion have to deal with humans; the units of these types of history are all the time humans, not artefacts. The artefacts may raise problems, but the answer to these problems lies on the human side, not in the artefacts themselves.

13

The Development and Diffusion of European Water Turbines, 1870–1920

Gunnar Nerheim

In his article on waterpower in the century of the steam engine Louis C. Hunter claimed that 'the role of steam power in early industrialisation has been exaggerated and the reasons for its preeminence misunderstood'. A common view, according to Hunter, sees the steam-engine as superior to water as a source of motive power. 'Yet much of British experience with stationary power was exceptional rather than typical, and thus in particularly striking contrast to the course of American industrial development prior to the 1860s.'[1] In most of the research done up to that time the role of water power in the industrialisation process had been underestimated. Hunter himself later wrote a splendid account of waterpower and the diffusion of water turbines in the United States in the nineteenth and early twentieth centuries.[2]

So far no similar study has been published on the development and diffusion of water turbines in Europe. One can find articles dealing with specific regions or countries in Europe. Some years ago the German historian of technology Hans-Joachim Braun wrote a stimulating article on the invention and diffusion of water turbines in the mid-nineteenth century. Lacking readily available statistical information, he made no effort to assess the diffusion of turbines in Germany. Having done little or no archival work on German turbine builders, Braun picked most of his 'empirical

---

1. L. C. Hunter, 'Waterpower in the Century of the Steam Engine', in B. Hindle (ed.), *America's Wooden Age: Aspects of its Early Technology*, Tarrytown, N.Y., 1975, p.160.
2. L. C. Hunter, *Waterpower: A History of Industrial Power in the United States, 1780–1930*, Charlottesville, 1979.

findings from secondary sources.[3] His generalisations were based on empirical research on the diffusion of turbines in Britain. This was very unfortunate. Even though he strongly insists that the diffusion process in Germany and other countries probably did not follow the British pattern, Braun nevertheless felt that the British experience could be used as a working hypothesis for more detailed regional studies on the diffusion of water turbines in Germany and elsewhere. Since the British market for turbines reached its peak in the mid-1850s and thereafter declined and levelled off, he infers that this was probably also the case elsewhere in Europe.[4]

After having pieced together empirical information on turbine-builders in different countries, I can safely conclude that Braun is definitely wrong in suggesting that the European market for turbines might have followed the British pattern. Certain industrialising regions in Europe never saw the steam engine in use. If a steam engine was bought, it was used only as a reserve prime mover.

The aim of this paper is to contribute to a more thorough understanding of the relative importance of water power in the industrialisation of different European regions outside Britain. My interest in the topic was excited in connection with research on technology transfer from Germany to Norway in the nineteenth and early twentieth centuries. In this broad context I was puzzled to see that while Norwegian machine shops were the main suppliers of water turbines to Norwegian industrial enterprises in the 1880s, with the coming of large hydroelectric power plants almost all of the larger Norwegian turbine orders went to German or Swiss firms. Why did the Norwegian producers lag behind?

I tried to answer this question in a comparative context, and found that it was not only the Norwegian turbine producers that came to lag behind. The same was true for machine shops in other countries - in Germany, Britain, France, Hungary, Sweden and Finland. Only a few firms were able to meet the new technological challenges in turbine-building set by electricity producers wanting to build ever larger generators. Turbine- building became specialised. In the 1890s even American turbine producers were surpassed by their European competitors. The first Niagara turbines were of European design.

In the following I shall try to identify some patterns in this process. Roughly speaking, I shall cover the period from when

3. H. J., Braun, 'Technische Neuerungen um die Mitte des 19. Jahrhunderts. Das Beispiel der Wasserturbinen', *Technik-geschichte*, no.4, 1979, p.291.
4. Braun, 'Technische Neuerungen', pp.296–7.

water turbines were an integrated part of complete machinery orders to the time when they were sold separately, and tailored to the specific site in question.

1 The National Period in Turbine-Building, or Water Turbines as an Integrated Part of Complete Machinery Orders

The most common turbines used in the second half of the nineteenth century were invented by Europeans. In 1827 the French engineer Benoît Fourneyron developed an inward-reaction turbine with much higher efficiency than the traditional water wheels. With his perfected Prony-brake he was able to demonstrate that his turbine had an efficiency approaching 80 per cent.[5] Ten years later the Henschel – Jonval axial-flow turbine was invented by Carl Anton Henschel in Kassel, and was used for the first time in Holzminden in 1841.[6] The French engineer Feu Jonval, from the firm of Koechlin in Mulhouse, saw the turbine in operation there. He later perfected it, and applied for a patent in France in 1843. It has since been associated with his name. This was the leading reaction turbine for low-pressure installations used in Europe in the nineteenth century. The third widely used turbine in Europe was the Girard turbine, introduced in the 1850s. This was an action turbine well suited for larger heads.

From the mid-nineteenth century water turbines increasingly began to supplant water wheels in central Europe, North America and Scandinavia in regions rich in water resources. The turbine, running faster than the water wheel, required less transmission and was easier to regulate. Above all the turbine had a higher efficiency, and produced more power from the water available. The turbine made it possible to use waterfalls with both higher heads and larger quantities of water.

The diffusion of turbines in Europe was primarily a function of the rate of industrialisation. Power needs combined with abundant water resources led many entrepreneurs to install water turbines instead of steam engines. Accordingly turbines were introduced early in certain regions of France, Germany and Switzerland. The industrialisation process in countries in eastern Europe, in Russia

---

5. K. Keller, 'Benoît Fourneyron', *Beiträge zur Geschichte der Technik und Industrie*, vol.4, 1912, pp.79–95.
6. *125 Jahre Henschel*, Kassel, 1935, pp.65–73.

and Scandinavia was some decades later in coming, and consequently the diffusion of turbines came later too. Around the mid-nineteenth century it seems that no special skills or expertise were required to build turbines other than those that most machine shops had at their disposal. The proliferation of machine shops trying their hands at designing turbines is a sign of this.

Discussing the diffusion of water turbines in Finland, Risto Keskinen classifies the period from 1870 to 1895 as the Finnish period in turbine design. Looking only at his own country he is right, but on the other hand this was the period when most European countries engaged in turbine-building. In Finland machine shops like Fiskars and Tampella were important suppliers of turbines. In Sweden Karlstads Mekaniska Verkstad (958 turbines delivered from the start until 1914), Nydqvist & Holm (250 delivered from 1848 until 1900), Arboga Mekaniska Verkstad (more than 1,800 turbines from the establishment in 1856 to 1921), Qvist & Gjers, and Aktiebolaget Finshyttan (first turbine delivered 1874 and the most important turbine manufacturer in Sweden in the last two decades in the nineteenth century) played a corresponding role. The two most important turbine manufacturers in Norway were Myrens Mekaniske Verksted (620 turbines between 1867 and 1900) and Kværner Brug (286 turbines between 1873 and 1900).

One of the early turbine builders in Germany in the nineteenth century was the company of Nagel & Kämp in Hamburg. Other suppliers worth mentioning are Lüneburger Eisenwerk of Lüneburg, the machine shop G. Luther of Brunswick and Amme, and Giesecke & Konegen of Brunswick. In southern Germany the Augsburg part of the firm later known as Maschinenfabrik Augsburg-Nürnberg (MAN) began manufacturing turbines in 1847, and continued with this up to the end of the century. Of the turbines exported many went to Russia.[7] At the end of the century, however, the two best-known manufacturers of turbines in Germany were probably the companies of Briegleb, Hansen & Co in Gotha and J. M. Voith in Heideheim. From the start to 1903 Briegleb, Hansen & Co produced more than 2,200 turbines, mostly for customers in central Europe. But one could also find their popular 'Knoop turbine' in or around Edinburgh, and, in Scandinavia, at Bergen, Moss, Odense, Malmø, Åbo and Tammerfors.[8]

---

7. F. Büchner, *Hundert Jahre Geschichte der Maschinenfabrik Augsburg-Nürnberg*, Frankfurt am Main, 1940, pp.42–3.
8. A characteristic of the Knoop turbine was that it worked as a pressure turbine, even if the running wheel was placed under water. See E. Reichel, 'Aus der

The company of J. M. Voith in Heidenheim, established in 1835, built its first turbine in 1867. Turbine no.100 was delivered in 1889, no.1,000 in 1901 and no.2,000 in 1905. At the turn of the century J. M. Voith had become one of the most important manufacturers of turbines worldwide. In Switzerland Escher Wyss & Co. was the predominant manufacturer. From 1840 to 1902 the company supplied more than 3,400 water turbines to customers all over Europe, but mainly in central Europe. Other Swiss machine shops early on showed a marked interest in designing and manufacturing turbines. The firm of Joh. Jak. Rieter & Co. in Winterthur took up the manufacture of turbines in 1853.[9] From the 1860s the firm of Theodor Bell & Co., Kriens-Lucerne, manufactured machine tools, steam engines, steam boilers, transmissions, water wheels and turbines. They delivered their first Jonval turbine to a silk mill in Zurich in 1859.[10] The company showed great inventiveness in the design and construction of turbines, and got first prizes for their turbines and turbine regulators both at the world exhibition in Geneva in 1896 and in Paris in 1900. The firm of Piccard & Pictet in Geneva became famous among engineers after it designed the first 5,000-hp turbines for the Niagara Power Co. in 1893, built by J. P. Morris of Philadelphia. The most important company in Italy was A. Riva, Monneret & Co. in Milan, and in eastern Europe the firm of Ganz & Cie in Budapest. This list of companies is representative, but far from complete.

Because of the ready availability of coal in Britain, attention was concentrated on developing steam engines rather than water turbines. Nevertheless, turbines in great numbers were designed and installed by British machine shops for both British and foreign customers. The principle of 'Barker's Mill', still used in lawn sprinklers, was used by the Scottish engineer James Whitelaw to design a reaction turbine in the 1830s, known as the Scotch turbine.[11] In 1850 the Irish engineer James Thomson was granted a patent on his 'Vortex turbine', an inward-flow reaction turbine with adjustable guide blades. This turbine, along with the Scotch

---

Geschichte der Wasserkraftmaschinen', *Beiträge zur Geschichte der Technik und Industrie*, vol.18, 1928, p.52.

9. B. Lincke, *Die schweizerische Maschinenindustrie und ihre Entwicklung in wirtschaftlicher Beziehung*, Frauenfeld, 1910, p.40.

10. M. Hattinger, *Geschichtliches aus der schweizerischen Metall-und Maschinenindustrie*, Frauenfeld, 1921, p.107.

11. P. N. Wilson, 'Early Water Turbines in the United Kingdom', *Transactions of the Newcomen Society*, vol.31, 1957–9, p.223.

type, was manufactured for decades by British machine shops. Williamson Bros. of Kendal began manufacturing the Thomson Vortex turbine in 1856, and by 1881 had delivered 439 turbines of this type. In 1881 Gilbert Gilkes bought out Williamson Bros., dropped the manufacture of miscellaneous agricultural machinery, and turned increasingly to water turbines. Between 1856 and 1881 Williamson's had averaged fewer than seventeen turbines per year: between 1881 and 1900 Gilkes averaged fifty-two, of substantially higher average power.[12] According to Hans-Joachim Braun, between 1852 and 1870 Williamson Bros. built 417 turbines, and Bryan Donkin & Co. of London 302.[13]

If Britain as a whole was looking to steam power, there were still districts, such as Cumbria and parts of Scotland and Ireland in the late nineteenth century, where water power could be more viable. For instance, evidence of more widespread use of turbines in Scotland comes after 1870, writes John Shaw in his book on water power in Scotland.[14]

As a rule most machine shops seem to have constructed and built turbines as part of larger machine orders for equipping whole mills or factories. Turbine building in the Swiss firm of Escher Wyss, for instance, grew out of the firm's main business as manufacturers of textile machinery. This machinery firm, founded in 1805 by eight merchants and bankers from old Zurich families, was originally a cotton spinning mill.[15] In 1810 the mill operated 5,232 spindles, the size originally planned. Some of the skilled workers were no longer needed by the firm, but instead of letting them go, Caspar Escher considered taking up the manufacture of spinning machines for sale to outside customers.[16]

Around 1830 the manufacture of textile machinery became as important to Escher Wyss & Cie as textile manufacture.[17] Escher's son, Albert Escher, was the driving force behind this change. Before joining the firm in 1826 he had worked for three years as an

---

12. Ibid., p.234.
13. H. J. Braun, 'Technische Neuerungen', p.296.
14. J. Shaw, *Water Power in Scotland 1550–1870*, Edinburgh, 1984, p.498.
15. *150 Jahre Escher Wyss 1805–1955*, Zurich, 1955, p.3.
16. H. Hofmann, *Die Anfänge der Maschinenindustrie in der deutschen Schweiz 1800–1875*, Zurich, 1962, p.37.
17. 'In den nächsten Jahren entwickelte sich nun neben dem Spinngeschäft auch die Werkstatt, so dass, als im Jahre 1831 Maschinfabrikation und Garnhandel von einander getrennt wurden, sich die Gewinne beider Produktionszweige die Waage hielten; ja, als die Trennung erfolgte war, verschob sie sich in den folgenden Jahren ganz bedeutend zugunsten des Maschinenbaues', Hattinger, *Metall- und Maschinenindustrie*, p.56.

apprentice with Georges Bodmer & Co. in Manchester. From 1826 to 1835 the number of employees in the workshop increased from sixteen to 400, and yearly turnover in this division of the company grew from fl.60,000 in 1830 to fl.375,000 in 1835.[18] The firm frequently received contracts for complete spinning mills, and because of this the product range was extended to include transmissions and prime movers, both steam-engines and water turbines. Most of the production was exported, especially to Austria and Italy.

To illustrate the general pattern in the development of turbines, we shall analyse more closely the development at Escher Wyss, Ganz & Cie in Budapest and J. M. Voith in Heidenheim, Germany, and compare it with the two most important Norwegian firms. Ganz & Cie became famous as producers of the new gradual-reduction all-roller mills. Like Escher Wyss they supplied their customers with prime movers, steam or water. J. M. Voith in Heidenheim specialised in the manufacture of paper and wood-grinding machines, and delivered turbines to drive them. The same was certainly true for the most important Norwegian manufacturer of turbines in the nineteenth century, the firm of J. & A. Jensen and Dahl, Myrens Mekaniske Verksted, Oslo. This company specialised in machinery for wood-grinding factories and sawmills. Consequently the customers for most of the turbines delivered were the same as for sawmilling, wood-grinding or pulp machinery. The power needs of textile mills, flour mills and wood-grinding factories therefore came to influence the average size of turbines delivered by the different manufacturers. This will become evident when looking more closely at the above mentioned firms.

The pioneering company in turbine-building among these was Escher Wyss. From th 1840s the company built Jonval turbines. Around 1850 one of the firm's leading engineers, Zuppinger, considerably improved the efficiency of the Poncelet or tangential wheel. The Zuppinger wheel, as it was called, was much used in small streams with low heads. The first Girard turbine was manufactured by the company as late as 1870. From 1844 to 1875 Escher Wyss delivered 229 tangential or Zuppinger wheels, 469 Jonval turbines and 103 Girard turbines, or a total of 801 turbines. Of these 103 were delivered to Swiss customers, while the remainder were exported.[19]

In the 1840s the average capacity of turbines delivered was

18. Hofmann, *Anfänge der Maschinenindustrie*, p.42.
19. Ibid., pp.84–5.

around 20 hp. The power needed to drive the fast-running mechanical spinning and weaving machines was not large. They required an even speed, however, and turbines were much easier to regulate than the traditional water wheels. They were early furnished with automatic governors of the type used on steam engines. This was one of the main reasons that turbines found an eager market among textile manufacturers in water-rich regions. Moreover, if a textile manufacturer wanted to extend the capacity of his mill without having to move to another site, he could install a turbine to get more power out of the water available. In the 1850s the average size of the turbines increased to a level between 40 and 50 hp. During the next two decades the capacity of turbines installed stayed much the same or increased a little. Not until 1889 did the average size of turbines manufactured by Escher Wyss pass the 100-hp mark.

When we look more closely at the period 1871–1902, two points are worth mentioning. In the 1890s the Francis turbine replaced the Jonval turbine as the dominant axial-flow turbine, and the Girard turbine was replaced by large-impulse turbines with spoon-shaped buckets. These new turbine types also represented a breakthrough in the average size of turbines from the late 1890s. We shall return to a more thorough discussion of this quantum leap later on.

In the last decades of the nineteenth century the machine shop of Ganz & Cie in Budapest emerged as one of the leading machine shops in central and eastern Europe. The firm manufactured steam engines, locomotives, all-roller mills, electrical machines[20] and many other products, among them water turbines. After having exhibited an all-roller grinder assembly with corrugated rollers of chilled iron at an international exhibition at Nuremberg in 1875 , Ganz & Cie experienced a flow of orders. Between 1874 and 1885 the firm sold 13,000 roller grinders all over Europe.[21] Their main customers were to be found in Hungary, Austria, Italy, Germany,

20. Thomas P. Hughes characterises Ganz & Cie as one of the leading electrical manufacturers in Europe in the early period. In 1878 the firm assigned part of its extensive engineering works to the manufacture of electric lighting apparatus in Austria-Hungary, and by 1883 had produced over fifty installations. The firm was a pioneer in the development of the alternating-current system. The company's engineers Blathy, Déri and Zipernowski invented a transformer in 1885. By 1890 the Ganz & Cie system had been widely adopted. See Thomas P. Hughes, *Networks of Power: Electrification in Western Society 1880–1930*, Baltimore, 1983, pp.96–7.

21. On the invention and diffusion of the all-roller, gradual reduction mill and the role of Ganz & Cie, see Gunnar Nerheim, 'Fra bekkekværn til automatiske møller', in Helge W. Nordvik, (ed.), *Rent mel i posen. Bjølsen Valsemølle A/S og mølleindustriens utvikling 1884–1984*, Stavanger, 1984, pp.71–7.

Serbia, Bulgaria, Romania and Russia, but Norwegian flour mills were also among the company's customers. Ganz & Cie listed a small Jonval turbine of 2.8 hp in 1866 as the first ever delivered by the company. The business of designing and producing turbines seems to have grown out of the success of marketing the company's all-roller gradual-reduction mills. Until 1880 the company had delivered only five turbines. Between 1881 and 1884 the company got orders for another twenty-six. The largest, with a capacity of 300 hp was installed at the 'Union' tinplate roller mill at Altsohl.[22] Until 1895 it is evident that all larger turbines were delivered either to iron and steelworks or wood-grinding factories, while turbines installed at flour mills usually had a capacity between 10 and 20 hp. Outstanding are the three 700-hp turbines delivered in 1892 to drive the electric generators for the lighting of Sofia, Bulgaria.[23]

When comparing the average size of turbines delivered by Escher Wyss and Ganz & Cie between 1885 and 1895 one can conclude that Ganz & Cie manufactured slightly larger turbines, at least up to the 1890s. The reason for this seems to be that two of the company's group of customers had fairly large power needs in their iron and steelworks and wood-grinding factories.

Let us then turn to the companies specialising in wood-grinding machinery and see how significant this was for the average capacity of turbines delivered. The firm of J. M. Voith traces its history back to the 1830s when Johann Matthäus Voith extended the family forge to a small general machine shop for repair of textile, paper-making machinery and the like. J. M. Voith was to play an important role in the development of the wood-grinder.

In the 1840s the Saxon master weaver Friedrich Gottlob Keller began experimenting with grinding wood to pulp on an ordinary grindstone. In June 1846 he signed a contract with Heinrich Voelter, the son of a paper maker in Heidenheim and managing director of a paper mill in Saxony. The two of them were to develop and exploit the commercial potential of the invention through patents. During the years 1846–8 patents on the process were granted in Württemberg, Hanover, France, Bavaria, Hesse, Austria, Prague, Saxony, Baden and Prussia.[24] Patent protection was important

22. Catalogue from Ganz & Cie, Budapest, 'Turbinen-Abteilung', Budapest, 1896, Ganz & Cie archives, Norwegian Technical Museum, Oslo, p.25.

23. Ibid., p.36.

24. A. Benedello (ed.), *Keller–Voelter. Die Einführung des Holzschliffs in der Papierindustrie*, Hagen-Kabel, 1957, pp.175–7.

enough, but much work was still to be done before the process could be an economically feasible part of the paper-making industry. Basically this work was performed by Voelter. After Voelter returned to Heidenheim in 1847, he was actively assisted in the development work by Johann Matthäus Voith. The first small wood pulp mill was delivered to a paper factory near Hanover in 1848. However, customers were few and far between in spite of a widely circulated prospectus describing the new invention. When Voelter finally began to receive orders in the mid-1850s, the mechanical workshop of J. M. Voith built the machines which Voelter delivered to his customers. In 1856 they signed a contract for six years, formalising their cooperation. Three years later J. M. Voith invented the refiner. Through the use of this machine the quality of the pulp was improved, and it also improved the profitability of wood-grinding. From 1852 to 1859 Voelter sold thirteen wood-grinders, while sixty complete wood-grinders of his construction were installed in Europe between 1860 and 1866.[25]

When Friedrich Voith took over the daily operation of his father's machine shop in 1867, the company had twenty-eight workers on the payroll. During the next forty years the number of employees grew a hundred times, to 3,000. After studying for four years at the polytechnic school in Stuttgart, Friedrich Voith worked as an engineer for one and a half years at Escher Wyss's subsidiary in Ravensburg. Here he became acquainted with both the design and manufacture of Zuppinger wheels and the Jonval turbines. This was followed by employment in the engineering bureau of Heinrich Voelter. The finishing touch to his practical education he got at the Henschel factory in Kassel.[26]

Friedrich Voith returned home with both theoretical knowledge and a wide range of engineering skills. In the first years after Friedrich Voith took over the running of the firm the company continued to serve a broad range of customers, weaving factories, dyehouses, breweries and the like. During the 1870s production was increasingly concentrated on machinery for wood pulp and paper factories and water turbines. At the Vienna Exhibition in 1873 J. M. Voith won a prize for a complete wood-grinding mill. The following year the company built its own testing and experimental mill for wood-grinding. By 1880 J. M. Voith had installed 100 complete wood-grinders. The market was mainly Württemberg,

25. Ibid., p.159.
26. *100 Jahre Voith*, Heidenheim, 1967, p.12.

Saxony and Bavaria. From 1881 to 1890 another eighty-seven were sold.[27]

Paper making is intimately connected with the manufacture of wood pulp, and in the course of the 1870s J. M. Voith extended his product range to include auxiliary machines for paper factories such as hollanders and calenders. In 1881 the company ventured to take the final step by building its first complete paper machine. Between 1881 and 1900 twenty-three paper-making machines left the factory, and the company had turned into one of the leading German manufacturers of such machines.[28]

The same was true of J. M. Voith as a builder of turbines. Turbine no. 1, a 100-hp Jonval turbine, was delivered to a customer in Bohemia in 1870, followed by five Girard turbines in 1872 and 1873. At first glance Friedrich Voith seems to have followed the standard pattern among European turbine-builders at the time. However, in 1873 the company deviated from this pattern. It built Germany's first 25-hp Francis turbine for the new textile mill of C. F. Ploucquet of Heidenheim.[29] Up to that time few if any European turbine-builders had shown any interest in the American mixed-flow turbine.[30]

During the planning and building of their new textile mill the firm of Ploucquet used the newly appointed professor of mechanical engineering at the Stuttgart polytechnical school, Wilhelm Kankelwitz, as their consultant. It was he who suggested that the mill should be powered by a Francis turbine instead of one of the traditional Jonval type. Friedrich Voith accepted the challenge and built the turbine. To regulate the flow of water into the turbine he integrated the principle of adjustable guide vanes as part of the construction, an idea put forward by Professor Fink some ten years earlier.

In 1875 Professor Kankelwitz recommended that the company hire the talented engineer Adolf Pfarr.[31] For the next twenty-two years he was to play an important role in the development of both turbines and paper-making and wood-grinding machinery. In 1897 he was appointed professor at the Technische Hochschule (TH),

27. 'Verzeichnis der jenigen Fabriken, welche seit dem Jahre 1880 Holzschleiferapparate von mir erhalten haben', listed in catalogue from J. M. Voith (1903), J. M. Voith archives, Heidenheim.

28. 'Empfängerliste', Papiermaschinen J. M. Voith, Maschinenfabriken und Giessereien, Heidenheim, undated catalogue (1917–18), J. M. Voith archives, Heidenheim.

29. *100 Jahre Voith*, p. 16.

30. Reichel, 'Aus der Geschichte der Wasserkraftmaschinen', p. 63.

31. *100 Jahre Voith*, p. 17.

Darmstadt, where he founded an institute for paper manufacture and wrote one of the leading textbooks on water turbines.

The number of turbines delivered per year by J. M. Voith between 1870 and 1890 is very modest compared with the equivalent figures for Escher Wyss and Ganz & Cie.[32] Because of Voith's many customers among central European wood pulp mills, however, the average hp per unit was higher, but not significantly so. What really looms large when comparing the three manufacturers is that J. M. Voith already from the mid-1870s delivered Francis turbines in all cases where other turbine-builders recommended Jonval turbines. With the coming of hydroelectricity in the 1890s, however, it became evident to all turbine builders that the Francis turbine best fulfilled the requirements set by the electrical manufacturers. J. M. Voith had by then more than two decades of experience in designing and manufacturing Francis turbines, and this contributed greatly to the future success of the company. Prior to 1893 no more than twenty turbines left the premises of J. M. Voith in any one year. Then orders began to flow in, and almost all of them were for Francis turbines. Between one-third and half of all orders came from hydroelectric power plants. At the turn of the century Voith could compete with Escher Wyss in the number of turbines delivered, but on the average Escher Wyss installed considerably more powerful turbines.

The largest turbine manufactured by Voith before 1890 was a 226-hp Francis installed at a wood pulp factory in Baden. This was surpassed by a 400-hp Francis and a 600-hp Francis for two different wood pulp mills in 1893 and 1894. Through the Electrizitäts-Aktien-Gesellschaft vorm. Schuckert & Co. of Nuremberg the Voith company in 1897 delivered three 618-hp turbines to an electrical power plant in Bergamo, Italy. Finally, the following year Voith passed the 1,000-hp mark with the delivery of two 2,000-hp turbines to a hydroelectric power plant in Milan. It was definitely the age of hydroelectricity that brought J. M. Voith to the forefront as an international manufacturer of turbines.

2 Wood-grinders and Turbines in Norway

Between 1840 and 1860 machine shops were established in most of the larger cities in Norway. The British influence on the first

---

32. Compiled from 'Verzeichnis der von J. M. Voith ausgefürten Turbinen', undated catalogue (1903) J. M. Voith archives, Heidenheim.

Norwegian machine shops was considerable. Most, if not all, of these firms operated initially as general machine shops, serving the needs of the business community in their respective cities. After some years, however, depending on skills and market demand, most shops specialised in the production of specific goods. Many of them branched out into shipbuilding. Others, like Myrens Verksted and Kværner Brug, specialised in machinery for the wood and paper industry. Myrens Verksted delivered its first steam-powered sawmill in 1859. By 1890 more than 400 planers had been delivered to customers in Norway, Sweden, Denmark, Finland, Russia, Germany, Austria, Switzerland, Holland, Belgium, France and Spain.[33]

Myrens Verksted built its first steam-engine and its first water turbine in 1850. The turbine was of the Scotch type, and the company applied for and was later granted a patent on the turbine regulator.[34] There is some doubt as to the number of turbines built by this company between 1850 and 1867, because in all official company catalogues a 55-hp Girard turbine delivered to the Eker Paper factory in 1867 is listed as turbine no.1.

In any case it is clear that only a small number of turbines were in operation in Norway before 1870. In the early years the machine shops were clearly influenced by the British development in the field. It was the new generation of engineers, educated in Germany and Switzerland, that was to make Fourneyron, Girard, Schwamkrug and Jonval turbines the standard types in Norway. Through their education they became familiar with the chapters on turbines in Ferdinand Redtenbacher's standard textbook, *Der Maschinenbau*, and some might even have read his *Theorie und Bau der Turbinen*. After his return from studies at the polytechnical school in Zurich, Antonius Jensen, the son of one of the founders of Myrens Verksted, specialised in the design of turbines.[35] It was hardly a coincidence that the first Girard turbine was constructed according to the Swiss 'Zuppinger system'. The first Norwegian-built Jonval turbine was delivered by Christiania Maskinverksted in 1868.[36] The manager of this shop, Chr. B. Mohn, was educated in Germany.

His shop was also one of the pioneering suppliers of machines to the fast-growing Norwegian wood and pulp industry. At the end

33. C. Gierløff, *Et brug ved Akerselven. Myrens Verksteds hundre års minne*, Oslo, 1948, p.143.
34. 'Bekjendtgjørelse fra Departementet for det Indre', *Polyteknisk Tidsskrift*, no.7, 1856, pp.104–8.
35. Gierløff, *Myrens verksted*, p.164.
36. *Aftenposten (Oslo)*, 2 November 1883.

of 1870 six mechanical wood pulp mills were in operation in Norway. By 1879 the number had risen to twenty-seven, and in 1890 the country had fifty-one such plants. The number of workers in the paper and pulp industry grew from 316 in 1870 to 7,606 in 1900. To put this growth in perspective, suffice it to say that the number of workers employed in Norwegian manufacturing industry in the same period grew from 34,597 to 79,635.[37]

Voelter's patent rights in Norway had expired by the time the demand for wood-grinders skyrocketed in Norway, and companies like the aforementioned Christiania Maskinverksted, Myrens Verksted and Kværner Brug all succeeded in designing and manufacturing wood-grinding machinery fulfilling the needs of the new domestic industry. Deliveries to the wood pulp industry constituted the backbone of Kværner Brug output between 1870 and 1890. Kværner delivered its first wood-grinder with a capacity of 130 hp to the Granfoss mill in 1873. The same year the company delivered its first turbine, a 230-hp Jonval turbine to Hoffsfos Træsliberi, another wood pulp mill.

At the turn of the century the number of pulp mills built by Myrens Verksted exceeded that of any other producer in Scandinavia. Myren started building wood-grinders in 1865, and up to 1905 had delivered 490 grinders of different sizes. Of these, 107 were hot grinders, introduced on the market in 1895. The power needs of the wood pulp industry were on a hitherto unknown scale. They could be met only by turbines. This new industry was without doubt the driving force behind the development of comparatively large turbines in Norway. However, the requirements for automatic regulation of these turbines were very modest.

In the two decades from 1870 to 1890 Myrens Verksted in Oslo delivered 267 turbines. One hundred and twenty-three of these, or 46 per cent of all turbines delivered, were installed to drive wood-grinders. However, if one looks at the total capacity of these turbines they accounted for 24,389 hp of a total of 33,070 hp, or 74 per cent of total delivery.[38] The single largest turbine was no.160, delivered to Skotfos Træsliberi in 1886 with a capacity of 370 hp. This record was not surpassed until 1895 when Myren delivered three 600-hp turbines to the wood-grinding factory of Hønefos

37. *Historical Statistics 1978*, Central Bureau of Statistics of Norway, Oslo, 1978, p.79.

38. J. A. Jensen and Dahl catalogue, 'Fortegnelse over leverede turbiner' (list of turbines delivered from Myren works), Kristiania, 1907, Myren archives, Norwegian Technical Museum, Oslo.

Brug. This coincided with the introduction of hot grinding, which led to a jump in the size of turbines delivered to such plants.

From 1873 to 1890 Kværner Brug delivered ninety-six turbines, and forty of these (or 42 per cent) went to wood pulp mills. Adding the number of turbines delivered to paper and pulp factories, 74 out of 96 (or 77 per cent) of all turbines manufactured went to the Norwegian pulp and paper industry.[39]

When comparing the Norwegian suppliers of turbines with J. M. Voith up to 1890, it seems that contracts for complete machinery deliveries to wood pulp mills were much more common in Norway than in Germany. It is also quite evident that the Norwegian turbine builders on the average manufactured turbines with a larger capacity than was the case with J. M. Voith, Escher Wyss and Ganz & Cie. The reason for this was that the Norwegian wood pulp mills with their large power needs were the dominating group of customers.

3 Hydroelectricity and the Quantum Leap in Turbine Building

In 1901 the new professor of turbine-building at the TH, Stuttgart, R. Thomann, maintained that turbine-building in Europe in principle had remained the same from the 1830s until the beginning of the 1890s. This changed drastically, however, after it had been proved at the Frankfurt Electrical Exhibition in 1891 that long-distance transport of electrical power was technically feasible. From now on turbine builders were continually challenged to build larger and more economic turbines.[40] Customers began formulating specifications concerning the efficiency and the speed of the turbines they ordered, with penalties on the part of the manufacturer if these specifications were not met. The fixed cost of a large electric power plant was on a scale hitherto unkown, and in the long run 2 per cent higher or lower efficiency could mean more in economic terms than 5 per cent higher construction costs.

It was the companies that could tailor site-specific turbines, and guarantee their performance that won the large contracts. Manufacturers who would not or could not guarantee the performance of

39. I am indebted to Helge Høye, Kværner Brug A/S, for supplying me with a copy of a handwritten list of turbines delivered by the company from 1873 to 1954.
40. R. Thomann, *Die Entwicklung des Turbinenbaues mit den Fortschritten der Elecktrotechnik*, Stuttgart, 1901, pp.4–6.

their turbines came to lag behind in the development of turbines in these crucial years. This was the case in both America and Europe.

Before the coming of the electric power station the efficiency of Jonval and Girard turbines at around 70 per cent had been sufficient for most customers. In some instances efficiencies of 80 per cent had been reached. Electric power plants and their consultants required that the turbine-builder guaranteed an efficiency of between 80 and 90 per cent or more. Moreover it became the rule to do away with belting and instead to couple the turbine directly to the generator. This in turn led to turbines with an increase in rpm (revolutions per minute). For medium and low heads this meant that the turbine manufacturer was forced to increase the speed of his turbine. The number of rpm for turbines had been more than doubled in the course of twenty years, reflected the Norwegian turbine-designer H. Stub in 1908.[41] Increased speeds reduced the fixed cost of the power plant both for turbine and generator. This requirement greatly favoured the Francis turbine.

At the turn of the century the Francis and the Pelton turbines emerged as the only turbines that could attain efficiencies around 90 per cent in conjunction with high specific speeds. The transition from tangential wheels to spoon-shaped buckets and cut-out sections for the jet in the Pelton turbine brought an increase in efficiency from 60 to more than 90 per cent. This represented 30 to 50 per cent more output from the same quantity of water. But larger turbines meant higher pressures and more complicated regulation of the turbines. Parallel to the quest for larger output the designers had to construct mechanisms that could keep the frequency of the turbine as constant as possible. In the field of automatic turbine regulators there was much inventive activity. It was clear that the traditional mechanical regulators were no longer suitable. The designer had to take into consideration the steady increase in the mass of parts to be accelerated. The regulating period should be as short as possible, and any unsatisfactory influence on the efficiency had to be prevented. Last but not least, the pressure in the pipelines leading to the power station must be as low as possible. In the course of the 1890s hydraulic regulators or servomotors came into general use, at first using water as the medium but later changing to oil regulators. The suppliers of the very large turbines led the way, and other manufacturers had to keep up if they wanted to compete in the field.

Let us, however, return to our main line of reasoning concerning

41. H. Stub, 'Turbiner og turbinanlæg', *Teknisk Ukeblad*, no.17, 1909.

the rapid increase in turbine output. Until 1893 the largest turbine built by Escher Wyss had a capacity of 600 hp, Ganz & Co. had delivered 700-hp turbines in 1892, J. M. Voith a 600-hp turbine to a wood pulp mill in 1894, and Myrens Verksted 600-hp turbines to a wood pulp mill in 1895. The new hydroelectric power plants wanted much larger units than this. In the course of a few years there ensued sweeping changes in the design and construction of water turbines. The traditional textbooks had given 500 hp as the limit in output for turbines. By 1908 H. Stub said that no one was now impressed to hear about turbines of 5,000 hp.

To illustrate the changes let us take a closer look at the development at Escher Wyss. In the 1890s the company tried to meet the requirements by developing very large, multi-stage Jonval turbines. The company passed the 1,000-hp mark in 1894 with the delivery of a 1,500-hp turbine of this type to the Chèvres power plant supplying electric power to the city of Geneva. This turbine achieved the very high specific speed of 360, while the normal specific speed for a Jonval turbine was 110. However, the increase in specific speed led to a drop in efficiency.[42] The company had similar experiences at other plants. Escher Wyss therefore abandoned the Jonval turbine completely and instead took up the production of Francis turbines.

Among competing firms it was above all J. M. Voith that had the longest experience in building Francis turbines. The demand for its turbines grew exceptionally fast around the turn of the century, and other manufacturers had to copy their constructions to keep up in the competition for orders. While J. M. Voith experienced an increase in the number of turbines delivered, other companies for a time led in the construction of the largest turbines.

The Geneva company Piccard & Pictet had designed 5,000-hp turbines for the Niagara Falls Power Co. in 1893, manufactured by the American firm J. P. Morris and installed in 1895. The Niagara project, according to Louis C. Hunter, 'revealed in a striking way the wide gap between the methods and achievements of the practical wheel-builders of this country and those of the engineering firms that led in the design and building of turbines in Europe'.[43] American turbine-builders, relying on traditional trial and error methods, were not able to design turbines to precise specifications of performance. The leaders of the Niagara project found that in so unprecedented an undertaking they had no alternative but to rely

42. *Escher Wyss. A Century of Water Turbines*, Zurich, 1940, pp.8–9.
43. Hunter, *Waterpower*, p.399.

on European methods of turbine design and construction. In Europe one rarely found two turbines alike. It had been customary to design each machine for the actual conditions of power, head and speed governing its installation. The construction of a turbine was the outcome of mathematical calculations and not trial and error. No American manufacturer was willing to underwrite the guarantees the Niagara Power Co. felt necessary. Instead, the design put forward by Piccard & Pictet of Geneva was accepted. J. P. Morris in Philadelphia built ten 5,000-hp Fourneyron turbines for powerhouse no.1 according to drawings and specifications supplied by the Swiss firm, giving the European company the final responsibility for the product.[44]

This first European success at Niagara Falls was followed by large contracts to Escher Wyss and J. M. Voith. For powerhouse no.2 Escher Wyss designed eleven 5,500-hp Francis turbines. These were built and delivered by Morris between 1900 and 1903 according to drawings and instructions delivered by Escher Wyss. Soon afterwards Escher Wyss got an order for three 10,250-hp units from the Canadian Niagara Co. These Francis turbines, delivered in 1903 and 1904, were the largest ever built up to that time.

J. M. Voith got the contract to supply power stations on the Canadian side of Niagara Falls with turbines. Four horizontal Francis 7,000-hp turbines for Dominion Power and Transmission Ltd were delivered in 1903. The same year the company got an order for several 12,000-hp Francis turbines to be installed at the power station of the Ontario Power Company on Niagara Falls.[45] The first turbine was delivered in 1904, followed by eight others in the following years. These turbines were at the time the largest in the world. Since Voith was able to meet all the customers' requirements, these deliveries contributed greatly to building up the reputation of J. M. Voith. After the turn of the century the German and Swiss turbine-builders were generally known as the most advanced in the world.

44. H. B. Taylor, 'Über die Entwicklung der Reaktions-Wasserturbine in Amerika', *Zeitschrift für das gesamte Turbinenwesen*, no.4, 1911, p.59. Article translated from *Engineering Magazine*, 1909–10, p.842.

45. 'Lieferliste Francis-Spiralturbinen über 750 kW', 1985, p.2, J. M. Voith archives, Heidenheim.

4   A Brisk Market for Turbines in Norway - the Domestic Manufacturers Lag Behind

Despite the abundance of water power, the first electric power plants of any size in Norway were driven by steam power. The market for very large turbines in power stations began to grow rapidly, however, after domestic and foreign venture capitalists began investing heavily in electrochemical and electrometallurgical plants from the late 1890s. Orders for the large turbines in these power plants went to central European suppliers. Only occasionally did Norwegian turbine-builders get a chance to prove their skills.[46] J. M. Voith got its first turbine delivery to Norway through the electrical manufacturer Elektrizitäts-Aktiengesellschaft vorm. W. Lahmeyer & Co., Frankfurt am Main. In 1899 the company supplied two 1,000-hp Francis turbines to power the generators of the carbide works at the Tinfos paper factory, Notodden. The same year A/S Glommens Træsliberi started construction work on the Kykkelsrud power plant, Glomma. J. M. Voith got the order for one of the 3,000-hp vertical Francis turbines through the electrical manufacturer Schuckert & Co. The second turbine was supplied by Escher Wyss. In 1899 large turbines were installed at the power plants at Hafslund and Borregaard. The six vertical Jonval turbines for the Hafslund plant were delivered by J. J. Rieter of Winterthur. The plant was extended in 1905 with four Francis turbines from Voith and two more Voith turbines of 3,750 hp in 1907. The two first turbines for the Borregaard plant were supplied by Escher Wyss and the generators by Brown Boveri & Co. The link between electrical manufacturer and turbine-builder obviously merits some further study.

In 1920 the journal *Engineering* published a table, based on information from the Institution of Mechanical Engineers, of some of the largest power plants in the world up to 1916. A remarkable number of Norwegian power plants were to be found on the list. Who supplied the turbines to these power plants?

Construction work on Svelgfoss power station, Notodden, started in the autumn of 1905. This station was to supply the necessary electric power to the Norsk Hydro plants, manufacturing synthetic nitrogenous fertilisers based on the electric arc method. At the beginning of 1906 an advertisement for bids was announced, and the result of this was that J. M. Voith got the order for four

---

46. N. de L. Kobberstad, 'Vandkraftmaskinens utvikling i vort land', *Teknisk Ukeblad*, 1924, p.406.

10,000-hp Francis turbines. The company was chosen both on the technical merit of the constructional solution submitted and on price.[47] In October 1907 the turbines were tested by a commission chaired by Professor Ernst Reichel of the TH, Charlottenburg. On the basis of the tests performed the commission concluded that the turbines were more powerful than required in the contract. At full water admission the turbines gave 11,750 hp, and at 10,000 hp the efficiency was 86.2 per cent.[48]

The test results were published both in *Teknisk Ukeblad* and *Zeitschrift des Vereins deutscher Ingenieure*, and were of great marketing value to J. M. Voith. When work commenced on the power plant for Rjukan I, J. M. Voith had excellent results to show for themselves. The same was the case with Escher Wyss, who had supplied the first six 4,600-hp Pelton turbines to the Tyssedal power plant in 1907.[49]

Rjukan I or the Vemork power station was built in the years 1908–11. The head from the top of the dam to the power station was 300 m. This head was at the time too high to use Francis turbines, and Pelton turbines were the natural choice. Ten 14,500-hp turbines were ordered. Up to that time the largest Pelton turbine delivered by J. M. Voith had a capacity of 4,000 hp and had been shipped to Santos, Brazil, in 1905. Through the Rjukan order J. M. Voith and Escher Wyss moved the borders of what was thought possible another big step forward. Each company delivered five of these large turbines. It is symbolic that the Norwegian firm Kværner Brug got an order for a 1,000-hp turbine to supply the plants with electric light and power.[50]

Work on Rjukan II or the Såheim power plant began in 1913. In the power station were to be installed nine 16,400-hp Pelton turbines for a head of 280 m. Besides two smaller turbines of 7,000 hp were ordered. J. M. Voith got the order for six of the largest turbines and the two small ones. Two were delivered by Piccard and Pictet of Geneva, and one by Myrens Verksted according to drawings and specifications supplied by Piccard and Pictet. To be able to compete on the Norwegian market Myrens Verksted had signed a contract with the Swiss firm concerning high-pressure Pelton turbines. In 1918 Myrens Verksted supplied a 12,300-hp turbine to Bjølvefossen power plant designed by the Swiss firm.

47. E. Reichel, 'Turbinerne. Norsk Hydro–elektrisk Kvælstofaktieselskaps Kraft-station ved Svælgfos, Notodden', *Teknisk Ukeblad*, 1908, p.425.
48. Ibid., p.429.
49. E. Dubislav, *Neuere Wasserkraftanlagen in Norwegen*, Munich, 1909, p.119.
50. 'Norsk Hydros kraftanlæg i Telemarken', *Elektroteknisk Tidsskrift*, 1918, p.135.

One of the two other turbines was delivered by Piccard & Pictet directly, and the third by Kværner Brug.[51] The head at Bjølvefossen is 843 m. J. M. Voith delivered two very large 27,500-hp Pelton turbines to the Glomfjord power plant in northern Norway in 1916, followed by a third of the same size in 1919. We can safely conclude that it was the foreign manufacturers that dominated the Norwegian market from the turn of the century until the end of the First World War.

How did domestic manufacturers like Myrens Verksted and Kværner Brug fare in this period? When evaluating delivery lists and the contemporary literature one gets the impression that the Norwegian manufacturers lost momentum at the turn of the century. At the end of the war, however, through investments in development work, they bridged the technological gap, but could not compete with foreign companies in price.

Norwegian manufacturers of turbines had always depended on the domestic market, even though they had delivered turbines to Sweden and Finland as part of turn-key contracts for wood-grinding machinery. In the mid-1890s there were no clear indications that the market for turbines for electric power plants would soon become the most important. Moreover the Norwegian companies were busy supplying new and more powerful turbines to wood pulp mills. They seem to have neglected the new market. The first time Myren delivered a turbine with a capacity of 1,000 hp was in 1900. It was not installed in an electric power station, but at Hønefos Brug, a wood-grinding mill.

Kværner Brug entered the power plant market in 1901, when two 1,000-hp double Francis turbines were shipped to Trondheim to be installed at the power station of the electricity works. The choice of turbine type was no coincidence. In 1900 Kværner Brug got a new managing director, the engineer Hans Gerhard Stub. He had introduced the Francis turbine in the 1890s during his time as manager of A/S Drammens Jernstøberi & mek. Værksted. From 1883 to 1886 he worked in America, and he had probably become familiar with the Francis turbine there. He designed and introduced the first hydraulic turbine regulator in Norway in 1896 and exhibited it at the World Exhibition in Paris in 1900.

The Norwegian companies were a few years later than their foreign competitors in changing to Francis and Pelton turbines, and this may be one of the reasons the Norwegian manufacturers came to lag behind. Kværner, however, responded much more aggressively

51. *Norske kraftverker*, vol.1 Oslo, 1954, p.283.

to the challenge than did Myrens Verksted. In the course of a few years Stub gathered a team of talented engineers specialising in turbines. After graduating from technical school at Christiania in 1898, Gudmund Sundby joined Kværner Brug. In 1903 he went to the USA for a year to become better acquainted with the state of the art. In 1905 he was appointed chief engineer of the company's turbine and pump division. In 1907 he got a fellowship to study high-pressure turbines installed at power plants in Switzerland and Italy. Later he was to be appointed professor of water turbines at Norway's Technical College in Trondheim.

The same year that Sundby was put in charge of the turbine division Paul Sagberg came on to the company payroll. After graduating from the Trondheim College in 1901, he studied at the Eidgenössische Technische Hochschule, Zurich. Sagberg concentrated on the development of turbine regulators and was after some years appointed the leader of this branch. Lorents Aksnes was another creative engineer in the service of the company. He graduated from Trondheim in 1900. Finding no suitable work in Norway, he worked at mechanical workshops in Germany until 1903. He then went to the USA, working in the construction departments of different engineering firms. The last years before returning to Norway and teaming up with Kværner in 1908, he was employed by the Westinghouse Machine Co. in Pittsburgh. For the next decade and longer these engineers and their colleagues were busy trying to bridge the gap between domestic and foreign manufacturers.

In the first years after 1900, however, the company got few contracts for large turbines. Worth mentioning is the 2,700-hp Francis turbine to the Borregaard plant in 1906. The largest turbine manufactured by Kværner Brug before 1910 was a 5,250-hp Francis delivered to Lienfoss power plant in late 1909. The three other turbines to this plant were supplied by J. M. Voith.

What was happening at Myrens Verksted in this period? In 1902 Myren delivered a 1,000-hp turbine to Kristiansands Electricity Works, and in 1906 a 2,300-hp turbine to the same company. English capital financed the development of water power at A/S Vigeland Brug in 1907–8. The new plant was to produce aluminium. Myrens Verksted got the contract for a 2,000-hp turbine for powerhouse no.1. But the large contract, the four 3,000-hp Francis turbines for powerhouse no.2, went to Th. Bell & Cie, Switzerland. Until 1911, when Myren delivered a 5,000-hp Pelton turbine to A/S Tyssefaldene, the company had only smaller orders.

It was at this time that Myrens Verksted signed the contract already mentioned with Piccard & Pictet, Geneva. The contract stipulated that the Swiss company was to supply the Norwegian firm with the necessary know-how to build large high-pressure turbines. Until 1910 Norwergian firms bidding for large contracts in Norway lost out in the competition with the foreign firms. Nevertheless the Norwegian companies improved their relative position in the years before the outbreak of the war. For both companies the year 1913 must have been experienced as a breakthrough. Myrens Verksted delivered all four 6,000-hp turbines to the Aarlifos power plant, and one of the 12,000-hp turbines to Vamma power plant. The second turbine installed in 1913 was delivered by Kværner. The latter company had already passed the 10,000-hp mark with a delivery to Svælgfoss the year before.

The war changed the competitive position in favour of the Norwegian companies. Norway was neutral and had import problems. It was hard to get enough coal and oil. As a consequence towns and municipalities all over the country involved themselves heavily in planning, financing and development of power plants and transmission lines to supply the necessary electric power. If the market for turbines had grown fast before the war, during the war years it grew even faster. Since the traditional German suppliers of both turbines and electro-technical equipment had problems supplying Norwegian customers, domestic firms got large deliveries. The profit from these deliveries enabled them to put more money into product development.

Most of the turbines delivered to power plants belonging to towns and municipalities were of medium size. However, the domestic manufacturers also got some of the large deliveries. Kværner Brug delivered another three 12,000-hp turbines to the Vamma power plant in 1914, 1915 and 1917. Myrens Verksted shipped a 13,000-hp turbine to the same plant in 1915, and a 15,000-hp turbine in 1919.

In the engineering community many felt that the final proof that Norwegian manufacturers could build as efficient turbines as their foreign competitors came with the delivery of seven Francis turbines to the Mørkfoss–Solbergfoss power plant in the years 1918–23. Kværner supplied four of the 12,000-hp turbines, the first delivered in 1918, and Myrens Verksted the three others. The runner wheel construction for these turbines was thoroughly tested at the new hydraulic laboratory at the Norges Technical College by Gudmund Sundby and his staff. According to the tests performed it was expected that the turbines would have an efficiency of 90 per cent. When tested after installation, the turbines had an efficiency of 94.3 per cent.

By that time, however, splendid test results did not help much. The Norwegian economy had gone into a depression. Many companies went into bankruptcy or were fighting heavy debts. Towns and municipalities had a hard time paying interest on the loans raised to finance the electric power plants during the war years. The consumption of electricity fell far behind expectations. There was a strong campaign to 'buy Norwegian', and the domestic companies certainly got most of the orders for turbines in the interwar years. Compared with the period 1900–20, however, the market for turbines had diminished considerably. In the long run there was no place for two large turbine manufacturers on the Norwegian market, and in 1928 the turbine division of Myrens Verksted was integrated as part of Kværner Brug.

As my findings show, there was a lively market for water turbines both in central Europe and the Nordic countries in the last decades of the nineteenth century. In most instances domestic machine shops supplied their customers with turbines. The coming of hydroelectricity, however, led to sweeping changes in the market for turbines. The market grew in importance, but only a few specialised manufacturers were able to meet the specifications set by the electro-technical industry. These came to dominate the international market for turbines in the twentieth century.

Appendix 13.1

Table 1 Turbines delivered by Escher Wyss & Co., Zurich & Ravensburg, 1845–1870

Year	Tang. wheels	Girard	Jonval	Total	Hp per turbine
1845			12	12	19.1
1846	11		9	20	23.6
1847	3		9	12	19.5
1848	2		4 ·	6	13.8
1849	3			3	56.0
1850	9		5	14	24.9
1851	5		9	14	31.2
1852	4		11	15	23.4
1853	6		14	20	47.7
1854	10		10	20	35.3
1855	5		9	14	42.1
1856	7		10	17	55.6
1857	14		4	18	47.7
1858	7		20	27	45.5
1859	7		7	14	51.1
1860	9		16	25	56.0
1861	19		27	46	51.7
1862	5		15	20	87.4
1863	4		23	27	58.6
1864	10		16	26	53.3
1865	5		14	19	73.2
1866	7		18	25	46.5
1867	5		19	24	66.5
1868	5		16	21	82.8
1869	6		35	41	53.5
1870	6	1	22	29	69.5

Source: Catalogue from Escher Wyss & Co., Zurich and Ravensburg, 1903

Table 2 Turbines delivered by Escher Wyss & Co., Zurich & Ravensburg, 1871–1902

Year	Tang. wheels	Jonval	Girard	Francis	Total	Hp per turbine
1871	4	25	8		37	62.5
1872	10	29	17		56	45.9
1873	4	19	25		48	72.7
1874	15	20	33		68	81.0
1875	22	21	22		65	48.1
1876	8	14	33	2	57	81.2
1877	2	17	29		48	62.7
1878	10	18	30	2	60	49.0
1879	11	18	30	1	60	59.4
1880	19	16	37		72	39.6
1881	9	28	44		81	78.5
1882	13	17	51		81	76.3
1883	31	17	41		89	51.9
1884	50	22	52		124	58.5
1885	26	23	25		74	46.5
1886	21	9	44		74	53.7
1887	17	18	49	1	85	79.0
1888	26	19	48		93	70.9
1889	16	45	72		133	104.2
1890	15	24	61		100	70.3
1891	30	18	39		87	61.2
1892	33	35	34		102	109.5
1893	32	23	39		94	59.7
1894	25	52	6	1	84	346.0
1895	46	16	30	4	96	134.0
1896	35	27	36	8	106	199.5
1897	45	8	12	13	78	148.3
1898	56	11	11	24	102	260.0
1899	50	10	10	49	119	358.0
1900	37		2	54	93	325.0
1901	63	6	5	95	169	284.0
1902	80		3	104	187	444.0

Source: Catalogue from Escher Wyss & Co., Zurich and Ravensburg, 1903

Table 3 Turbines delivered by Ganz & Co., Eisengiesserei– und Maschinenfabriks–Actiengesellschaft, Budapest

Year	No. of turbines	Hp per turbine
1885	43	73
1886	21	66
1887	20	101
1888	30	89
1889	49	170
1890	36	66
1891	43	84
1892	61	69
1893	59	110
1894	67	98
1895	79	114

Source: Catalogue from Ganz & Cie, Budapest, 1896

Table 4 Turbines delivered by J. M. Voith, Heidenheim

Year	Jonval	Girard	Francis	Total	Average hp
1870	1			1	100
1872		3		3	98
1873		2	1	3	59
1874		3		3	83
1875			2	2	42
1876			2	2	113
1877			2	2	40
1878		1	3	4	65
1879		2		2	15
1880	2		4	6	57
1881		1	5	6	83
1882		1	3	4	94
1883		2	1	3	87
1884		1	4	5	77
1885		2	1	3	75
1886		5	10	15	109
1887		5	5	10	140
1888			11	11	114
1889		11	8	19	143
1890	Schwamkrug		6	6	63
1891		1	6	7	130

continued on page 360

Table 4 *continued*

Year	Jonval	Girard	Francis	Total	Average hp
1892	2	1	8	11	96
1893	2	1	24	27	92
1894	2		19	21	123
1895	1	1	38	40	109
1896	1		74	75	110
1897			84	84	134
1898	4	1	132	137	161
1899	10		129	139	141
1900	13		179	192	123
1901	9		170	179	149
1902	13		139	152	121

Source: 'Verzeichnis der von J. M. Voith ausgeführten Turbinen', Heidenheim, 1903

Table 5 Turbines delivered by J. & A. Jensen and Dahl

Year	No. of turbines	Average hp	To wood pulp mills	
			No.	Average hp
1867	1	55		
1869	2	70	1	60
1871	1	180	1	180
1872	5	35		
1873	16	122	7	200
1874	10	85	2	126
1875	10	37		
1876	9	88		
1877	15	42		
1878	7	87	1	220
1879	5	53		
1880	12	105	8	130
1881	19	126	13	154
1882	8	100	1	210
1883	11	86	3	112
1884	14	141	6	237
1885	12	84	4	130
1886	31	139	18	195
1887	24	167	17	213
1888	36	178	27	213

Table 5 *continued*

Year	No. of turbines	Average hp	To wood pulp mills	
			No.	Average hp
1889	19	217	15	260
1890	41	89	3	200
1891	17	126		
1892	21	94	4	203
1893	23	104	8	215
1894	45	131	33	152
1895	33	171	27	283
1896	39	225	25	378
1897	42	142	11	290
1898	19	98	6	188
1899	39	165	5	309
1900	34	250	17	333
1901	44	288	26	372
1902	49	292	27	411
1903	4	215	4	215
1904	24	217	7	421
1905	32	373	21	496

Source: Catalogue of J. A. Jensen and Dahl, Kristiania, 1907

Table 6 Turbines delivered by Kværner Brug, 1873–1903

Year	Jonval	Girard	Grenz-t	Tang. wheel	Total	Average hp
1873	1	1			2	135
1874	3				3	180
1875	2				2	260
1876					0	
1877	3				3	40
1878	1				1	60
1879	2				2	85
1880	9				9	171
1881	3				3	157
1882	9	2			11	122
1883	2	2	1		5	56
1884	2				2	68
1885	3	1			4	152
1886	3	6			9	216

continued on page 362

Table 6 *continued*

Year	Jonval	Girard	Grenz-t	Tang. wheel	Total	Average hp
1887	7	4			11	115
1888	8	1	1		10	146
1889	7	5		4	16	116
1890	1	3	2		6	96
1891	3	3	1		14	64
1892	4	2	4		10	144
1893	8	11	9		28	131
1894	24	1	2		27	210
1895	10	2	5		20	115

Source: Kværner Brug, Oslo

14

Norway's Entry into the Age of Paper: The Development of the Pulp and Paper Industry in the Drammen District

Eli Moen

Throughout the nineteenth century the manufacture and use of paper increased enormously in Europe. In Britain, the leading paper manufacturing country, the production of paper multiplied about fifty times from 1800 to 1900. Growth was spread throughout the period, but was particularly strong in the last quarter of the century.[1] Paper was a sign of modernisation, and it was said that the cultural level of a country could be measured by its use of paper; on contemporaries the use and influence of paper made such an impression that it was declared to be 'one of the leading characteristics of the nineteenth century', and towards the end of the century, people were talking about the 'Age of Paper'.[2]

The growth and development of the paper industry have often been explained as a result of increasing demand for paper. A rising population and the modernisation of society have given grounds for such a view. On the other hand, if we consider the supply side, there is no doubt that the industry itself also gave strength to the diffusion and use of paper. New raw materials, product and process innovations resulted in new and better paper qualities – and further innovations. Optimistic inventors started to make use of pulp in an infinite variety of ways. In Germany, railway wheels were made of paper, and people were offered keys, plates, cups, inner soles,

---

1. D. C. Coleman, *The British Paper Industry 1495–1860*, Oxford, 1958, p.201; A. Shorter, *Paper Making in the British Isles*, Newton Abbot, 1971, p.142.
2. A. Dykes Spicer: *The Paper Trade*, London, 1907, p.2; O. Winckler, *Der Papierkenner. Ein Handbuch und Rathgeber für Papier-Käufer und Verkäufer, technishce Lehranstalten etc.*, Leipzig, 1887, p.1.

jewellery and so on, of paper.[3] Shorthly after the end of the nineteenth century, A. Dykes Spicer summarised the development in this way: 'Catering for modern demands has become a science.'[4] At the beginning of the nineteenth century newspapers were a rarity; by its end they had become the possession of ordinary people. The background for this development was the technical changes which took place in paper manufacture during the century.

The most important developments towards a modern, industrialised paper industry were completed by about 1890, and the technological change in paper manufacturing brought about new opportunities to countries with forestry resources. From a modest start around 1870, pulp and paper had, within a period of three decades, become one of Norway's foremost industries. In 1900 the earnings from forestry products accounted for about 40 per cent of the value of the country's exports. While in 1875 '98 per cent of the export earnings from forestry products originated with sawn and planed wood, and only 2 per cent with pulp and paper', we find, twenty years later, that export earnings were equally divided between the two.[5]

This chapter will try to trace the origins of the pulp and paper industry in the Drammen district, south-west of Oslo, the district which took the lead in the development, and then discuss the motive forces in the development. In this process, the transfer and adaptation of new technology were crucial, and we shall see that our study reflects and confirms international developments. Norwegian manufacturers were quick to import and adapt wood pulp technology. On the other hand it has been maintained that they were late starters in adopting advanced technologies within paper and cellulose.[6] If this is true, it must be an appropriate question why some technological capabilities were more easily acquired than others. When we ask such a question, it is necessary to bear in mind that it is difficult to say what is slow and what is fast in the diffusion of technology, and it remains a problem how to measure rates of technological diffusion. In this chapter we shall start by looking at the transfer mechanisms. What mechanisms were involved in the technology transfer that took place? Before approaching the prob-

---

3. Ibid., p.2.
4. Spicer, *The Paper Trade*, p.13.
5. T. Bergh et al., *Growth and Development: The Norwegian Experience 1830–1980*, Oslo, 1981, p.96.
6. The modern pulp and paper industry consists of mechanical wood pulp, cellulose and paper.

lem, it is, however, necessary to describe briefly the technical and economic development of the industry.

1 The Development of the Modern Pulp and Paper Industry

At the beginning of the nineteenth century, paper was made of cloth and by hand. Towards the end of the century, paper making had turned into large-scale industry, and paper was to a great extent made of wood fibres. This change came about gradually, but was speeded up in the last quarter of the century. Three innovations were crucial in this development: the paper-making machine, mechanical wood pulping, and cellulose.

The paper-making machine was invented in France in 1799.[7] The invention was, however, significantly developed and improved in England, where in particular Bryan Donkin made important contributions. Donkin was also the first to establish works for building paper machines.[8] By 1830, the development of the three main parts of the machine – the wet and the dry parts, and the reels – were accomplished. Apart from size and complexity, the basic principle remains the same today: refining the paper mass by passing it over a wire mesh or grate in constant motion. The essence of the process consists in freeing the cellulose fibres from the surrounding substances, because it is only cellulose which has the property of forming the bonds in the paper. The macerated fibres are suspended in water in such a way that, by the use of a sieve (the wire), the fibres can subsequently be lifted from the water in the form of a thin layer, which is paper.

Paper making consists of two main processes: the preparation of the raw material, and the forming of the paper. Before the pulp is allowed to flow into the paper-making machine, it is treated in a beating machine (hollander) and other materials such as china clay and dyes added.[9] The shaping of the paper takes place on the paper-making machine. From a chest box the pulp is conveyed on

7. During the French Revolution, Nicolas Louis Robert, who managed a paper mill in Essonnes, invented the paper-making machine in order to replace striking workers. R. H. Clapperton, *The Paper-making Machine: Its Invention, Evolution and Development*, Oxford, 1967.
8. R. L. Hills, *Papermaking in Britain 1488–1988*, London 1988, p.175.
9. China clay or kaolin was added in order to give body and weight to the finished sheet.

to the endless wire, and from the wire the paper is passed between two couch rolls covered with felt in order to squeeze out the water. Next it is passed on to drying cylinders, heated with steam, and pressed by endless felts. Leaving the drying cylinders, the paper passes through a set of rollers, or calenders, which impart the final treatment. Lastly, it is reeled and cut into sheets of various sizes.

The introduction of the paper-making machine made possible an increase in productivity, and most of the machine-made paper was also of better quality than the hand-made. In spite of these advantages, the machine diffused slowly. The reason for this is to be found in a type of technical imbalance. In paper making the supply of raw materials had been a problem for a long time, and this bottle-neck was felt more strongly after the introduction of the paper-making machine. It was not until after 1850, when substitutes for textile fibres were successfully brought into use, that more widespread mechanisation could take place. Accordingly, the development of the paper industry gives a good demonstration of innovation induced by technical imbalance, which in general occurs, according to N. Rosenberg, because 'the relationship among components was usually such that some imbalance *had* to be corrected before an initial innovation could be fully exploited.' The history of technology is replete with such examples, and inducement mechanisms due to imbalances have been regarded as an enormously important source of technical change.[10] As for the development of the paper industry, D. C. Coleman emphasises in particular this factor: 'Outstanding amongst the forces influencing supply conditions were those which flowed from mechanization.'[11]

By 1850 the supply situation in paper making had attracted general interest. Many efforts were made to find substitutes for cloth, and a great number of different materials were proposed: artichokes, coconuts, rhubarb, thistles, nettles and turf. In the 1850s a process of boiling straw in soda was patented. The invention created optimism, and many – 'even intelligent' – manufacturers maintained that straw would become the sole substitute for cloth in the future.[12] Paper made from straw was produced in France, and to some extent in Germany. In Britain, however, paper makers came to rely on a different material, namely esparto grass.[13] Paper made of esparto grass was patented in 1856. As a substitute for

10. N. Rosenberg, *Perspectives on Technology*, Cambridge, 1976, p.117.
11. *The British Paper Industry 1495–1860*, Oxford, 1958, p.212.
12. *Polyteknisk Tidsskrift*, 1859, pp.92ff.
13. A sort of grass, generally found in Spain and North Africa.

cloth, esparto satisfied many technical requirements, and during the 1860s was widely adopted by British manufacturers.

In 1845 paper made with an admixture of mechanical wood pulp was patented by F. G. Keller, a German weaver.[14] The patent was taken over by H. Voelter, a paper manufacturer, who started developing machinery for the production of mechanical wood pulp.[15] Developing machinery suitable for grinding wood was a demanding and time-consuming process, and it was not until the Paris exhibition in 1867 that a fully equipped pulp mill was presented. Initially, mechanical wood pulp was met with scepticism, and its diffusion has to a large degree been ascribed to the indefatigable work of Voelter. The Paris exhibition led to Voelter's final breakthrough, and in the following years wood pulp became generally accepted.[16] Its cheapness and its good printing qualities were important characteristics for the successful diffusion.[17] These qualities proved to be decisive for newsprint, and wood pulp thereby contributed to the spread of newspapers.[18]

About 1870, straw, esparto and wood pulp had improved the supply situation to such an extent that there was no longer any hindrance to the diffusion of the paper-making. Paper making was heavily mechanised in the following years, and most of the vat paper mills, based on manual production, had to shut down. Increasing mechanisation was accompanied by an increased number of engineering workshops entering into paper-machinery production. In Great Britain, the Bryan Donkin Company was seriously challenged by Bertram & Sons in Edinburgh and Bentley & Jackson in Glasgow.[19] In Germany the machine shop of J. M. Voith, among others, was playing an important role in the production of machines for the paper industry.

Right from the start it was obvious that mechanical pulp could not completely replace cloth in paper making. The grindstones cut the wood fibres too small, and accordingly paper made of mechanical

---

14. G. Bayerl and K. Pichol, *Papier. Produkt aus Lumpen, Holz und Wasser*, Hamburg, 1986, p.125.

15. The development of the machinery was carried out in cooperation with the mechanician Johann Voith.

16. A. Benedello, *Keller–Voelter, Die Einführung des Holzschliffes in der Papierindustri*, Hagen-Kabel, 1957, p.160.

17. E. Lange has shown that Norwegian manufacturers had to make great efforts to sell their wood pulp. E. Lange, 'Om å "gjøre sig en hel Del Umage". Norsk tremasse paæ det britiske marked', *Historisk Tidsskrift*, no.3, 1988. See chapter 15.

18. J. A. McGaw, *Most Wonderful Machine,*, Princeton, 1987 p.203. Hills, *Papermaking in Britain 1488–1988*, p.146.

19. A. H. Shorter, *Paper Making in the British Isles*, Newton Abbot, 1971, p.127.

pulp was too crisp and broke easily. Consequently, ground wood paper had to contain an admixture of rag fibres. The hunt for alternative processes therefore went on. The search was concentrated on chemical processes, and in the 1870s a sulphite as well as a sulphate process was developed. The former became of greater importance. In 1872 C. D. Ekman in Sweden and A. Mitscherlich had succeeded in producing sulphite cellulose. After a period of trial and error, paper makers were finally offered a raw material which in every respect could substitute for cloth, and which was sufficiently cheap. It is still a question what mechanisms were involved in the diffusion process. The results of the achievements were, however, soon to be seen. Within a few years cellulose became the dominant raw material. Thereby the development of a modern paper industry was completed, and paper making developed into large-scale industry.

2 New Opportunities and Old Traditions

As paper making was developing into a modern industry, Norwegian manufacturers joined the trade, in the first place as suppliers of raw material. Mechanical wood pulp was exported for the first time in 1870. In 1880 the volume of exports had increased to 100,000 tons, and in 1895 to 200,000, an increase that gives an annual growth rate of about 15 per cent between 1880 and 1895.[20] In the 1890s the picture changed. The growth rate of mechanical pulp decreased. Instead the production of cellulose and paper improved. (See figure 14.1) From 1890 to 1910 cellulose experienced an annual growth rate of about 10 per cent, while growth rate in the export volume of wood pulp had fallen to less than 7 per cent. The strongest growth rate, however, took place in paper. The growth rate of this industry reached almost 19 per cent in this period.[21] At this point in time, the founding period of the Norwegian pulp and paper industry was almost completed. As the figures indicate, the period was marked by different stages, each with its special characteristics. How did the development come about, and what factors influenced the course of events?

The first attempt to manufacture mechanical wood pulp in Norway was made in 1863 at Bentse brug, an old paper mill close to Christiania (now Oslo). If we look at Sweden, we find that the

20. F. Hodne, *Norges økonomiske historie 1815–1970*, Oslo, 1981, p.90.
21. Ibid, p.95.

Figure 14.1 The export value of mechanical wood pulp, cellulose and paper from Norway 1870–98

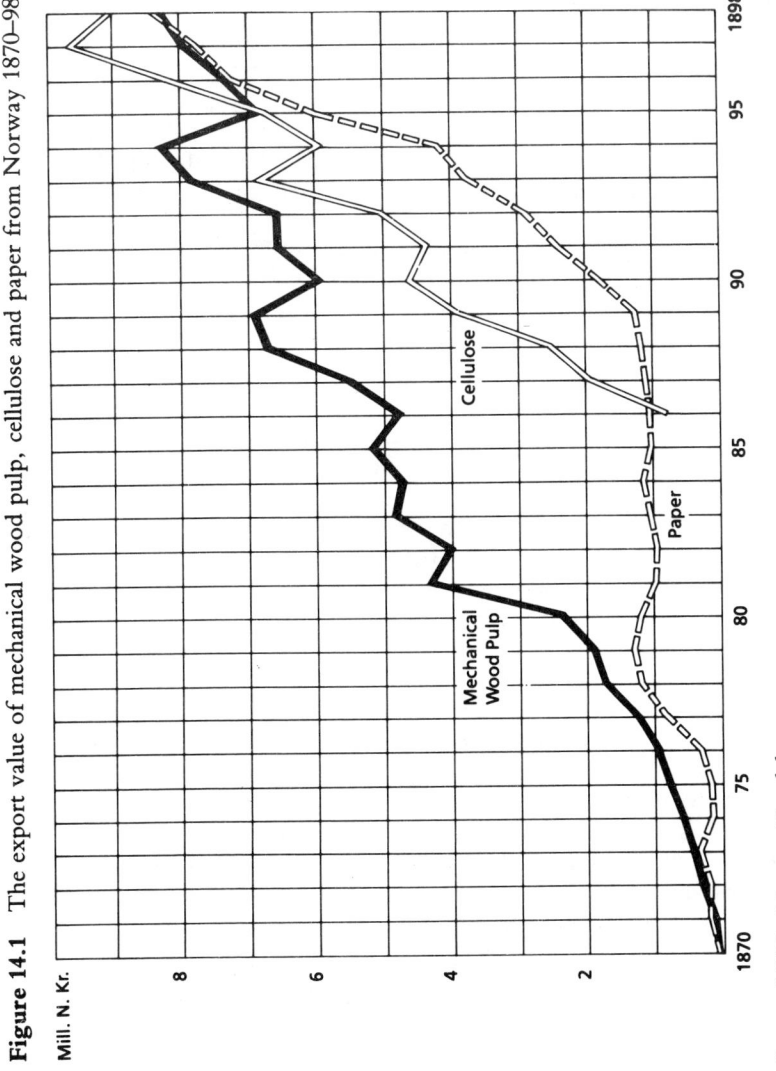

Mill. N. Kr.

Cellulose

Mechanical
Wood Pulp

Paper

Source: NOS, Norges Handel

first pulp mill had already been founded in 1857. It took ten years, however, before the pioneering enterprise was followed by others.[22] When first introduced, the new technology spread more rapidly in Norway. By 1870 ten small pulp mills had been erected, and the demand for pulp and paper created by the Franco–German war in 1870–1, accelerated the process. Within a few years twenty-five new pulp plants were established. While the pace slackened in the second part of the 1870s, in the 1880s thirty new pulp plants were established. By this time Norway was the most important supplier of pulp on the world market. In the course of the process the Drammen district played a leading role. From 1870 to 1895 the district contributed almost one half of Norway's exports of wood pulp.[23]

Exploiting natural resources such as timber and water was not a new activity in the Drammen district. Timber had for centuries been a major Norwegian export commodity, and the Drammen district had long traditions in this trade. Drammen had taken a leading position in the late eighteenth century as a port for timber exports.[24] After the Napoleonic Wars the town suffered a heavy setback, but it recovered and experienced another flourishing period in the 1830s and 1840s. Towards the mid-nineteenth century a restructuring of the sawmill industry took place. In order to mini-mise transportation costs, the sawmills were concentrated at the waterfalls in the lower part of the district and production units were at the same time enlarged. Ninety per cent of timber export came under the control of a group of ten men.[25] Some of these timber merchants were also engaged in shipping, and shipping was as important in the economic life of the town as the timber trade. The culmination of these traditional trades occurred in the period pre-ceding 1875. In these years the export of timber reached its peak, and the town's fleet ranked as the third largest in Norway.[26]

By 1875 the Great Depression brought an end to the flourishing times. This time the setback was definitive. The cause of the permanent decline of the timber trade has been found by many historians, within internal factors. Because of the extensive trade, the supply of timber is thought to have decreased, especially in the

22. B. Rudén, *Papperets historia*, Stockholm, 1987, p.129.
23. P. Mikkelsen, *Norsk tremasseindustri 1863–1895. En studie i et industrielt gjen-nombrudd*, Oslo, 1975, p.185.
24. Kr. A. Olsen, *Skipsfarts- og trelastbyen Drammen*, Drammen, 1959, p.10.
25. Ibid., p.459.
26. O. Thorson, *Drammen. En norsk østlandsbys utviklingshistorie*, Drammen, 1972, vol.3, p.461. Olsen, *Skipsfarts- og trelastbyen Drammen*, p.16.

larger dimensions. In pursuing this line of thought, it has also been maintained that prices were higher in the Drammen district than in other large river systems. Consequently, the sawmill industry of the Drammen district has been considered as lagging behind in the competition.[27] To continue in the business seemed to be no real alternative. In order to survive, the timber merchants had, according to the Norwegian economic historian F. Sejersted, either to switch to the production of wood pulp, or move east to the Glomma river system on the opposite side of the Oslo fjord and continue sawmilling there, where timber dimensions were larger.[28]

If this presumption is correct then we would expect to find local merchants or forest owners among the founders of the new industry. Some examples have been found, but there is no obvious connection between the regression in the traditional trade and the new industry. ' The complicated reality is constantly slipping away from clear generalisations', Sejersted considers, and thinks the explanation must be looked for in a combination of various factors. Technical enthusiasts and reckless investors, wherever they came from, were met with approval in this district, because the need for reorientation was felt most strongly there.

If, however, we take a closer look at the picture, we find it even more complex. The 1870s in Drammen have been characterised as the most extraordinary period of the town's modern economic history. The timber trade increased enormously during the first part of the decade, and optimism reigned. In 1871 there was money in abundance, and 1873 in particular was a marvellous year. Later on, townspeople were sadly dwelling upon those years.[29] The 'golden years' were to a large extent due to the rise of steam sawmills in the area. With the repeal of the law on sawmill regulations in 1860, trade was set free, and men from different parts of the district poured into business. The total number of sawmillers increased considerably. In the 1860s the number of timber merchants was about thirty, by 1875 it had risen to more than fifty.[30] Several of the newcomers succeeded in making substantial fortunes during the boom of the early 1870s.[31] Nearly all the people were more or less engaged in the timber trade, wrote a local author; it

27. F. Sejersted, 'Teknisk utvikling i sagbruks- og treforedlingsindustrien under krisen i 1880-årene' in F. Sejersted (ed.), *Vekst gjennom krise*, Oslo, 1982.

28. Ibid, p.91.

29. S. Eier, *Fløting og trelasthandel. Drammensvassdragets fellesfløtingsforening 1807–1957*, Drammen, 1956, p.185.

30. Ibid., p.184.

31. Thorson, *Drammen*, vol.3, p.473.

was, after all, the town's principal industry. For those who wanted to enter the trade, there was still money to be earned in the Drammen district. Even if the 'golden years' were over in 1874, it took some time before the timber trade had had its day. Even in 1889 only a quarter of the rafted timber went to the pulp industry.

Considering these facts, it is hard to find support for the view that the start of the pulp and paper industry was induced by a crisis in the traditional trade. Apparently, other forces must have been at work. In what follows we shall approach the problem by taking a closer look at the founders of the pulp and paper industry. Who were exploiting the new opportunities?

3 The Start of the Pulp and Paper Industry in the Drammen District: The Pulp Period

The first paper mill in Norway, Bentse Brug, was founded in 1698. It was not, however, until 100 years later that a second paper mill was established. Pre-industrial paper making was, in other words, limited in Norway. In contrast to Sweden, where there were ninety-one paper mills in 1835, there were only seven in Norway.[32] A reason for this has been seen in the mercantilistic policy of the political authorities.[33] Paper making was encouraged by the monarchy, but privileges were for the most part granted to paper mills in the Danish part of the joint Danish–Norwegian kingdom. Thanks to these measures, considerable progress was made in Danish paper making throughout the eighteenth century.[34]

In 1802 Eker Papirmølle was founded by Hans Nielsen Hauge. This was the only paper mill in the Drammen district before modern paper making started. In the late eighteenth century Hauge had started a strong revivalist movement, and in order to encourage industriousness and to provide work for poor people, he also founded enterprises of various kinds. Eker Papirmølle was started against this background, and its situation at the waterfall of Vestfossen was chosen because of its proximity to the Kongsberg Silvermines. Cordage from the mining was used as raw material.

32. Ruden, *Papperets historia*, p.56; H. Fiskaa, *Norske Papirmøller og deres vannmerker*, Oslo, 1973, p.101.

33. G. C. Wasberg and A. Strømme Svendsen, *Industriens historie i Norge*, Oslo, 1969, p.69.

34. O. Nordstrand, 'Some Notes on the Earliest Paper-machines in Denmark', in J. S. G. Simmons (ed.), *VII International Congress of Paper Historians*, Oxford, 1967, p.95.

In 1856 Eker Papirmølle was taken over by Harald Lyche and Erland Kjösterud from Drammen. Lyche was a man of enterpreneurial talents, and the take-over of the old mill was apparently part of his expanding ambitions. Five years earlier he had established himself as a paper stationer in the town. Links between marketing and the modernisation of paper making have also been identified in Britain; R. L. Hills writes that it was the stationers that were the first people to introduce paper machines because they were interested in selling paper of high quality.[35] Characteristically enough, Lyche's enterpreneurial activities did not stop with the acquisition of the paper mill. A couple of years later, his business was extended to comprise printing works, and by 1870 it consisted of a lithographic studio, a bindery, a bag manufacture, a toy factory and whip factory.[36]

The new owners realised that modernisation was necessary if the running of the mill was to continue. To achieve this aim they also went abroad to study paper making.[37] The introduction of wood pulp as raw material became crucial to their future plans. There was timber in abundance in the area, and the use of this material would make possible an extension and mechanisation of the mill. In 1867 they installed a water turbine in the old paper mill – the first in the area – and also ordered a paper-making machine. The same year the first wood-grinding plant in the area was established at a nearby waterfall, erected by a local mechanic. The following year a second pulp mill was established at the same waterfall. Both pulp mills supplied Eker Papirmølle.

In 1870 the idea of producing mechanical wood pulp had spread to another part of the district. A little further north, at Modum, the third pulp mill was erected on the grounds of the old Kongssagene sawmills, which recently had been taken over by new owners. This was the beginning of a wave of wood pulp establishments in the district; within three years another nine mills had been founded, and the Drammen district from now on played a leading role in the development of this industry.[38]

A characteristic feature of this development was that the new industry did not emanate from the traditional sawmill milieu in the

35. Hills, *Papermaking in Britain 1488–1988*, p.174.
36. O. Alsvik, *Bokhandelen i Drammen*, Drammen, 1952, pp.135, 139.
37. Fiskaa, *Norske Papirmøller og deres vannmerker*, p.74.
38. The mills were: Eker, Vestfossen (Borch), Kongssagene, Embretsfos, Labro, Vittingfos, Aadal, Hofsfos, Follum, Veme and Land. H. Fiskaa, 'Treforedlingsindustrien i Buskerud fylke', in J. Sætherskar, *Det norske næringsliv, Buskerud Fylkesleksikon*, Bergen, 1951, p.148.

area. After the first impetus, which came from Eker Papirmølle, it was engineers and industrialists from various parts of the country who arrived in order to exploit the opportunities brought about by new technology.[39] Consequently the transformation, induced by positive 'pressure', was promoted by the so-called optimistic technicians and reckless investors.[40] The same phenomenon has been observed in other countries as well during processes of technical change. In the richly forested district of northern Sweden the pulp and paper industry was started by entrepreneurs who came from other parts of Sweden – and even from Norway.[41] As mentioned, it was, according to R. L. Hills, the stationers who were the first to introduce paper machines in Britain.[42] Writing about the early period of mechanisation in south-west Germany, F. Schmidt writes that the first paper-making machine was introduced by local businessmen.[43] During the period of mechanisation of paper making in Berkshire, USA, J. McGaw found that this process called forth a new breed of mill owner.[44]

The picture is, perhaps, more complex than this. It seems to be the case that a number of various forces were at work in these processes of technical change. Evidence from the Swedish industry suggests that enterprises founded during periods of prosperity had a tendency to resist changes when external conditions were altered. Glete has used theories about institutionalising and inertia within organisations to explain resistance to change. Introducing new technology demands different qualities, because uncertain technological and economic conditions favour those 'who call on others for advice and assistance', writes McGaw.[45] In her study, McGaw concludes that the mill owners who survived the process of mechanisation were those with an extended social network, and she maintains that this factor also determined the pace of industrialis-

---

39. Embretsfos, Labro, Aadal, Hofsfos and Follum were started by engineers, J. F. Lied, Steenstrup, O. Tobiesen and O. E. Gröndahl, and Vittingfos by mill owner A. O. Haneborg. Eker, Kongssagene and Veme were founded by locals. Borch's pulp plant was the only one that was started by a sawmill owner. Source: Fiskaa, 'Treforedlingsindustrien i Buskerud fylke', p.148 (see note 38).
40. In 1973 F. Sejersted put forward a theory about the engineers as the prime movers of the industrialisation in this period in *En teori om den økonomiske utvikling i Norge i det 19. århundre*, Oslo, 1973.
41. J. Glete, *Ägande och industriell omvandling*, Stockholm, 1987, pp.201ff.
42. Hills, *Papermaking in Britain 1488–1988*, p.174.
43. F. Schmidt, *Von der Mühle zur Fabrik. Geschichte der Papierherstellung in der württembergischen und badischen Frühindustrialisierung*, Stuttgart, 1988.
44. McGaw, *Most Wonderful Machine*, p.130.
45. Ibid.

ation as 'the co-operative network of paper mill owners explains a great deal about the rapid and successful mechanization of the local industry.'[46] Glete, too, has emphasised the network factor as a strategically important element in the growth of firms.[47] As to the entrepreneurs of the Drammen district, a characteristic for the group of men who founded the new industry was their many-sided activities and connections. They all had widespread involvement in the financial and economic life of the south-eastern part of Norway, and some of them were bound together by family ties as well as by business. They were all men with an extensive network.

When the Drammen district came to prevail in the development of the Norwegian pulp and paper industry, it was to a large extent due to the district's comparative advantages. Grinding wood is highly power-intensive, and the course of Drammen is replete with middle-sized waterfalls which were well suited for the dimension of the water turbine at that time. For paper making, trees are divided into two main classes, softwoods and hardwoods. With regard to softwoods, such as pines, spruces, firs and larches, 95 per cent of the volume of the fibres can be used. Among these, the Norway spruce is particularly well suited because of the length of its fibres.[48] In the Drammen district, spruce happened to be the dominant type of wood.

In addition to the comparative advantages in the form of natural resources, the district was on the brink of industrialisation and in possession of modern means of transport. The modernisation of transportation is vital for industrial development, and from the course of events in Berkshire, McGaw writes that 'new mill construction followed closely the opening of new rail lines'.[49] These factors were closely linked in the Drammen district as well. The building of the railway was accomplished at the very moment when the new industry was emerging. The main line was opened in 1868, and the side-lines in 1871 and 1872.[50] Most of the pulp mills were directly linked up with the railway. To summarise, the Drammen district offered exceptionally good conditions for the development of the mechnical wood pulp industry, and this circumstance to a large degree explains the growth of the industry in that region.

46. Ibid., p.157.
47. Glete, *Ägande och industriell omvandling*, p.29.
48. Hills, *Papermaking in Britain 1488–1988*, p.144.
49. McGaw, *Most Wonderful Machine*, p.119
50. E. Østvedt, *De norske jernbaners historie*, Oslo, 1954, pp.197, 204 and 206.

The wood pulp industry continued growing in the 1880s. In 1880 the export volume from the district was about 13,000 tons, and in 1890 more than 90,000.[51] The growth was furthered by various factors. From 1880 to 1894, thirteen new enterprises were founded in the district. An examination of the founders suggests that the search for profits played an important role in the establishment of a number of the enterprises. The mills were to a great extent founded by the saw-mill entrepreneurs who had made fortunes in the early seventies. Wood pulp was in heavy demand, and the trade highly profitable. During this period the wood pulp industry yielded exceptionally high returns. Thus, the industry attracted a lot of capital.[52]

Growth was also due to mechanisms within industry itself. Throughout the seventies a number of technical improvements were introduced, and the process continued in the following decade. During the 1880s production costs were reduced by one-third, a change necessitated by fierce competition.[53] In order to reduce costs, technical improvements were of great importance, but the use of economies of scale was of equal importance in this process. As a result, production capacity was considerably increased. The pulp plants founded in the early 1870s were constructed for an annual capacity of less than 2,000 tons; in the 1880s the average annual capacity was augmented by several thousand, and the biggest units were extended to such a degree that they were capable of producing 20,000 tons. In this way, Norwegian manufacturers succeeded in standing their ground on the world market. However, in the 1890s this state of affairs changed.

4 The Paper Industry: The Case of Drammenselven Paper Mills

Nineteenth-century Norway has been characterised as a country exhibiting several traits of backwardness, in particular as regards industry and economic life. The successful adoption of mechanical pulp production has usually been explained by historians by the fact that wood-grinding demanded little technical skill. In accordance with this view, it has been enmphasised that the technological level

51. Mikkelsen, *Norsk tremasseindustri 1863–1894*, p.185.
52. Ibid. One of the foremost men in Norwegian business life in the last decades of the nineteenth century, Chr. Christophersen, made his investments mainly in the pulp and paper industry.
53. Ibid.

of the country was too low for the development of more complex industries like paper and cellulose.

The wave of enterprise formation which took place during the early 1870s was not, however, restricted to pulp plants but included the establishment of several paper mills. In the Drammen district the owners of Eker Papirmølle started making plans for extending the production capacity of the mill. Further expansion of the old mill was impossible, as the head of water was too small. A new site for the establishment of a paper as well as a pulp mill was located at the Geithus waterfall at Modum, some miles north of Eker. As the new project required a considerable amount of capital, a limited company was founded in 1873. Approximately 40 per cent of the capital was raised among the owners' families and relatives.

Regular operation was achieved in 1878. By that time the boom had turned into a slump, and paper prices were falling. Several of the newly established mills were facing serious difficulties, and some were shut down in about 1880, for example Otterelven (Hunsfos) and Union C. Drammenselven Paper Mills, however, managed to weather the storm. From 1880 to 1888 production almost doubled, from 1,00 tons to about 2,000 a year. During the 1890s further expansion took place, so that by the turn of the century production capacity had reached 7,000 tons. Why did Drammenselven succeed where others failed?

5 A German Offer and Fruit Trade in South America

In the last quarter of the nineteenth century most Norwegian wood-grinders were delivered from Myren's mechanical workshop in Christiania. Soon after the incorporation on 19 March 1873, Drammenselven's board of directors asked Myren for a tender, and Myren worked out a plan for a wood-grinding plant. This was not readily accepted; instead the chairman of the board, Jörgen Meinich, travelled to Germany along with the company's technical manager, a Dane called Michael Hansen who had been trained as a pharmacist and had experience from Danish paper mills. Germany was in the process of developing an important paper industry, and was known as the mother country of wood-grinding. A couple of years earlier, the Norwegian technical journal *Polyteknisk Tidskrift* had recommended Heinrich Voelter's machinery as the best available.[54] Heinrich Voelter strenuously attempted to spread the new paper

54. *Polyteknisk Tidskrift*, 1870, p.83.

technology not only in Germany but also in most European countries as well as in the USA. In 1858 he had founded an engineering firm for paper manufacturing, and he co-operated with several machine shops in the production of the various paper-making machinery in Germany and abroad.[55] In his pursuits Voelter managed to establish a worldwide organisation.[56]

Voelter was also aware of the value of publicity and therefore participated at several exhibitions from 1854 to 1873, in Munich, Paris, London and Vienna. The 1867 exhibition in Paris led to his first order from Sweden. The indefatigable innovator also had booklets printed in order to spread knowledge about his products.[57] When Meinich and Hansen called on Voelter in Heidenheim, it was not by mere accident, as Voelter at that time was one of the best-known men in paper manufacturing.

In Heidenheim, the two Norwegians and their project were met with interest on Voelter's side. He worked out a plan for an integrated plant for the production of both wood pulp and paper. In addition he offered a licence for the use of Oswald Meyh's process for the manufacture of brown paper.[58] The process consisted in boiling the logs in steam before they were ground. Pulp treated in this way gave a strong, leathery, brown-coloured paper which could not be subjected to dyeing. As wrapping-paper, however, it was excellent, and it was entirely made of wood fibres.[59] This paper was shown for the first time at the Vienna world exhibition in 1873.[60]

Voelter succeeded in convincing the Norwegians about the advantages of his proposal, and they returned to Norway full of enthusiasm and recommended it to the company. In spite of the fact that such a plant far exceeded the expenditure the company had had in mind, the plan was accepted. In order to implement the project, the capital stock was raised from 400,000 kr. to one

55. E. Raithelhuber, 'Heinrich Voelter', in M. Miller and R. Uhland (eds), *Lebensbilder aus Schwaben und Franken*, Stuttgart, 1966, p.403.
56. Voelter was not, however, the first to pursue an active policy for the diffusion of paper-making machinery. In Heilbronn, not far from Voelter's native place, Bryan Donkin had erected the first paper-making machine in south-western Germany in the 1820s. Donkin was in opposition to the British prohibition of machine exports, and his consistent activity in expanding the market has been regarded as of great importance for the development of the paper industry on the Continent. Schmidt, *Von der Mühle zur Fabrik*.
57. A. Benedello, *Keller–Voelter*, p.159ff.
58. Bayerl and Pichol, *Papier*, p.129.
59. Ibid.
60. Raithelhuber, 'Heinrich Voelter', p.409.

million, at that time a substantial amount of money. Union Co. and Otterelven, both founded at the same time and both for the purpose of producing wood pulp and paper, started with capital stocks of 320,000 and 480,000 kr. respectively.[61] By investing in Voelter's proposal, the company chose a course outside the main stream, as most of the Norwegian pulp and paper mills went in for Norwegian grinding plants and British paper-making machines. As to the choice of products, the company also differed from other paper mills, as these, to a great extent, started producing printing-paper.[62]

In due time, E. Kjösterud went to London, Paris and Hamburg to find outlets for the products. In Hamburg he succeeded in recruiting a well-established company as their agent. According to the agreement, Kratzenstein and Co. was to take charge of the sales of Drammenelven's products for ten years. The first deliveries met with approval, and the company was optimistic about the future prospects although prices were falling. From 1879 to 1880 the export price of paper fell by almost 17 per cent, and under these circumstances several mills succumbed. Drammenselven, on the contrary, was increasing its exports year after year.

The company's success was closely connected with the kind of paper quality they had decided to produce. Through Kratzenstein and Co., Drammenselven succeeded in getting a foothold in the overseas market. Wrapping-paper was heavily in demand in the booming fruit trade in South America,[63] and brown paper was well suited for this purpose. It was cheap, and sufficiently strong. Every week, cargoes containing small bales of brown paper were shipped to Hamburg for further shipment. During the Atlantic crossing these bales were used as stowage on big ships. Freights were thereby reduced, and Norwegian paper was accordingly competitive in spite of the long transportation. Drammenselven was a pioneer in this trade, and it attracted much attention. It was not long before other manufacturers were imitating Drammenselven's course of action. Within a short span of time, several mills were producing brown paper. Some mills were technically integrated,

---

61. A. Strømme Svendsen, *Union 1873–1973*, Oslo, 1973, p.45; S. Tveite, *Vennesla*, Kristiansand, 1986, p.216.

62. *Papir-Journalen*, no.9–10, 1940, p.88. It is noted that most of the machines until 1890 were delivered either by Bentley and Jackson or by Bertram and Sons in Scotland.

63. In 1885 paper products to the amount of about 6.5 million marks was shipped from Hamburg and Bremen to South America, which was the most important of the overseas markets. Winckler, *Der Papierkenner*, p.221.

including a pulp plant, as was the case of Böhnsdalen, Funnefos, Granfos and Tinfos, whilst others, such as Stray and Holmen, were established as pure paper mills.[64] Adopting brown paper production was relatively easy, since all that was needed was a boiler for the steaming of the logs in addition to ordinary paper-making and wood-grinding equipment.

The immediate success of the brown paper mills is evident in the export statistics. From 1881 to 1891 the volume of Norwegian paper exports rose by 260 per cent. The main share of this exports (80–90 per cent) consisted of brown paper.[65] As prices at the same time were constantly falling, the value of paper exports remained at the same level throughout this period. It is partly due to this fact that the initial period of the Norwegian paper industry has often been overlooked. This is, however, not a period devoid of interest, but provides us with valuable insight into which factors were particularly important in the early adoption of new technology. The progress made in Norwegian paper making was, however, noticed by contemporaries. Most of the mills participated at the exhibition in Christiania in 1883. The part of the exhibition dedicated to paper was praised in the journal *Teknisk Ukeblad*, and some of the mills were awarded medals for their products, among others, Böhnsdalen and Drammenselven.[66] Owing to the success of the brown wrapping-paper, the period from about 1880 to 1990 has been called 'the brown paper period' in the Norwegian paper industry.

It seems to be the case that Norway's alleged backwardness proved to be no hindrance for the adoption of more complex technologies. Furthermore comparative advantages apparently played an important role in the success of the paper industry. Norway offered favourable conditions for the production of brown paper; water power and spruce were to be found in abundance. The fierce competition, however, did not allow the manufacturers to rest on their laurels.

6 Competitive Pressures

According to A. D. Spicer it was 'only with the advent of new mechanical devices [that] the truth slowly dawned upon the

64. *Papir-Journalen*, no.9–10, 1940, p.88.
65. NOS, *Norges Handel*. In 1881 the volume of paper exports was 3,189 tons; in 1891 it had increased to 11,551.
66. *Papir-Journalen*, no.11–12, 1940, p.100.

Anglo-Saxon mind that the old mercantile system encouraging exports and discouraging imports was unsound, and that markets were won by selling cheap'.[67] This was in any case true of the paper industry. As early as 1832 American paper makers predicted that 'competition in our land will reduce the prices *as low* as the article can be manufactured: and *improvements in the mechanic arts* will lend a helping hand'.[68] Technology was to solve the manufacturers' problem and J. A. McGaw writes that 'Investing to reduce production costs was the millowner's principal competitive strategy'.[69]

Right from the start, such considerations played an important role in the strategy of the Drammenselven Paper Mills. Initially the mill was equipped with only one paper-making machine. The buildings, however, were constructed for two machines. After one year of full production, calculations for the purchase of another were made, in order, it was stated, to bring production capacity into line with the capital already invested. The cost of this investment was regarded by the company as rather low in comparison with the potential production increase.

In 1880 a new paper-making machine was ordered, this time from Bertram & Sons in Edinburgh. At the end of the first year the second machine was put into operation and production rose by 35 per cent. By continuously improving and extending the machinery, manufacturing capacity was further augmented by 40 per cent during the period 1881–8. In this way the enterprise managed to keep pace with constantly falling prices. In 1889 the factory was damaged by fire, and all the equipment had to be replaced. The rebuilding was accomplished in one year and the mill re-entered business with two paper-making machines from Bertram & Sons. At the same time production capacity was further extended. This time, however, the measures taken did not suffice to cope with the fierce competition. In spite of all kinds of rationalisation, prices fell faster than the reductions in costs.

In 1894, therefore, it was decided to extend production even further, and it was under these circumstances that the company ordered a so-called 'high-speeder' from an American machine shop. Again, the company was taking a pioneering step, as the 'high-speeder' was the first of its kind in Norway, and it was not until the turn of the century that the first American multi-cylinder

67. Spicer, *The Paper Trade*, p.132.

68. J. A. McGaw, 'Business Practice in Berkshire County Paper Industry', *Technology and Culture*, vol.26, 1985, p.710.

69. Ibid.

machine was imported to England.[70] Why did so much innovative activity take place at Drammenselven Papermills, and what made the company capable of keeping up with international development?

7   A Matter of How: Ethics and Travels

In the case of Drammenselven Papermills there were two factors which contributed significantly to the successful running of the enterprise: professional management[71] and flexibility.[72] Both factors were closely linked with the qualifications of the owners and their purpose in establishing the enterprise. H. Lyche and E. Kjösterud belonged to families that for generations had been engaged in the economic life of the area. The bourgeoisie of which these families were part was influenced by the revivalism of Hans Nielsen Hauge. Lyche himself was a religious man, and this engagement was reflected in the way he directed his business.[73]

Work, industriousness and endurance were highly appreciated in this community, and hence order and seriousness were the guiding principles for their business activity. In the development of Berkshire paper making McGaw has emphasized the importance of the local revival movement, as it 'encourages manufacturers and workers to value discipline, punctuality, and temperance'.[74] In McGaw's reasoning, regular habits were important preconditions for the mechanisation of paper making.[75] Qualities like endurance and serious engagement were none the less essential for developing a paper industry that to a high degree demanded long-term capital investments.

One of the first steps that was taken after the incorporation, was to establish a professional management. This meant the engagement of two full-time managers, one responsible for the technical department, and the other the commercial. Several of the technical managers were recruited from Sweden, the business managers, on

70. Shorter, *Paper making in the British Isles*, p.138.
71. According to E. Penrose the firm's size is determined by its management's potential for growing with the firm. E. Penrose, *The Theory of the Growth of Firm*, Oxford, 1959.
72. In his book, J. Glete draws a dividing-line between the power to change and the power to conserve. In enterprises where the power to change prevailed access to information and its effective handling were crucial. Glete, *Ägande och industriell omvandling*, pp.16ff.
73. H. Alsvik, *Drammens handelsstands forening 1847–1947*, Drammen, 1947, p.36.
74. McGaw, *Most Wonderful Machine*, p.81.
75. Ibid., p.89.

the other hand, were Norwegians.[76] Both the technical and the business managers had their offices at the plant.[77] The meetings of the board of directors also took place there. Thus the management was closely related to the daily operations at the factory. This factor has often been emphasised as important for a firm's innovative capability. At that time, this arrangement was rather unusual. For many Norwegian enterprises within the pulp and paper industry, the managerial functions were often geographically separated from the mills. Many enterprises were directed from Christiania. As sales of the products from the pulp and paper industry were usually entrusted to agents, little attention was paid to developing efficient management, as was the case of Embretsfos pulp mill, where no business manager was engaged.[78] Clearly, the successful running of Drammenselven was closely connected with the organisation of the firm and the qualities of the directors.

As important as a firm's organisation, however, was access to information, since paper makers were operating in a highly competitive environment. Through what channels did they obtain information? During a journey to Massachusetts in 1890 the young engineer Mohn[79] happened to hear about a paper-making machine that could run at the incredible speed of 120 m per minute. The usual speed at that time was 50–60 m. Initially this was not believed at home, and Mohn was suspected of telling stories.[80] Drammenselven, which was about to buy two additional machines, came to hear these rumours. The technical manager was immediately sent to the USA to make further inquiries into the matter. The rumours proved to be true, and on his return the company decided, as mentioned above, to buy a 'high-speeder' from the Pusey & Jones Co.

The case reveals essential features about the information situation. To Norwegian paper manufacturers, information diffused accidentally by way of personal contact and travels. In the late nineteenth century the channels of information were to a great extent the same as they had been at the beginning of the century.

76. Several of the technical managers were Swedish because Sweden had had a stronger tradition in paper making than Norway.
77. N. Rosenberg has emphasised the importance of the management's daily confrontation with the existing range of productive activities. N. Rosenberg, *Perspectives on Technology*, pp.110ff.
78. The Embretsfos factory was headed by Chr. Christophersen. In 1899, however, he went bankrupt.
79. Victor Mohn was son of the director of Christiania Mekaniske Verksted.
80. *Papir-Journalen*, no.11–12, 1940, p.98.

N. Rosenberg has emphasised the role of institutionalised innovation in industrialisation processes.[81] The experience of the Norwegian paper industry suggests that in the first decades of its development this industry had to manage differently. It was not until 1889 that this situation changed. In that year G. Hartmann, a graduate from the Technical High School in Munich, established himself in Christiania as a consulting engineer, specialising in machinery for paper, cellulose and pulp mills. Hartman established his firm on the advice of a man in the trade, Carl Christensen.[82] He acted as an agent for various machine shops, among others, H. Füllner in Warmbrunn, Silesia. The influence of his activity is to be seen in the fact that Füllner supplied a number of Norwegian mills with machinery and equipment.

Throughout the nineteenth century, Norwegian paper makers had largely to manage on their own in the search for information. As we have seen in the case of Drammenselven, foreign travel was of great importance in this respect. Such travel was sometimes undertaken for a definite purpose, e.g. in order to buy new machinery, as was the case in 1873, 1889 and 1896, or more generally to keep up with technical developments. From such travels we learn not only how technology was diffused, but they also provide us with valuable information about the mechanisms involved in the diffusion process.

8 New Products and a New Development

In 1881 Scheide, the manager at Drammenselven, was sent to Germany to study technical development within the paper industry, and in particular Mitscherlich's process for the production of sulphite cellulose. As a guest at the Second Congress of the German paper makers' association, he expressed himself content with his studies, and stated that the company would probably erect a factory for the production of cellulose.[83] Drammenselven experimented with the new raw material, but gave it up as it was found too expensive. The same happened at Embretsfos; the establishment of a cellulose plant was considered, but the project was rejected, as the price of the product was considered to be too high.

81. Rosenberg, *Perspectives on Technology*, p.152.
82. E. Sundt, *Norges handel og industri*, Christiania, 1907, p.481.
83. *Festschrift zum 50-jährigen Bestehen des Vereins deutscher Zellstoff-fabrikanten*, Berlin, 1930, p.10.

The first attempt to produce sulphite cellulose was made in 1872. In 1881 the first sulphite plant was erected in Norway. However, it was not until about 1890 that his industry became of importance, a fact which explains why Norwegian manufacturers have been regarded as late starters in this industry. A closer investigation into the development of the cellulose technology, and the timing of its diffusion, suggest a different picture. Cellulose technology was still in its childhood in the 1870s and early 1880s, and during its period of maturity the cellulose product was too expensive in relation to paper prices.[84] Because of these circumstances the breakthrough of the cellulose industry did not occur until the second half of the 1880s. Compared to developments in other countries, this suggests that the Norwegians to a great extent were following the general course of events. In Germany the first sulphite plant to apply Mitscherlich's method was built in 1879–80, only one year prior to the establishment of a similar plant in Norway.[85] A general breakthrough did not take place until the late 1880s and early 1890s. As regards the German development, the repeal of Mitscherlich's monopoly in 1884 has been seen as important for the spread of cellulose – allegedly because of threatening competition from Swedish, Norwegian and Austrian manufacturers.[86]

Viewed against this background, it appears that Norwegian manufacturers were, on the contrary, quick in adopting the new technology. Thanks to the innovative activity of the Swedish entrepreneurs, Scandinavia came to play a leading role in the development of the cellulose industry.[87] The general breakthrough of this industry occurred in the mid-1880s, as the new product by that time had reached a stage of commercial profitability. When first accepted, the spread of cellulose was rapid. Cellulose was for the first time separately classified in British import statistics in 1887. In 1893, it had already outdistanced other sorts of raw material in paper making.[88] Simultaneously cellulose appeared for the first time in Norwegian export statistics, and during the 1890s the export of cellulose quintupled.[89]

84. Bayerl and Pichol, *Papier*, pp.142ff.
85. Der Verein Deutscher Papierfabrikanten, *Festschrift zum 50 jährigen Jubiläum des Vereins*, Berlin, 1922, p.199.
86. *Festschrift zum 50 jährigen Bestehen des Vereins deutscher Zellstoff-fabrikanten*, p.37.
87. E. Kirchner, *Das Papier. III Teil. Die Halbstofflehre der Papierindustsrie*, Chemnitz, 1907, p.18.
88. Spicer, *The Paper Trade*, appendix 1.
89. NOS, *Norges Handel*.

Cellulose proved to be the perfect substitute for textile fibres, and within a few years became the dominating raw material in paper making. The diminished cost of making paper near the source of supply gave countries like Germany, Norway and Sweden an advantage in paper production. This opportunity resulted in a second spurt for the Norwegian paper industry, and the export of paper also quintupled during the 1890s.[90] This time, however, it was another paper quality that came to form the basis for the growth and development of the industry: newsprint. As this paper quality consists of a mixture of cellulose and wood pulp, but mostly wood pulp, Norway offered favourable conditions for the production of this quality. By exploiting their comparative advantages – by investing in power-intensive production with a comparatively low use of timber per ton finished product[91] – Norwegian paper makers managed to assert themselves in a highly competitive environment. This structure still forms the basis of the Norwegian pulp and paper industry.

10 Conclusion

In this paper I have tried to focus on the factors influencing the start of the pulp and paper industry in Norway. Decisive for industrialisation was the development within the modern paper industry; in this process technological transfer was crucial. Observations made on international developments, along with evidence from the industry in the Drammen district, suggest that Norwegian manufacturers were quick in adopting the new technology at every stage during the foundation period 1870–1900. It was also of importance that there were local groups of people who understood how to exploit the new opportunities. The structure of the industry, however, was shaped in the intersection between the technical development and the country's comparative advantages.

90. NOS, *Norges Handel.*
91. Wood pulp uses far less timber per ton of finished product than cellulose.

15

'To Take Great Pains': Norwegian Wood Pulp on the British Market in the 1870s

Even Lange

The development of a modern wood-processing industry during the last three decades of the nineteenth century is central to the economic history of Norway: wood pulp production constituted Norway's most important growth sector during the period when industrialisation made its breakthrough. Accordingly, interpretations of how this industry first was established have a significant influence on our understanding of Norwegian industrial development in general.

We are dealing with problems of great complexity, which can be treated from many angles, but one of the basic questions concerns the relative importance of supply and demand forces in the process: Was Norwegian industrial development in the decisive phase towards the turn of the century mainly precipitated by increased demand on the world market, or should the development be explained mainly by conditions on the production side in Norway, as a result of measures implemented by the local businessmen? Were the decisive impulses provided by supply or demand factors?

There is general agreement that an important characteristic of the industrial breakthrough from the 1870s onwards was production for export markets. The export of goods was more than doubled up to the end of the nineteenth century, and branches such as the textile, metal and chemical industries crossed the threshold from the national to the international market.

Wood processing played a central role in this export-oriented industrial growth. The forest sector had for a long time been the strongest pillar of Norway's foreign trade, and in 1870 delivered close on 40 per cent of exported products. Its share rose to almost 50 per cent in 1900. It was modern wood-processing, primarily wood pulp production, which was in the forefront of the forest sector's expansion in export markets. Whereas the production and sale of timber stagnated, pulp production increased dramatically from c.500 tons in 1870 to nearly 10,000 in 1875 and over 30,000 in 1880. Practically all this was exported to Britain, but in the course of the 1880s France also became an important market.[1] The growth of the pulp industry formed the prelude to a modern, broadly based development of forestry. From a zero-point in 1870, wood processing came to represent more than half of the forest sector's export at the turn of the century, an export which again comprised over half of the country's strongly increasing total exports.

The pulp industry's breakthrough thus provides a fine example of the opening phase of modern industrial growth, and of the transition to an industrial economy in Norway. The answer to the question regarding how the pulp industry developed will help us to understand the central phase of the transition to an industrial economy in Norway.

In an otherwise rather sparse literature dealing with Norwegian economic history, the breakthrough in the export of wood pulp has been the subject of much attention and discussion. Several dissertations on the development of the pulp industry during this period have been written; Per R. Mikkelsen's of 1975 is the most important. Francis Sejersted has discussed the subject in a number of connections, both in specialist articles and as a section in larger presentations. Fritz Hodne has given the wood-processing industry exhaustive treatment in his survey of Norwegian economic development. In addition, we can find histories of branches of the industry and a comparatively copious literature of jubilee books published by the many industrial concerns which were established during this period.[2]

According to the prevailing opinion in this literature, the most important starting-points for the Norwegian pulp industry's growth during the 1870s were an increasing demand for paper on the world market and a lack of alternative raw materials. These conditions

---

1. NOS: *Historisk Statistikk 1968*, table 161; P. E. Mikkelsen, *Norsk Tremasse-industri 1863–1895*, Hovedfagsoppgave, University of Oslo, Oslo, 1975, appendix 1, table A.

2. F. Hodne, *Norges Økonomiske Historie 1815–1970*, Oslo, 1981; Mikkelsen, *Norsk Tremasse*; F. Sejersted, 'Teknisk utvikling i sagbruks- og treforedlingsindustrien under krisen i 1880-årene', in F. Sejersted (ed.), *Fra Linderud til Eidsvold Værk*, vol.3, Oslo, 1979. See also K. Anker Olsen, *Follum gjennom 75 år*, Oslo, 1948; A. Strømme Svendsen, *Union 1873–1973*, Oslo, 1973.

especially applied in Britain, which became Norway's most important market.
Fritz Hodne is perhaps the most incisive exponent of this view.
He writes:

> When the pulp mills took the stage as a growth industry in the Norwegian
> economy in the 1870s, it can be worth taking a glance at the international
> market situation. There was an increasing lack of raw materials for paper in
> the world, and paper prices were going up. The time was ripe. The
> world's paper consumers were waiting for an innovation.

Hodne considers that such a demand–dictated development is the
main feature of the Norwegian economic growth process. The
pulp industry's growth fits, according to his presentation, beauti-
fully into the 'theory of export–led growth', which advances the
'the external or exogenous impulse' as 'the essential factor trigger-
ing off the growth' to follow Hodne's formulations.[4]

Per R. Mikkelsen, who has studied the pulp branch most thoroughly
and represents an important source for Hodne in this connection,
maintains that the development of the pulp industry was in all
essentials precipitated by pressure of demand. This applies to both
the early and later phases. Mikkelsen points to the fact that the first
Norwegian mills in the 1860s were established by paper manufacturers who wished to expand production, but who could not obtain
enough rags as raw material. Every establishment of wood pulp
mills up to 1870 had, according to Mikkelsen, the aim of meeting
the demand caused by the precarious raw material situation on the
domestic market.[5]

Mikkelsen asserts that the first Norwegian mills demonstrated
that conditions were favourable for producing a paper raw material
which was also internationally in demand. As a result, from the
1870s onwards a number of pulp mills were established outside the
original centres of paper production in Norway, the target being
the large foreign markets.

These mills were for the most part established at the beginning of
the 1870s under the influence of the boom and good market
prospects, Mikkelsen states. Of the 25 Norwegian pulp mills which
started to operate in the 1870s, eighteen were built in the years
1870–3. The export offensive for Norwegian pulp in the 1870s was

3. Hodne, *Økonomiske Historie*, p.87.
4. Ibid. p.19.
5. Mikkelsen, *Norsk Tremasseindustri*, pp.20ff.

thus, according to Mikkelsen, set in motion under the strong pressure of demand from abroad. He carries the argument forward into the 1880s and 1890s, and arrives at the result that 80 per cent of the mills up to 1895 were established in situations with strong foreign demand.

As soon as the normal state of the market changed, and especially in the 1880s, Mikkelsen also emphasises the fall in prices and the difficulties within the traditional sawmilling industry as contributory factors to the expansion of the mills. Declining profits and lack of opportunities for expansion in the timber trade resulted in many sawmills switching to pulp production, especially in the Drammen River district. Nevertheless Mikkelsen supports in the main Hodne's argumentation, inasmuch as he regards these factors to be of secondary importance, and considers that the demand for pulp was the main driving force behind the expansion in the 1880s and 1890s.[6]

Francis Sejersted has strongly rejected such an interpretation as far as concerns the 1880s and 1890s. He has in more general terms maintained that there was a transitional period from demand-directed to supply-directed industrial growth during the depression of the last part of the 1870s. As regards the development in the forestry sector, he has also attached decisive importance to the role of the supply situation. Sejersted considers that technological change lay behind the expansion in the 1880s and 1890s.[7]

Even though Sejersted is in basic disagreement with Mikkelsen and Hodne regarding the driving forces behind the development in the 1880s and 1890s, he shares their most important suppositions concerning the expansion of the pulp industry at the beginning of the 1870s. He writes about 'a strongly increasing consumption of paper in the world as the nineteenth century progressed' and states that 'the increased consumption led to an intense pressure on the supplies of raw materials', which again 'gave rise to tremendous efforts to find alternative fibres from which to make paper'. 'There was an early focus on wood fibres', Sejersted says, and he describes the development as 'an obvious example of how market impulses give rise to efforts leading to technological innovations'.[8]

The impulse arising from demand was passed on through the paper manufacturers, first at home, then from Britain. The first phase is described by Sejersted as 'an optimistic time of general

6. Ibid. pp.19, 68ff.
7. Sejersted, 'Teknisk utvikling', p.82.
8. Ibid. p.81.

prosperity' where production, as he says, 'at any rate partially' was set in motion 'as a result of pressure from the paper manufacturers'.[9]

In Sejersted's presentation the whole process is started through pressure of demand, and the background for the development is increasing paper consumption and lack of raw materials abroad. More specifically he asserts that 'the impulse of demand from England' seems 'to have played an important role' at the beginning of the 1870s, and mentions that H. C. Mathiesen could inform his cousin Christian Anker that John Neck, the English agent, urged him to 'build a factory', because he promised easy sale of a considerable quantity of wood pulp.[10]

The basic conditions for the growth of the Norwegian pulp industry in the 1870s, according to the above authoritative versions, can be summed up as follows:

1. There existed a general shortage of raw materials for paper because the supply of rags and cloth, which constituted the traditional raw materials, was limited.
2. The problem of shortage of raw materials was solved by developing methods of manufacturing paper from wood material.
3. Strong demand for wood pulp, caused by rapidly increasing paper consumption on the British market, was the driving force behind the establishment of the wood pulp mills, directed at the export market, in Norway at the beginning of the 1870s.

In the light of British accounts of the development of the paper industry and new source material relating to the market conditions which Norwegian mills encountered in Britain, I shall argue that the above picture of the growth of the Norwegian pulp industry should be corrected in several important respects.

The increase in production and the raw material problems which plagued the world's paper industry from about 1850 were real enough. From early in the nineteenth century, when hand-made paper was gradually replaced by machine-produced paper, production had increased considerably. In Britain, which was by far the world's largest paper market, production was doubled seven to eight times from 1800 to 1860, when it reached c.100,000 tons per annum.[11]

The supply of rags and cloth did not manage to keep pace, *inter*

9. Ibid. p.88.
10. Ibid. p.83.
11. D. C. Coleman, *The British Paper Industry, 1495–1860*, Oxford, 1958, figure 7.

alia because people changed their clothing habits from primarily cotton to material with a wool mixture, which was unsuitable for paper. The price of rags rose by nearly 30 per cent during the 1850s, and as they represented c.50 per cent of the paper's cost, this was a serious problem.[12]

The search for alternative raw materials was intense. In 1854 *The Times* offered a prize of £1,000 to the person who could find a substitute for rags. By 1861 120 patents were taken out for this purpose, ninety-two of them after 1850. They included almost every imaginable material, such as rope, turf, coconut and rhubarb, but only two had reached a commercial stage by 1861, when the paper tax, the last so-called 'tax on knowledge', was abolished. The first to be commercialised was cotton and flax waste material from the spinning mills in Lancashire, which was used by the local paper factories, and the second was straw, which was used with varying degrees of success wherever it could be obtained.

None of these raw materials, however, represented real substitutes in the sense that supply, quality and price could measure up to the rags. A prominent British paper manufacturer stated in 1861 that he took a gloomy view of the situation: 'I think there is no probability of any substitute for rags being discovered'. The prize offered by *The Times* was not awarded either.[13]

However, in Germany experiments had been made with wood as raw material for paper since the 1840s, and around 1850 an engineer named Voelter introduced a method of grinding up wood pulp which gave a usable raw material when it was mixed with the traditional rags. Norwegian paper manufacturers, who for a long time had had severe raw material problems and had been dependent on a considerable rag import, naturally welcomed the new material.

As I have mentioned, several of these paper manufacturers set up pulp mills attached to their factories in the 1860s, and the thrust into the export markets was started in the next decade. The first purely export-oriented mill was Granfos, with O. M. Tobiesen, the founding engineer, as the driving force behind it. Swedish manufacturers had entered the field a little earlier and, together with some Germans, established a certain amount of export to Britain. Here, in the world's largest market, the Norwegian timber merchants really did have good contacts, and, as we have mentioned above, in 1870 H. C. Mathiesen was urged by his agent John Neck to set up a wood pulp factory for export purposes.

12. Ibid. p.338.
13. Ibid. p.344.

He was not the only one to receive such signals. His cousin Nils Anker, a timber merchant in Fredrikshald (Halden today), could report similar inquiries. Together with his brother Christian, he had started up a little pulp mill at Skaaningsfoss in Tistedalen during the autumn of 1870, and he wrote at this time that 'from many different quarters' he had 'demands and exhortations to send wood pulp'.[14] The pressure of demand from the British market as the starting-point for concentrated efforts in Norway to produce pulp in the early 1870s thus seems well documented.

The subsequent experiences of the Anker brothers as regards this market oblige us, even so, to place a big question mark on such an interpretation. When production had started up properly, it turned out to be extremely difficult to sell the wood pulp. The agent, who also in the Anker brothers' case was Neck, did not manage to speed up the sale. In May 1871 stocks were so large (and at risk, because pulp decomposes in warm weather) that Nils Anker had to go to Britain himself to straighten things out.

His letters home to his younger brother Christian give a completely different picture of the British market's demand for Norwegian wood pulp from what we have previously surmised. After travelling widely round Scotland, where he first landed, and visiting nearly fifty factories, he had to admit that he was 'most dissatisfied' with his trip, and continued, 'I have often been depressed about the prospects for our Skaaningsfos factory.' There existed, as far as he could see, 'only two manufacturers in the whole of Scotland who use pulp according to Voelter's method', and they were already well supplied.[15]

The Norwegian raw material supplier even had great difficulty in getting interviews with the manufacturers. Anker counted himself lucky that he had a local contact 'without whose help I would never have been granted an audience with the haughty and rich paper manufacturers', as he put it. Obviously Scottish manufacturers were not short of raw materials for paper at this juncture, or at any rate not after wood pulp.

This did not mean that the factories had not tried it. Even when he tried to hand out samples, Nils Anker met rejection: 'Unfortunately they had all without exception experimented with pulp, and received instructions from the Swedes in its use.' All except for two, 'declared that it was unusable, even for newspaper'.[16]

14. Mikkelsen, *Norsk Tremasseindustri*, p.98.
15. Letter, Nils Anker to Christian Anker, 28 May 1871, Anker Archives, Rød Herregård, Halden.
16. Ibid.

The situation in Scotland in the early summer of 1871 was thus as follows. Pulp was little used, and the paper factories were not very interested, but were well informed. There was no sign of any great demand for war materials, but pulp had definitely been tried out and found unsuitable as raw material for paper by most users. In other parts of Britain the situation was not much better. In London, through the help of one of Borregaard's directors, Anker established contact with somebody he called 'a good agent', who was 'much respected by the manufacturers'. The Borregaard owner, Mr Godman, had himself 'speculated' by establishing a pulp mill at Borregaard. However, the agent would 'have nothing to do with wood pulp', and prophesied the 'early demise' of wood pulp. It was, as Anker wrote home, 'an embarrassing reception' and Godman was 'completely cured of the idea of building a pulp mill'. Other agents came with similar warnings. A senior partner in the reputable firm of Clarkson, whom Anker considered 'well informed about everything . . . predicted a gloomy future for pulp'.[17]

The picture presented by this correspondence of Nils Anker is thus completely different from the situation described by the rest of the Norwegian literature at the beginning of the 1870s. There is no trace of any pressure of demand. Pulp was, as Anker concluded after he had travelled round the London area too, 'hardly used at all anywhere'. It was almost only those who manufactured low quality products of secondary importance such as paper hangings, 'long elephants', millboard, and 'cartridges', i.e. wallpaper and cheap cardboard, who could use pulp. According to Anker, 'the Germans had been especially quick off the mark in delivering pulp, and so had the Swedes', so that the market was saturated.[18]

In general Nils Anker characterised the state of the British pulp market as follows after his month's stay and travel in all the paper districts: 'The business competition is tremendous, because the manufacturers are swamped with samples and commercial travellers, and the production of pulp is certainly twice as much as consumption.'[19]

Even if this is an exaggeration and Anker's impression more generally could have been reinforced by seasonal variations, his market analyses seem so thorough that they must be given weight. As so many of his assessments are of a long-term nature, his reports can hardly be dismissed as a reflection of a temporary swing

17. Letter, Nils Anker to Christian Anker, 4 June 1871, Anker Archives, Rød Herregård, Halden.
18. Letter, Nils Anker to Christian Anker, 9 June 1871.
19. Letter, Nils Anker to Christian Anker, 18 June 1871.

towards a buyers' market. They seem to support the conclusion that the theory of over-demand for pulp on the British market in this vital early phase of the 1870s must be revised.

The reason for the weak interest in the new raw material was not stagnating paper production, as one might naturally assume from an economic line of argument. On the contrary, paper production increased more rapidly than before. The rates of growth, which had fallen somewhat up to the middle of the 1860s, recovered strongly in the early 1870s, corresponding with the general economic trend. From 1870–1 there is a leap in British paper production from 117,000 to 155, 000 tons, i.e. about 30 per cent.[20] Even so the manufacturers were not, as far as we can understand, hard up for raw materials at this time.

Such a strong expansion in paper manufacturing at the same time as Anker finds an over-saturated market for raw materials is puzzling. The explanation must be that the basic raw material situation for British paper manufacturers in fact was different from the one presented in traditional Norwegian historiography. The precarious raw material situation in the 1850s no longer applied.

The most important reason why demand for pulp was so weak twenty years later quite simply is that the paper factories in the mean time had found a perfectly good substitute for rags. The raw material problem in paper production was solved before wood pulp entered the British market.

During the course of the 1860s British manufacturers had changed over on a grand scale to a new raw material for paper with which they were satisfied and which could be supplied in large quantities. The raw material was a type of grass called esparto. It played a decisive role in the important growth phase for the British paper industry between 1860 and the end of the 1880s.

As an element of the intensive efforts to find a substitute for rags, Thomas Routledge, the paper engineer, had developed a process of boiling esparto grass in caustic soda with subsequent bleaching, following the same process as used with rags. He arrived at a viable method which he patented in 1856, and in 1860 he patented improvements which made the process commercially competitive. At the world exhibition in London in 1862 Routledge was awarded a medal for paper made solely from esparto, with the jury's verdict 'very satisfactory for common or ordinary printings'.[21]

This raw material thus had great success in the 1860s. It was most

---

20. A. D. Spicer, *The Paper Trade*, London, 1907, appendix IX.
21. Coleman, *Paper Industry*, pp.340ff.

popular in Scotland, *inter alia* because the process was dependent on a steady supply of chemicals and soft water. For the British paper industry as a whole, the use of esparto grass signified, according to D. C. Coleman, that 'a new contentment' spread from the 1860s 'to gladden the hearts of paper manufacturers'. A. H. Shorter, in his latest survey of the British paper industry, makes similar remarks: 'The anxiety of the paper-makers in the 1850s and 60s was largely relieved by the development of the use of esparto grass, on which Britain has since relied to a considerable extent.'[22]

Esparto paper turned out, as the jury's verdict indicates, to be especially suitable as raw material for the fastest growing product: printing paper for newspapers and books. Esparto gave a homogeneous paper which absorbed printer's ink quickly and easily without making too strong demands on the type.[23]

Already by 1863 the British were importing 50,000 tons of esparto, and the volume more than doubled during the following decade. The importance of esparto is clearly shown by the development in the market for the old main raw material, rags. About 1860 the import of rags into Britain reached a level of c.15,000 tons yearly, and the price was pressed up to £28 per ton. However, during that decade both the price and the import volume of rags fell by one-third and one-quarter respectively. In 1870 Great Britain had in fact become a net exporter of rags.[24]

Esparto had thus solved the British paper factories' raw material problem before Norwegian pulp appeared on the scene, and the 1870s turned out to be the golden era of esparto. The grass was first imported from Spain and Algeria, and in the 1870s rich new sources of supply materialised from Tunis and Libya. Large paper manufacturers like Edward Lloyd were backers of this raw material. Lloyd had his own enterprise for publishing popular literature and newspapers – potential consumers of his paper. He secured for himself a safe raw material source through plantations and the right to harvest esparto grass over large areas of Spain and North Africa. Lloyd alone controlled more than 400,000 acres of esparto grass plains. Tens of thousands were engaged locally in the seasonal harvesting and sorting work.[25]

This, then, was the marketing situation for raw materials facing the Norwegian pulp producers who wished to enter the British market in the 1870s. They had to compete not with a scarce,

22. A. H. Shorter, *Paper Making in the British Isles*, Newton Abbot, 1971, p.139.
23. Ibid. p.141.
24. Spicer, *Paper Trade*, appendices I, II and IX.
25. W. J. Reader, *Bowater: A History*, Cambridge, 1981, p.11.

expensive and declining old raw material, but with a new and expanding one which had rich supply sources and strongly established interests backing it.

In addition, esparto produced a paper of much higher quality than wood pulp. This must surely be the main explanation of why Nils Anker was cold-shouldered by the 'haughty rich manufacturers'. He was informed by the paper manufacturers that they might consider changing over to pulp 'when they could get as good quality paper as esparto made'. But they could not get this in 1871.

Throughout the 1870s British paper production increased more than fourfold, and in 1883 had reached 330,000 tons. Increased supply of esparto provided the bulk of the raw material of this strongest growth phase in the British paper industry during the nineteenth century. Parallel with paper production the import of esparto also increased more than fourfold during the 1870s, and in 1883 was well over 200,000 tons. By then, however, the most intensive growth phase was on the wane, and the rates of growth fell quickly to one-third of what they had been.[26]

Even though the starting point seemed pretty hopeless for Nils Anker, pulp managed to get a foothold in the British market during the 1870s, although in modest quantities compared with esparto. Later, at the end of the 1880s and beginning of the 1890s, wood pulp and chemical pulp were to take over as the main raw material for paper, but the ground was prepared in the 1870s, and it is this breakthrough which needs an explanation.

According to British import statistics, 12,000 tons of the product 'pulp of rags and wood' were imported in 1873. Since two-thirds came from Sweden and Norway, most of this must have been wood pulp. Ten years later the import of wood pulp had risen to c.70,000 tons, and Norway alone provided some two-thirds of the total.[27]

How then are we to explain the entry of Norwegian wood pulp into the British market at the beginning of the 1870s if the old theories about lack of raw materials and pressure of demand do not hold water?

Let us return to Nils Anker to examine the question more closely. We do not know for sure why he and his brother Christian started their pulp mill in 1870, and similarly the motivation for establishing most of the other early pulp mills is obscure. Nevertheless, Per R. Mikkelsen, who otherwise attaches decisive signifi-

---

26. Spicer, *Paper Trade*, appendices I, II, and IX.
27. Ibid., appendices I and II.

cance to the factors of demand, considers that the founders' 'interest in the new wood conversion product' in many of the early cases 'was due to lack of alternative utilisation of their resources'. Mikkelsen is also of the opinion that this was 'possibly the reason for the building of Skaaningfos', Anker's mill.[28]

Untapped waterfall power and a surplus of small timber must have seemed a tempting opportunity for many timber merchants, especially for entrepreneurs with a pioneer spirit, who had not only plentiful energy but also the compulsive need to be successful at something new and the will to take incalculable risks. Christian Anker was such a man, and it must have been he who managed to persuade his more careful brother Nils to join him in establishing the mill. As Francis Sejersted has pointed out, entrepreneurial engineers such as Tobiesen in Christiania were behind a large number of the early wood pulp mills. These two impulses from the supply side in combination – the surplus resources and the entrepreneurial zeal – provide a sufficient explanation of why the early Norwegian pulp mills were established.

But what about the signals from the demand side of the market? Both H. C. Mathiesen and Nils Anker received reports from their agent that the product was in demand. These signals cannot, in my opinion, be regarded as any pressure caused by demand. They were rather signs of market opportunities, which could be tempting given an entrepreneurial spirit and a suitable resource basis.

It was the underlying uncertainty, not buying pressure of any permanence, which characterised the market situation for wood pulp in the 1870s. When Nils Anker came to London in 1871, he discovered that the established, large agents who were willing to become involved in wood pulp at all regarded it as an object for speculation. His agent Neck, who had so completely overestimated his capabilities of disposing of Anker's pulp, would not even divulge to whom he had sold it. When Anker sought help from Reid, the other main agent for the Norwegian producers who had sold pulp for Tobiesen, he did not get much further.

In Anker's opinion Reid appeared to be 'a very smart fellow, but otherwise a brigand'. Reid made it quite clear that he did not sell pulp 'as an agent, but wanted to buy it for the purpose of speculating'. With reference to both Reid and Neck, Anker had to finally declare: 'It's just nonsense dealing with them.'[29]

28. Mikkelsen, *Norsk Tremasseindustri*, p.97.
29. Letter, Nils Anker to Christian Anker, 4 June 1971, Anker Archives, Rød Herregård, Halden.

Uncertainty in connection with the Franco-Prussian War made the transport of esparto much more expensive. Paper manufacturers feared they would be completely cut off from supplies of esparto. For a short period this enabled agents like Reid and Neck to sell some consignments of wood pulp at a 25–30 per cent net profit, but during the summer of 1871 the uncertainty ended and the paper manufacturers lost most of their interest in the new raw material.[30]

Most of their interest, but not all. Right at the end of his stay of more than one month in England, Nils Anker assessed his chances of opening up a market for his wood pulp in Britain more optimistically. After he had made contact with some of the biggest paper manufacturers in Lancashire, he was convinced that pulp 'had selling possibilities', as he wrote, 'when one sells it cheaply and takes great pains'.[31] The way Nils Anker tackled difficulties, both by his own actions and by his advice and instructions to his brother Christian, indicates what had to be done to give Norwegian wood pulp a footing in the critical and fastidious British market in the 1870s.

The first part of the 'pains' he had to take was a comprehensive marketing survey, which was what his journey turned out to be. At a bookshop in London he procured a catalogue of all the paper manufacturers in Britain, with details of what kind of paper they made. After he had studied it carefully he sent a 'handsome circular' to about 200 producers, asking them whether they used wood pulp as raw material, and enclosed a sample with quotation. The percentage of replies was not impressive, but of the twenty-five who answered, ten gave the information that they used pulp, and he arranged to call on two of them. He travelled round Lancashire and got access to the largest factories he 'had ever seen'.[32]

Even though the supply of wood pulp far exceeded the demand, he could see big opportunities for the future and wrote home, 'If one could only persuade the Lancashire people, who are the most skilled factory workers in England, to use my pulp, it would sell like hot cakes.' The premium for 'taking great pains' was thus considerable. The problem was that the paper manufacturers had to be 'persuaded' to use wood pulp, and Nils Anker suggested measures on several levels.

30. Ibid.; Spicer, *Paper Trade*, p.37.
31. Letter, Nils Anker to Christian Anker, 18 June 1871, Anker Archives, Rød Herregård, Halden.
32. Ibid.

In the first place, it was absolutely necessary to establish a network of 'local agents who could visit and pester the manufacturers regularly', as Anker put it. Otherwise, in his opinion, it would be 'impossible to get the pulp sold'. These agents were also needed to collect the money and settle disputes etc. with the manufacturers. Unlike the timber buyers, the paper manufacturers did not accept bills of exchange, and were in Anker's opinion 'as stubborn as mules'.[33] The frequent disputes probably resulted from the fact that the customers were not familiar with wood pulp. It was difficult to get them to accept that the pulp did not contain more than 50 per cent water, which was the standard specification for wet pulp. The manufacturers considered it almost impossible 'to press pulp into 50-per cent dry material, and they felt they had been cheated', Anker wrote. Mutual agreement on existing specifications for the pulp had to be established.[34]

Aggressive selling and pressure were needed to get the product accepted as a raw material in principle, and in addition the introduction on the British market necessitated an improvement in the quality of the product. The suppliers had to establish new standards of quality in line with the needs of the manufacturers.

This primarily meant being able to deliver a 'uniform pulp free of knots', as Anker wrote on several occasions. Knots or blemishes were a nightmare for the producers, since they 'could completely spoil a whole brewing of paper, worth £50–100'.[35] Secondly, the paper factories had problems dissolving the pulp before production. At the time it was pressed down in sacks and packed in boxes; it therefore often became lumpy, and as a result 'it was almost impossible to dissolve it without a lot of work'. The paper manufacturers wanted it to come in thin sheets which dissolved more easily.

To solve these two main problems, Nils Anker suggested to his brother that they should buy 'a machine to remove the knots' which he had heard of, and he asked whether it was not an idea 'perhaps to deliver cardboard'. A year later it was in fact a type of cardboard-making machine which Anker's mills started using to squeeze out the water from the wood pulp before it was packed as a bale consisting of many sheets. A sieve to remove knots was successfully introduced to most pulp mills in the first half of the

33. Ibid.
34. Letter, Nils Anker to Christian Anker, 4 June 1871.
35. Letter, Nils Anker to Christian Anker, 18 June 1871, Anker Archives, Rød Herregård, Halden.

1870s.[36] Through practical measures of this nature on the marketing and production side, as well as a number of practical improvements in dispatch 'so that the paper manufacturers don't have a lot of bother', as he put it, Nils Anker considered it possible to overcome the paper factories' opposition to wood pulp as raw material, and perhaps 'persuade them to use a great quantity of pulp'.

They had to be enticed, of course. Aggressive marketing and improvements in quality were not enough: the sales price had to be considerably lowered. To dispose of the wood pulp, it had to 'be sold cheap'. In his circular to the manufacturers Anker had given a quotation for his pulp at £7 10s–£8 per ton, which was the price he had obtained before. The small consignment he had reached agreement on delivering was sold at a 25 per cent discount. He wrote to his brother that even if the prices were not much to shout about, he was 'glad that he had managed to sell the pulp'. He was prepared for the necessity of further reductions to achieve a breakthrough, but thought that 'there would be a much increased use when the prices went down'. If the quality problems were solved and the manufacturers were sufficiently badgered, then the economic mechanisms in an elastic market would apply. That was Anker's recipe for putting wood pulp as raw material on to the British market.

As we know, both the Anker brothers and others in Norway succeeded in carrying out such a strategy in the long term. By the beginning of the 1890s Norway had become the world's leading wood pulp producer and the largest supplier of paper raw material to the British market.[37]

This development was not triggered by demand for the new raw material, as has been the general opinion until now. Rather it was an uphill battle. In 1874 Nils Anker was still complaining how difficult it was to get the pulp accepted. He wrote: 'If the English had not been so extremely conservative, England could have used many times what we produce here, to great advantage. But the English think it is cheating to use wood pulp in paper.' In fact the Norwegian pulp industry as a whole ran at a considerable loss each year until 1879.[38]

The competition with esparto (not to mention the competition inside the wood pulp branch itself) was extremely fierce all through,

36. I. Kittelsen, *Trekk fra skogens og treforedlingens historie i Norge*, Oslo, 1938, pp. 72ff.

37. Mikkelsen, *Norsk Tremasseindustri*, p.13; Spicer, *Paper Trade*, appendix II.

38. Mikkelsen, *Norsk Tremasseindustri*, pp.104ff., appendix IX.

and the price fell dramatically for both types of raw material. The price of esparto fell by 25 per cent in the space of ten years from the early 1870s, while the wood pulp price was cut by over 40 per cent during the same period. The fall in prices continued at a similar rate in the 1880s.[39]

Competition between the raw material suppliers cut prices down to a level which in the 1880s made it possible to sell newspapers for a penny a piece. The result was mass circulation and a strong expansion of the demand for paper. Lord Northcliffe, the legendary founder of the *Daily Mail* and a prominent figure in the development of the British popular press, described the process as follows: 'From 1875 to 1885 paper cheapened rapidly, and it has been estimated that the introduction of wood pulp trebled the circulation of newspapers in England.'[40]

In his discussion of the development of the Norwegian wood-processing industry, Francis Sejersted has considered the impulses from the supply side as being the decisive factor, especially during the 1880s, although this view has still not won general acceptance. The argument I have put forward in this article is that these impulses from the supply side were also decisive in the important introductory phase during the first half of the 1870s, and that the prevailing view concerning the reason why wood pulp started to be utilised as raw material for paper must be revised.

My conclusions can be summed up in three points, which correspond to the outline I gave of the currently established view:

1. There was no strong demand for wood pulp on the British market. The reasons for the launching of Norwegian wood pulp export were surplus supply and entrepreneurial skill.

2. The problem of getting a suitable raw material for modern paper production was solved before wood pulp came into use. There was abundance and not scarcity of raw material in Britain after 1870, caused by the rich supply of esparto.

3. Increasing consumption of paper linked to rising demand for newspapers was not the driving force behind the Norwegian wood-processing industry in the last three decades or so of the nineteenth century. The reverse was the case: the Norwegian wood-processing industry played an important part in making it possible for newspapers to increase their circulation up to the million mark by its ability to cut the price of wood pulp radically.

39. Spicer, *Paper Trade*, p.33.
40. Reader, *Bowater*, p.5.

The main basis for the Norwegian wood-processing industry was laid at the beginning of the 1870s. It was during this critical phase that the new raw material was introduced. This process was not advanced by any international pressure of demand, but through individuals such as Nils and Christian Anker finding it necessary 'to take great pains'.

Index

Index

Index